Endorsements

Physics manifests itself in our lives, like the metamorphoses of a kaleidoscope. The book of famous physicists — Attilio Rigamonti, Andrey Varlamov and Jacques Villain — carries us away from the very first chapter. The curious reader will find out why the sky and the sea are blue (but not only), why the rivers are meandering, how natural waveguides work, why the climate changes, with what forces tides associated…

Furthermore, the path of knowledge leads from the beauties and uncovered mysteries of nature to the physics of everyday life — from the train in the gallery to the secrets of a glass harmonica and a good violin. Over the aperitif, the authors ask questions about the secrets of bubbles of champagne and wine tears on the walls of the glass, knowing the answers to which will help you to be known as an experienced sommelier. Then you enter the kitchen: the physical phenomena around ovens and stoves are endless, from microwaves to cooking, with corresponding phase transformations. Making pizza requires deep thought, not to mention pasta, where Italian and German philosophies collide. The physics behind making good (or bad) coffee is very interesting, and you can't go wrong.

After an excursion to the kitchen, the authors invite us to a strange quantum world. The quality of the text and the splendour of the illustrations convey the aesthetics of the physics of the atomic and subatomic world. The chapter entitled Physics, Geometry and Beauty deals with fullerenes, Leonardo, Piero della Francesca and Luca Pacioli. Even though our senses are already heightened by wine, cooking and good coffee, perhaps it is in the field of quantum physics that two cultures, once united, and then separated by Aristotelian constructions, again merge into world harmony.

Giorgio Benedek, Member of the Lombard Institute Academy of Science and Letters, and of the Italian Academy of Science "Dei Lincei"

This book talks about physics and its role in the world around us. It was written by professional scientists who have devoted their entire lives to finding answers to the riddles posed by Nature. Riddles that authors find in a seemingly mundane world, and riddles of the quantum world, which they manage to penetrate, continuing the path of many generations of scientists.

The book has an unusual history. It originated back in the 1980s on the pages of the "Kvant" magazine of the Academy of Sciences of the USSR for schoolchildren, widely known in those years.

Then, together with its first author, the famous theoretical physicist and popularizer of science Andrey Varlamov, the book moved to Italy where it was gradually enriched with its culinary part and other chapters. This happened thanks to his many years of collaboration with another author of the book — the remarkable experimental physicist Attilio Rigamonti.

Scientific meetings and joint work of the authors at the Lombard Academy of Literature and Science in Milan with its foreign member, the French theoretical physicist Jacques Villain, ended with the enrichment of the Italian "Il Magico Caleidoscopio della Fisica" with new chapters and ideas and, most importantly, an appeal to a much wider audience. In 2014, it was published as Le Kaleidoscope de la Physique by the publishing house Belin. The following year it received the Roberval prize, an international award for the best popular science book of the year in French. Its Russian language version, published in 2020 was recognised among the best scientific popular books in the same year and awarded the Diploma of *Russian Academy of Sciences*. Today, thanks to the efforts of the World Scientific Publishing Company, the book, having been considerably expanded, becomes available to the English-speaking reader.

Lev Pitaevskii, Member of Russian Academy of Sciences

I wish to tell you how much I had appreciated *Le Kaléidoscope de la Physique* which I discovered when I was a member of the jury of the Prix Roberval (sponsored by the Académie des Sciences). This yearly prize is awarded to books in French devoted to technology and science, and one of them is addressed to a wide audience. The large variety of subjects of your book, which stimulate the curiosity of the readers, the spirit in which they are presented, and the pleasure that anyone can take in discovering each topic, the nice illustrations, had led the jury, a few years ago, to rapidly focus on your book in the discussions and to readily select it. This is remarkable as I was the only physicist in the jury, which contained scientists and engineers from all specialities and from all French-speaking countries.

Roger Balian, Member of the French Academy of Sciences

THE KALEIDOSCOPE OF PHYSICS

FROM SOAP BUBBLES TO QUANTUM TECHNOLOGIES

THE KALEIDOSCOPE OF PHYSICS

FROM SOAP BUBBLES TO QUANTUM TECHNOLOGIES

Attilio Rigamonti
University of Pavia, Italy

Andrey Varlamov
SPIN-CNR, Italy

Jacques Villain
Academy of Sciences of France, France

World Scientific

NEW JERSEY • LONDON • SINGAPORE • BEIJING • SHANGHAI • HONG KONG • TAIPEI • CHENNAI • TOKYO

Published by

World Scientific Publishing Co. Pte. Ltd.
5 Toh Tuck Link, Singapore 596224
USA office: 27 Warren Street, Suite 401-402, Hackensack, NJ 07601
UK office: 57 Shelton Street, Covent Garden, London WC2H 9HE

Library of Congress Cataloging-in-Publication Data
Names: Rigamonti, A., author. | Varlamov, Andrey, author. | Villain, Jacques, author.
Title: The kaleidoscope of physics : from soap bubbles to quantum technologies /
Attilio Rigamonti, University of Pavia, Italy, Andrey Varlamov, SPIN-CNR, Italy,
Jacques Villain, Academy of Sciences of France, France.
Description: New Jersey : World Scientific Publishing Co. Pte. Ltd., [2023] | Includes index.
Identifiers: LCCN 2022056885 | ISBN 9789811265242 (hardcover) |
ISBN 9789811265259 (ebook for institutions) | ISBN 9789811265266 (ebook for individuals)
Subjects: LCSH: Physics--Popular works.
Classification: LCC QC24.5 .R55 2023 | DDC 530--dc23/eng20230302
LC record available at https://lccn.loc.gov/2022056885

British Library Cataloguing-in-Publication Data
A catalogue record for this book is available from the British Library.

First published 2023 (hardcover)
Reprinted 2026 (in paperback edition)
ISBN 9789819833320 (pbk)

For any available supplementary material, please visit
https://www.worldscientific.com/worldscibooks/10.1142/13111#t=suppl

Desk Editors: Aanand Jayaraman/Adam Binnie/Shi Ying Koe

Typeset by Stallion Press
Email: enquiries@stallionpress.com

About the Authors

Attilio Rigamonti (born in 1937) is an experimental physicist, professor, member of the Lombard Institute Academy of Science and Letters (Italy), of the G. Cardano Institute, and of the Mediterranean Institute of Fundamental Physics. Presently he is Emeritus Professor for the discipline Structure of Matter. Since 1976, he held the chair "Structure of Matter" at the Department of Physics "A. Volta" of the University of Pavia, teaching as well at various universities of Europe and the USA. He was Director (2007–2012) of the Doctorate School of Science and Technology at Pavia University. He is the author of more than 250 scientific articles dealing with various aspects of solid-state physics (nuclear magnetic and quadrupolar resonances, liquids, polymers, phase transitions, incommensurate magnetic systems, and superconductivity). He is also the co-author of institutional books (one in collaboration with the Nobel Laureate K.A. Muller) and popular science books.

Andrey Varlamov (born in 1954) is a theoretical physicist, professor, member of the Lombard Institute Academy of Science and Letters (Italy) and of the Mediterranean Institute of Fundamental Physics, as well as a popularizer of science. After graduating with honours from Landau Institute for Theoretical Physics (1977), he worked as a researcher, then as a professor at the Department of Theoretical Physics of the Moscow Institute of Steel and Alloys (1980–1999), in the theoretical group of condensed matter physics of the Argonne National Laboratory, USA (1993). Since 1999 he has been a leading researcher at the Institute of Superconductivity, New Materials and Devices (SPIN-CNR, Italy). His research focuses on the theory of superconductivity and condensed matter physics. He is the author of more than 170 scientific articles, co-author of the monograph *Theory of Fluctuations in Superconductors*, the book *The Wonders of Physics*, and other scientific publications.

Jacques Villain (born in 1934) is a theoretical physicist, professor, member of the Academy of Sciences of France, foreign member of the Lombard Institute Academy of Science and Letters (Italy), and a popularizer of science. He worked as a researcher at the Nuclear Research Centers of Saclay and Grenoble (1961–2010) and was engaged in research work at the European Synchrotron Radiation Research Center (Grenoble) and at the Institute of Solid-State Physics in Jülich (Germany). His research focuses mainly on condensed matter physics (magnetism, surfaces and crystal growth) and statistical physics. He is the author of 170 scientific articles; he is also co-author of two monographs *Physics of Crystal Growth* and *Molecular Nanomagnets*.

Acknowledgements

This book is dedicated to the authors' grandchildren, who are just entering a world that physicists have completely transfigured, a world considerably different from the one their grandfathers met when they were kids. A world that may be more comfortable, where it is difficult to get bored, but not necessarily where it is easier to find one's own place and serenity:

to Luca and Margherita Strozzi Rigamonti
to Michail Varlamov
to Mathilde and Lucas Schmitt

We express our deep gratitude to our editor, Dr. Keith Mansfield; our colleague Dr. Rufus Boyak, who carefully read the manuscript and greatly improved its literary merit; and Alexandra Fridberg for her help with drawings. Finally, the authors wish to thank colleagues and friends who kindly assisted in revising parts of the book, helping the authors to avoid being blinded by the brilliance of the kaleidoscope and to find the proper paths along the lonely spots of a complex science, of which even the professionals do not know all its deep meanings.

In particular: Alain Aspect, Giuseppe Balestrino, Sébastien Balibar, Giorgio Benedek, Claude Berthier, Marc de Boissieu, Hélène Bouchiat, Patrick Bruno, Alexandre Buzdin, Giancarlo Campagnoli, Christiane Caroli, Anne Marie Cazabat, Alexei Chepelanskii, Robert Dautray, Guy Dellaval, Daniel Estève, Stephan Fauve, Yuri Galperin, Etienne Ghys, Maurice Goldman, Denis Gratias, Isabelle Grillo, Sylvie Guérin, Jacques Haissinski, Catherine Hill, Jean-Pierre Hulin, Jacques Lambert, Dmitri Livanov, Yu Lu, William-Luise Mame, Valérie

Masson Delmotte, Keizo Murata, Philippe Nozières, Sergei Parnovsky, Jean-Louis Pautrat, Raymond Pierrehumbert, Olivier Pierre-Louis, Pierre Pugnat, David Quéré, Y. Saito, Sergei Salikhov, Irina Seregina, Sergei Sharapov, Stefania Tentoni, Hervé This, Igor Tralle, Alexandre Valance, and Nicolas Villain.

Contents

Introduction

Physics touches several aspects of our lives in ways that appear like the changes in a kaleidoscope. This book is for readers wishing to understand some of the strange and astonishing phenomena they commonly see. Why the sky is blue? How do rainbows form? What is the driving force of the tides? Why do we see a head on top of a glass of beer or a frothy white mousse over a creek? These are some of the questions we shall answer and explain in the first part of the book.

We as humans have considerably increased the number of the physical phenomena that are nowadays sources of wonder: how do microwave ovens work in comparison to induction plates? How is electricity generated? These and other devices we have created ourselves will be the topics addressed in the second part of the book.

The third part is more relaxed. It has been written mostly for the friends who like cooking. If one turkey is twice as big as another, should we double the time it spends in the oven? We have no intention to teach the art of cuisine but simply to answer some of the more scientific questions that might arise when cooking. At the very least, we shall offer some recreation while fine dishes are produced by high-class cooks.

The fourth and final part of the book is the most ambitious. We wish to describe some of the mysterious aspects of modern science, in order to emphasize how our view of Nature has been drastically changed by the new area of physics known as quantum mechanics. We explore how, around the turn of the millennium, some revolutionary ideas have been illustrated and experimentally supported. We hope we are not being too ambitious, considering the first three parts will have provided the reader with the drive and the courage required to attack the fourth.

We shall assume some basic scientific knowledge, say at high-school level. However, we shall try to avoid complexity by using simple language. Correspondingly, the number of equations has been reduced to the bare minimum, leaving only those that should favour real understanding in a simple way. Even though many equations are undesirable, a few may occasionally help break up the text and might even stimulate a reader's attention when appropriate. In comparison to a high-school graduate, the general reader might have the advantage of being able to skip steps considered too hard, too high level or too elementary, instead referring to other chapters or to other sections they may enjoy more. In such places, we have suggested a few exercises to stimulate our readers' attention. They are straightforward equations to write and solve or very simple experiments.

This book is largely a translation of *Le kaléidoscope de la Physique* published in the French language by Belin (Paris) in 2014. That book was itself based on previous editions of articles, papers or books published by one or more of the authors. In particular, some chapters are largely based on *Magico Caleidoscopio della Fisica* published in 2007 by La Goliardica Pavese or from *Wonders of Physics* published in 2012 by World Scientific. All chapters have been extended and updated.

Part 1

Open-Air Physics

According to etymology, Physics is the science of Nature (in ancient Greek φ. σις) or at least of natural phenomena. Indeed, it is the very spectacle of Nature that offers human beings their first physics problems, for example, the tides, that astonished the sailor Pytheas when navigating his familiar Mediterranean sea;

the rainbow, that Descartes was able to explain; the apparent motions of the planets in the sky, that Aristotle wrongly explained and which caused troubles to Galileo; and the shape of bubbles or droplets, that worried Pierre-Simon Laplace and Thomas Young.

Many of these problems, apparently simple, could not be properly explained until the 19th century. More recently, Albert Einstein was still asking himself about the shape of the meanders in a river. We are going to devote the first part of the book to some of those phenomena.

Chapter 1

Rivers, Meanders and Lakes

A river is a complex system that has evolved in a complex environment. Although science cannot explain all the details of its movement, still it gives some clues that allow us to understand several properties of this phenomenon.

Often, when walking beside a creek or a river, we might ask ourselves why the flow of the water does not follow the most direct path (namely a straight line), instead making a series of meanders. Of course, some parts of the river do follow a straight path, when the local topography or the work of humans dictates this. However, when the river is free to choose its path on an almost flat area, it proceeds through zigzags. The curves follow one after the other with a certain regularity (Fig. 1). How do we explain the meanders?

Some Tea Leaves in a Cup…

One of the first scientists who tried to explain the reason for the formations of these curves in a river was Albert Einstein in 1926. Temporarily leaving behind the difficult mathematical physics of general relativity and its curved space–time, at the meeting of the Prussian Academy of Sciences Einstein presented a note without any equations, entitled *Causes of the meanders formation in the stream of the river banks and about the Baer law.* What is this famous Baer law? Based on the observations by outstanding 19th-century geographers, naturalist Karl Baer concluded that in the Northern hemisphere, in the absence of mountain reliefs, the right-hand bank of a river has the tendency to become increasingly eroded compared with the left bank; the opposite happens in the Southern hemisphere.

Before dealing with the sinuosity of rivers and the shape of their banks, Einstein proposed a simple experiment, reproducing a common everyday

Figure 1. Meanders of Siberian River from aircraft porthole.
Source: Prof. Keizo Murata.

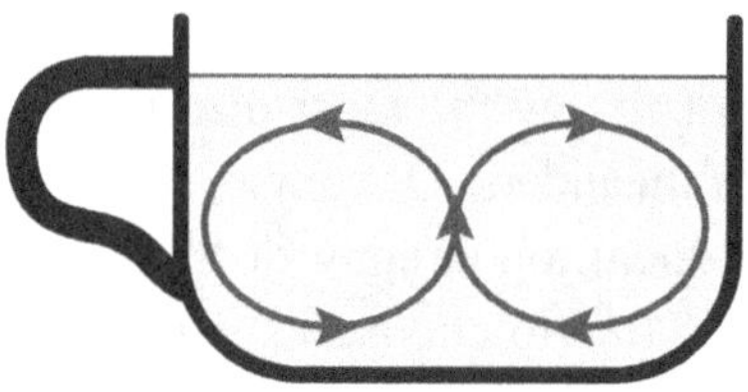

Figure 2. When the water in the cup is moved by the little spoon, vertical vortices are induced in the liquid.

movement: when with the swirl of a small spoon we encourage sugar to dissolve in a cup of tea. What Einstein was interested in was a phenomenon scarcely intuitive at first glance: the rotation of the liquid induced by the teaspoon and the subsequent creation of vertical vortices (Fig. 2).

In order to give evidence for the vortices, Einstein dispersed some fragments of tea leaves. When the rotation of the liquid is instigated by means of the spoon, the leaves start to gather in the centre of the bottom of the cup (Fig. 3). Readers can practise the experiment themselves.

Einstein's explanation was as follows: as a consequence of the rotation, a centrifugal force acts on the liquid, towards the exterior of the trajectory. This force increases with the increase of the velocity (see Chapter 4). Within the boundary of the cup, the liquid is slowed down by friction and therefore rotates more slowly than in the centre of the cup. In particular, Einstein added, the angular

(a) (b) (c)

Figure 3. Einstein's experiment. Fragments of tea leaves are forced to rotate by the little spoon (a). Quickly the fragments gather at the center of the glass (b) and then they deposit at the bottom (c). This motion is evidence of the counterintuitive occurrence.

velocity and then the centrifugal force are slower at the bottom compared with the layers higher up. In this way, the circulation of liquid indicated in Fig. 2 arises, dragging the fragments of the tea leaves towards the centre.

How the Riverbed Evolves

We are going to examine the motion of the water when the river reaches a bend. Here, the analogous motion inside the cup occurs, as Einstein remarked. As in the experiment, the river water is slowed down by friction at the walls and the velocity of the water flow decreases near the bottom: the centrifugal force, directed towards the outside of the bend, is lower in the vicinity of the riverbed. Thus a vertical circulation is induced towards the outside of the meander, close to the surface, and instead towards the inner curve close to the bed (Fig. 4).

This circulation leads to some transfer of ground and stones, pulled out from the external bank towards the interior of the curve. A deposit of materials at the interior of the curve occurs, just as there was a deposit of tea leaves at the centre of the bottom of the cup in the experiment described above. In both cases, when the water rises, precipitation occurs, the liquid depositing the heavier material due to gravity. The erosion of the external bank and the deposit on the internal bank induces an elbow in which the exterior is rougher while the interior is smoother.

Sometimes, after continued erosion, the riverbed can be so dramatically affected at the entrance and exit of the meander that the ground is wiped out and an island is formed (Figs. 5 and 6).

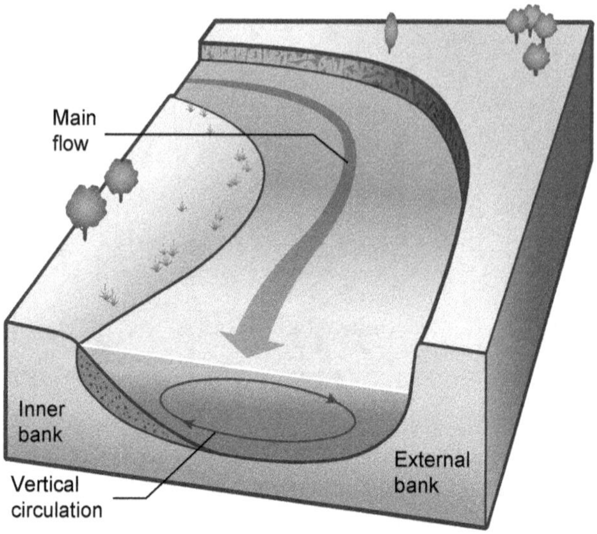

Figure 4. Water circulation in a river with a curve, from Einstein. A centrifugal force from the internal towards the external bank is acting on every part of the liquid. Due to friction, the force decreases in the vicinity of the bottom and a vertical circulation is superimposed on the main flow. The material pulled out from the external bank is carried towards the interior of the curve.

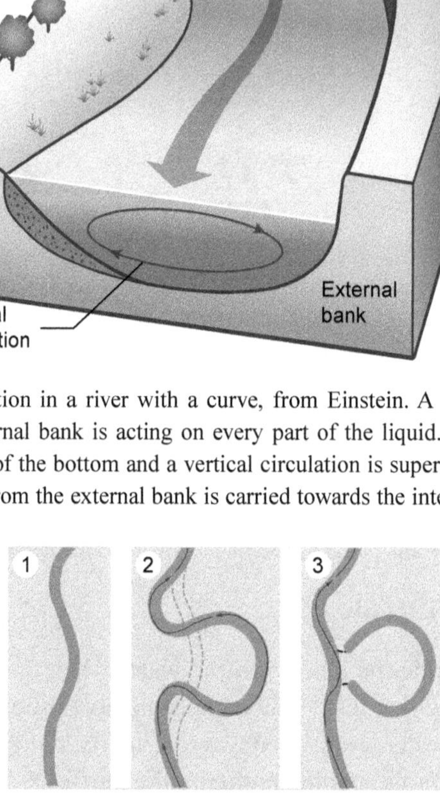

Figure 5. The sinuosity of a river, first moderate (1), progressively increases due to the deposit of material on the internal bank (2) until finally it can lead to the origination of an island (3).

The considerations we have developed cannot explain everything. For instance, how do we explain the Baer law in regards to the difference between the two banks, even when the river is not affected by any sinuosity? And how to justify the observations of the geographers describing the way the effect is inverted between the Northern and Southern hemispheres? The reader may already guess that the rotation of the Earth is somehow involved: we shall return to this in Chapter 4.

What Is the Shape of the Meanders?

The path of a riverbed is strongly related to the relief of the area in which it runs. Where the ground is irregular, the river proceeds in such a way as to avoid

Figure 6. Meander of the river Seine at Andelys and the island, as seen from the Gaillard castle. The external bank is steep while the internal bank is smooth.

asperities, and it takes the path characterised by the pronounced slope. When the ground takes the form of a plane, a straight path is not so stable. Any little avalanche, or the fall of a tree over the bank, forces the flow into a curve that is progressively increased so that a meander is induced, as explained in the previous section.

More generally, what is the shape that characterises the sinuosity of a river flowing on flat ground? In the 1960s, geologists argued that the sinuosity tends to take a particular shape: the one of a rod that is curved when its two extremities are brought near (Fig. 7). The shape assumed by the rod is a curve described by the Swiss mathematician Leonhard Euler (1707–1783), who studied its properties. The analysis by Euler is still widely reported in texts dealing with buckling when the weight being borne is increased by too much (see Panel on page 9). That hypothesis is in good agreement with experiments carried out in laboratories by simulating the flow of a river and studying the changeover of its bed. Initially straight, in ideal conditions, some meanders quickly appeared; the final distribution was in good agreement with Euler's law (Fig. 8(a)). We remark that real rivers do not show the regularities appearing in the lab (for instance, because of irregularities in the ground). Several times, similar patterns, approximately periodic, spontaneously occur when the river runs across flat land (Fig. 8(b)). In general, the sinuosity has long periods when the river is large.

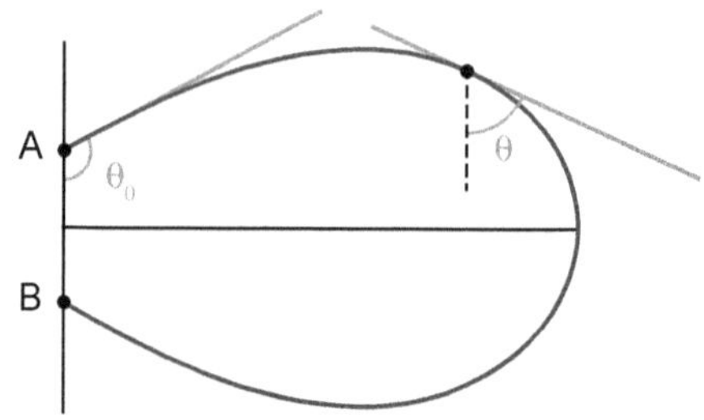

Figure 7. The shape assumed by an elastic rod at fixed extremities is called the Euler curve. The angle between the tangent and the line AB allows one to define the curvature $d\theta/ds$, namely the derivative of the angle with respect to the line. The Euler curve minimises the mean square curvature of the line: in other words, it minimises the integral $\int(d\theta/ds)^2ds$, where θ is the angle of the tangent with respect to a fixed direction.

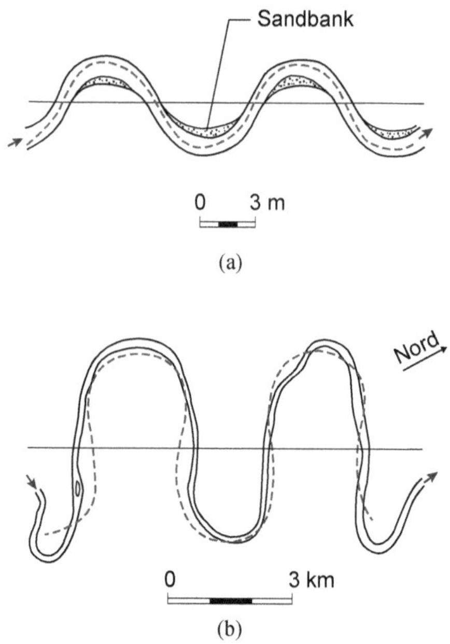

Figure 8. (a) Meanders observed in a water flow in the laboratory. In a homogenous environment, the flow is initially straight. After some hours, due to the erosion, curves occur in the channel, and the meanders appear. The Euler curve is given by the dotted line. (b) Sketches for a real river (the Potomac at Paw in USA) and the Euler curve (dotted line) approximately fitting it. From L. B. Leopold and W. B. Langbein, (1966) "River Meanders", *Scientific American*, 214, 60–70. http://dx.doi.org/10.1038/scientificamerican0666-60.

An especially winding river is found in Turkey, the Büyük Menderes that is at the origin of the term meander for the sinuosity of the rivers: its name comes from the Greek "Maiandros". However, the name meander also applies to the sinuosity of those streams that arise at the surfaces of the glaciers as well as to ocean

An experiment about buckling

Let us place a plastic ruler on a table, keep it vertical and press on its top. When we do not push too severely, the ruler remains vertical and straight. When a certain force is reached, the ruler starts to form a curve: this is the phenomenon of buckling. It occurs when the force acting at the extremities has reached a certain value, below which the ruler maintains the straight shape. Architects and builders are careful to not exceed that limit, for instance, when a terrace is supported by metal columns.

In our experiment, the ruler bends towards the right-hand side when the force is increased above that limit. In the same way, the ruler could obviously bend towards the left-hand side! One may compare this situation to that of a traveller when a bifurcation appears in front of them. Bifurcation is just the name mathematicians give to the physical phenomena that can occur in two equivalent ways under the variation of a certain parameter.

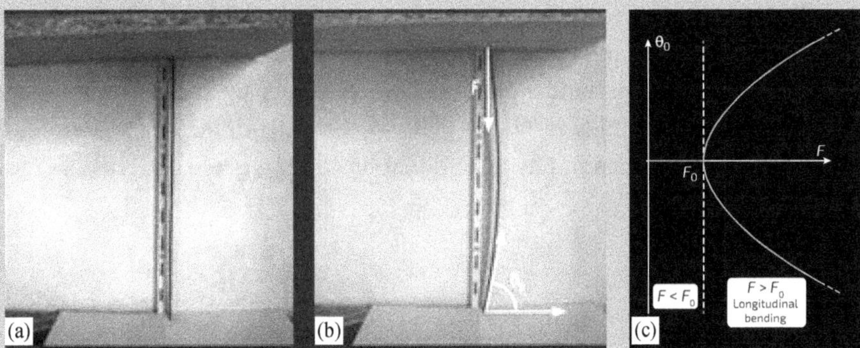

The phenomenon of buckling: (a) When the vertical force F is smaller than a certain value F_0, the ruler does not bend. (b) When the force is above F_0, the ruler bends and could even break if the force is increased. (c) Variation of the angle θ_0 between the ruler and a vertical plane as a function of the value of the force F. When the force F_0 is increased, the ruler bends to the right ($\theta_0 < 90°$) or to the left ($\theta_0 > 90°$): the curve of the variation of the angle exhibits two branches that display a bifurcation.

currents such as the Gulf Stream. For all these phenomena occurring in homogeneous environments, random processes contribute to the formation of an approximately periodic sinuosity; other causes may play a role.

A Lake, One River

Usually, several rivers are tributaries of a large lake. For instance, Lake Geneva receives the Rhone but also other small rivers such as the Dranse from South and the Veveyse from North. However, only one river, the Rhone, is effluent. More generally, independent of the number of water flows entering a lake, only one comes out! How do we explain this?

The reason is that the water from the lake flows out along the deepest channel it finds. Excluding exceptional cases of floods, the water surface in the lake is usually at a lower level than all other possibilities of channel formation, therefore only one stream emerges from it. Even if more than one flow is temporarily possible, this situation is unstable, and we could only possibly find it for lakes recently formed. In fact, the flow of water at the lowest level and with the strongest current will cause an increase in the erosion. Thus a decrease in the level of the lake will occur. The amount of water contributed by other rivers will progressively decrease, just to fill the bottom. Therefore, only one flow of water, the deepest of the flows, will survive.

Rivers are characterised by analogous properties. It is well known that there are confluences, while very seldom they spontaneously split. The flows generally follow the steepest line, and there is little chance that this line will split. An exception is represented by a particular situation: where rivers are about to enter the sea, they split to create a delta (Fig. 9). In fact, when far from the sea, water courses have to travel across pleats in the ground that has a long geological history, going

Figure 9. The delta of the Rhone. On approaching the Mediterranean Sea, the river splits into several branches.

back several millions of years. On the contrary, at the delta, the river makes the banks by depositing flood materials in a shallow sea.

This chapter is at the end since our river has now arrived at the sea. We shall find the sea again in Chapter 5 when describing the astonishing physical phenomena represented by the tides.

Chapter 2

Artificial and Natural Waveguides

How can a sound that originates on the Australian coast arrive at the Bermuda islands, tens of thousands of kilometres away? In order to understand why, we are going to make an analogy between sound propagation and light propagation, for the latter, the concept of "ray" being more familiar. We will go deep in the ocean, searching the mysterious waveguide that is capable of propagating the sound over very long distances.

For about the last 20 years, an enormous and increasing amount of data has been transmitted from continent to continent, thanks to the optical fibres that crisscross the ocean (Fig. 1). These fibres drive light waves which carry information over

Figure 1. Bunch of optical fibres inside a protection sheath. The single fibre, in glass or in plastic material, measures 125 μm in diameter, including its core, cladding, and outer jacket.

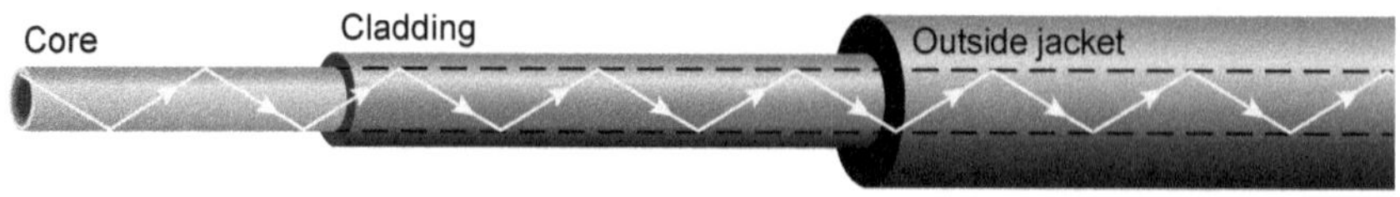

Figure 2. Trajectory of a light ray inside an optical fibre. The propagating ray is totally reflected at the interface between the heart and the sheath, thus being driven along the fibre. The data are codified by means of variation in the intensity of the light.

large distances (Fig. 2). This is the way in which a message instigated by your computer or telephone is transmitted to a colleague in the USA or Japan. Obviously, those light waves are dampened as they travel along their path, but the reduction is relatively weak, and only a few relays are required.

The Propagation of Sound Waves

Without the addition of technology, the ocean is capable of working as a sound waveguide. This is the astonishing phenomenon was discovered by Russian and American scientists around 1940 and 1950, respectively; sound waves propagated in the ocean and could sometimes be detected at a distance of tens of thousands of kilometres from the source! During one of the most spectacular experiments, the sound due to an explosion in the deep sea near the Australian coast could make half a circuit around the Earth and be detected at the Bermuda islands, an archipelago in the Atlantic Ocean. Thus, the sound signal could propagate underwater for more than 19,600 km. This is indeed a record!

The intensity of a sound signal inexorably decreases on moving away from the source. In fact, the energy emitted by the source has to divide approximately along all the possible directions. By neglecting the frictional damping, the total energy of the sound wave is conserved during the propagation: at the distance R, the energy is distributed over a surface proportional to R^2. Therefore, the intensity of the sound decreases as $1/R^2$ on moving away from the source (Fig. 3). And this is without taking into account the dissipation, absorption, and scattering of the sound from the medium where the sound is propagating! Therefore, in order for the explosion near Australia to reach Bermuda, with the intensity of the sound wave still being sufficiently strong, it had to be the case that the energy was driven towards the archipelago without scattering along other directions (Fig. 4). That means that the ocean includes a kind of acoustical waveguide, namely, a channel along which the propagation can occur with weak attenuation. For the waveguide to work in a proper way, it is essential that the walls are perfectly reflective: not permeable or absorbent.

Figure 3. The intensity of sound emitted by a speaker (assumed as a point source and propagating in all the directions) decreases as $1/R^2$ on leaving the source, in the absence of guides as well as of obstacles. The total energy remains the same, but is distributed over the surface of the sphere of radius R, given by $4\pi R^2$.

Figure 4. An example of directional propagation of acoustic waves in air. The two young girls are exchanging secrets. While placing the hand in the vicinity of the mouth, the sound signal cannot longer propagate along all directions.

What is the principle that supports the existence of an "acoustic waveguide" inside the ocean? We can guess that it is a principle similar to the one for optical waveguides, namely, based on the total internal reflection of the waves from the walls (see Panel on page 16). Could the total reflection of the acoustic waves occur at the interface of water and air? No! The speed of sound is greater in water than in air (in the cold Greenland Sea, it averages 1,411 m s^{-1}, in the warm Mediterranean Sea, it is 1,554 m s^{-1}, while the speed of sound in the air under normal conditions is 335 m s^{-1}). This means that water is a much less "dense" environment for sound than air — the situation is exactly the opposite of the case of light propagation. Therefore, the conditions for the total reflection cannot be met for a sound wave passing from the water to the air. When a sound wave coming from the deep sea arrives at the surface, there is still a reflected and a transmitted wave. Another related consequence is that instead of departing from the vertical direction in the case of the acoustic wave, it approaches the vertical.

Do we have to give up on the idea that the surface of the ocean can be reflective? Not so fast. In fact, the fraction of the energy that is reflected at the water–air interface strongly depends on the angle of incidence and on the ratio of the corresponding velocities. When the two velocities are markedly different, as in the present case, the refracted (outgoing into the air) wave intensity is indeed small for any angle of incidence. Thus, the reflection is almost total at the surface anyway: it can be proved that no more than 1% of the intensity of the sound wave propagating almost horizontally can pass from the water to the air. Therefore, in principle the ocean surface can be capable of reflecting the sound coming from the deep.

So, do we have an explanation of why the sound can propagate for such a long distance within the ocean? No, and for two reasons. First, a certain amount of energy is lost anyway when the wave arrives at the surface. Second, the

Reflection and refraction of the light waves

Let us recall the properties of reflection and refraction in optics. When a light ray propagating in a medium 1 hits an interface (assumed a plane) with a medium 2, part of the light is reflected back towards medium 1, while part enters medium 2 (figure): it is the phenomenon of refraction. The angle of refraction α_2 *is related to the angle of incidence* α_1 *by the Snell–Descartes law:*

$$\frac{\sin\alpha_2}{c_2} = \frac{\sin\alpha_1}{c_1},$$

where c_1 *and* c_2 *are the velocities of the light in the media 1 and 2.*

One can also write this law by resorting to the refraction indexes of the two media, respectively, $n_1 = c/c_1$ *and* $n_2 = c/c_2$*, where c is the velocity of the light in the vacuum.*

(*Continued*)

(*Continued*)

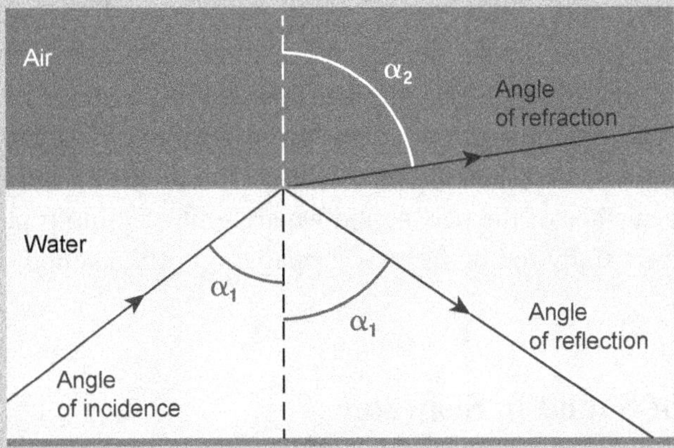

When the ray arrives at the interface with an incidence angle sufficiently high, the term $\frac{c_2}{c_1}\sin\alpha_1$ *becomes larger than 1 and no value of* α_2 *can satisfy the equation given above. In this case, we no longer have any refraction, and the internal reflection is said to be* ***total*** *(see figure). This is the phenomenon we take advantage of in optical waveguides: the light beam undergoes a series of total reflections inside the guide, and thus the light propagates with minimum loss. Optical fibre is the most popular example of optical waveguide.*

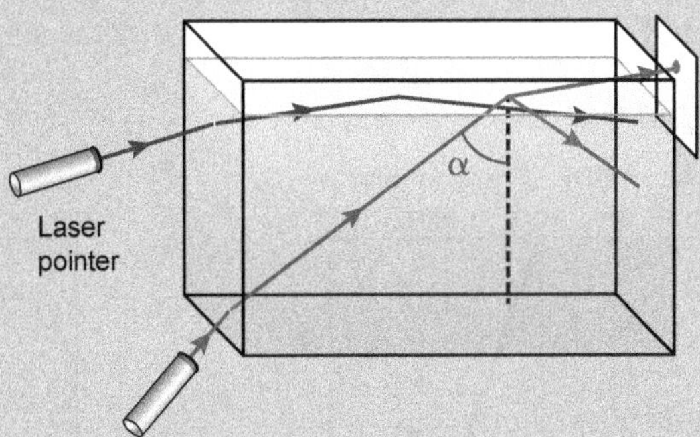

Similar to light waves, even sound waves can undergo reflection and refraction. If we consider a "sound ray" (the ray corresponds to the direction of the propagation of the energy), the Snell–Descartes equation is again valid, where evidently c_1 *and* c_2 *represent the speed of sound in the different media rather than that of light.*

surface of the sea is practically always affected by fluctuations, and thus the reflection of the waves is perturbed. In conclusion, the surface of the sea does not offer the upper part of the oceanic waveguide we are looking for, unless the sea is really very calm. On the other hand, the bottom of the ocean is even less capable of being the lower part of the waveguide. The bottom sediments do not reflect the sound; on the contrary, they more frequently absorb it. Thus, the oceanic waveguide we are looking for has to be found elsewhere, in between the bottom and the surface of the ocean, and we are going to find it in what follows. To that end, let us study the process of propagation of the sound in the ocean in more detail.

The Speed of Sound in Seawater

In a liquid such as seawater, the speed of sound is a function of its properties, which are not the same in every point in the ocean. This is the key to the problem! Depending on the amount of salt, temperature, and pressure, the speed of sound in the water can vary from 1,450 m s^{-1} to 1,540 m s^{-1}. For instance, the pressure increases with increasing depth, and therefore the sound travels faster. In the same way, the speed of sound increases with increasing temperature. Cold water, being more dense, is trapped at the bottom of the ocean. These two competing effects mean that the speed at which sound can travel is dependent on the ocean's depth, as depicted in Fig. 5. In the immediate vicinity of the surface, a sharp drop

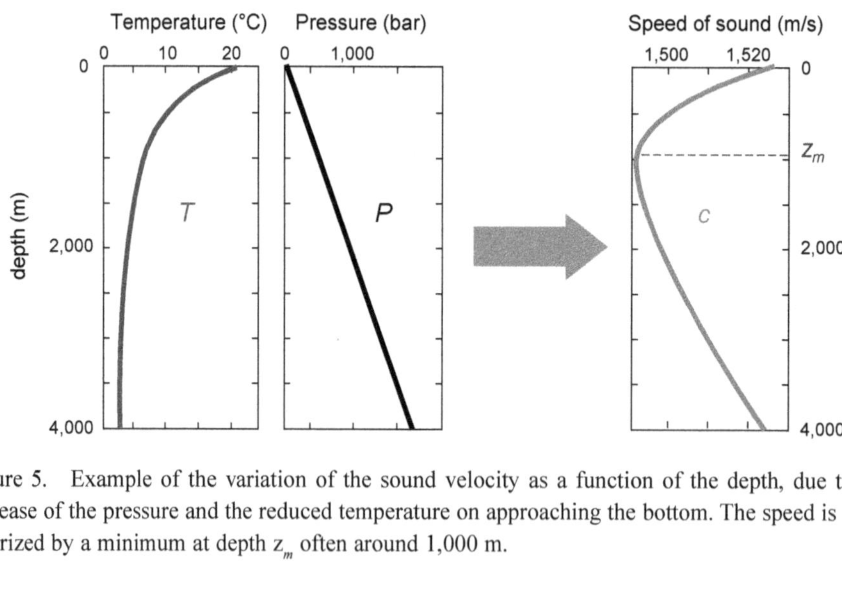

Figure 5. Example of the variation of the sound velocity as a function of the depth, due to the increase of the pressure and the reduced temperature on approaching the bottom. The speed is characterized by a minimum at depth z_m often around 1,000 m.

Sound waveguides due to man

The sound propagation in a fluid is characterized by a perturbation temporarily affecting the density of the particles microscopically representing the fluid itself. Locally, a given layer of the fluid experiences a periodic succession of compressions and of dilatations. In solids, the sound is transmitted through a succession of local vibrations that propagate.

In general, the speed of sound is greater in solids and liquids in comparison with that in gases. This is not surprising, considering that the particle density in a gas is somewhat in between that in condensed matter and in a vacuum (where it is zero), if one takes into account that in a vacuum sound is not transmitted at all. Furthermore, when the speed of sound is very different in two media, then the transmission of the sound from one to the other is rather hampered at the interface. This phenomenon is something we see at work in a stethoscope, the instrument used by the medical doctors in order to auscultate the sound effects coming from the chest of the patient. Long ago, the stethoscope was simply a long wooden funnel.

Other waveguides are based on the phenomenon of total reflection, occurring when a sound has to pass from air into a solid or a liquid. For example, one can mention the ancient way of transmitting words from different parts of a ship by resorting to a long pipe. Usually, this was made of copper or brass and used by the commander to send orders from the stateroom to the engine department. In such a device, the sound propagation is almost mono-dimensional, meaning that the intensity of the sound wave remains almost constant along the pipe, regardless of the distance from the source. The attenuation of the sound is indeed very weak, and it would be possible to make a pipe 750 km long without almost any absorption from the walls, so that a telephone call from Paris to Marseille could be achieved. Unfortunately, the speed of the sound in air being 340 m s^{-1}, the words would take about half an hour to travel from the source to the end of the pipe, and thus communication would be rather difficult!

in temperature first leads to a gradual decrease in the speed of sound $c(z)$. At great depths, the temperature change is not so noticeable, the effect of increasing pressure dominates, and this leads to an increase in c as we approach the bottom. The velocity reaches a minimum at a depth z_m, often found between 1,000 and 1,200 m, but it can go down as far as 2,000 m at low latitudes, where the water is warm even at relatively large depths. On the contrary, z_m can be at 500 or 200 m or even less in the region of the poles. The variation in the salt concentration with depth in general is small and results in practically no effect.

When Sound Propagates in Zigzags

Consider now a sound ray whose source is at a depth of z_m. Regardless of whether it goes up or down, in the area in which it finds itself, the speed of sound is greater than at origin. Thus, as a result of the successive passage of layers of water along its path, the sound ray is gradually bent, up to a grazing incidence for which total reflection occurs (see Panel on page 16). Then, it begins to bend in the direction of increasing (or decreasing) depth, until it reaches the depth z_m again, where the change in the speed of sound changes sign. Thus, the ray moves along a zigzag path between two planes (Fig. 6). The reader, if they have some knowledge of differential equations, can find the equation of the curve from the Snell–Descartes relation (see Panel on page 16), which can be written as

$$\frac{\cos\alpha(z)}{c(z)} = \text{const.}$$

These two planes are the equivalent of the upper and the lower walls of a waveguide, still lacking the lateral walls to be really complete. However, thanks to the phenomenon that we have described, the sound can propagate over very large distances in the ocean. Our study is complete!

Efficiency of the oceanic waveguide

Not all the sound rays emitted by a given source can take the "oceanic waveguide". A source initially emits in all directions, and the destination of the sound rays depends on the angle that they form with the vertical. When this angle is sufficiently large, the acoustic ray propagates indefinitely. If the angle is small, then the ray can reach either the bottom or the surface. The bottom of the ocean is

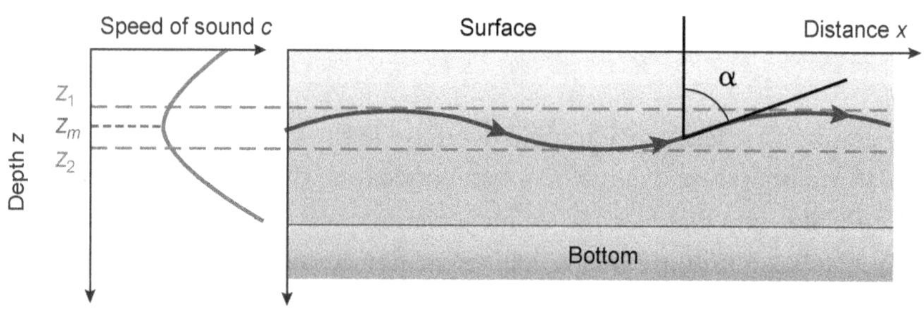

Figure 6. The acoustic ray emitted at depth z_m propagates in between two planes where total internal reflection occurs. The dependence of the speed of the sound in the ocean versus distance from the surface $c(z)$ is shown by the dark curve. The values z_1 and z_2 (assuming zero at the level of the surface) depend on the angle of incidence at depth z_m and are given by the Snell–Descartes law.

rough, and the sound is dispersed, and in general, this is also the case for the surface, unless the sea is exceptionally calm and flat. Thus, in general, the sea by itself cannot work as a waveguide. In practice, it has been verified that there are acoustic channels where the sound is driven at long distances, or other regions where the sound does not arrive at all.

The study of sound propagation in the oceans was the cause of serious worries for American and British scientists during the Second World War. The problem was to be able to detect the presence of German submarines before they could arrive at a distance sufficient for the launch of their torpedoes. The submarine detection by acoustic means, such as sonar, played a major role during the Battle of the Atlantic: in 1943, after heavy losses, the Allies could finally destroy a significant number of German U-boats, thus sounding the death knell for their supremacy over the seas.

A simple model

It is interesting to consider the case when the sound velocity is a simple function of the depth z. The simplest function granting a minimum at z_m is

$$c(z) = c(z_m) + k(z - z_m)^2,$$

where k is a constant.

The curve yielding the dependence of the speed of sound as a function of the depth (see curves in Figs. 5 and 6) is then a parabola. In practice, this is valid for a depth not too far from z_m. Thus, for an acoustic ray that is propagating along a direction not so different from the horizontal one, it is found that the propagation takes a sinusoidal path, with a period independent of the inclination. Thus, all the sound rays lying on a given vertical plane converge to points of the axis of the waveguide $z = z_m$ (Fig. 7). These points are analogues of the focuses of optical instruments such as lenses, points where the incident optical rays converge: this is

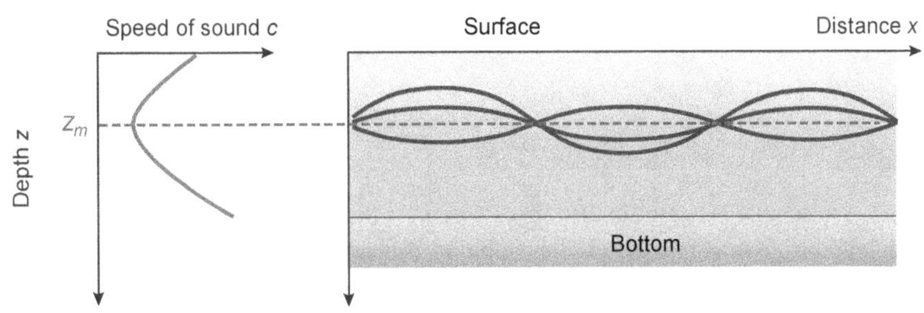

Figure 7. The phenomenon of the focalization of acoustic rays.

the phenomenon of the acoustic wave focalization. The parabolic form describes pretty well the variation of the sound velocity as a function of the depth in the deep ocean. However, the parabolic form $c(z)$ is not always strictly valid, and in these cases, the focalization cannot be perfect.

Conclusion

When a sound is emitted at a particular depth in the sea, a relevant part of the acoustic energy remains inside an "acoustic channel". Is this sufficient to explain how a sound can reach Bermuda after being emitted from Australia? One can ask that. The mechanism we have described certainly confines the sound in the vertical plane, but still other directions of propagation remain possible. A sound wave emitted in a given place in the ocean at a time t can propagate a distance R of the order of ct, where c is the average speed of the sound in the water, say 1,500 m s^{-1}. Even given the assumption that there are no losses, the energy of the sound wave has to be distributed approximately on a surface of a cylinder, namely, $2\pi Rh$, h being the height between the upper and the lower walls of the acoustic channel that

Figure 8. An example of mirage in the Libyan desert. As the light rays get closer to the ground and approach layers of ever warmer air (and therefore of increasingly small refraction index), they are progressively deviated, in a way similar to the sound rays in Fig. 7, until they are reflected. Then, an observer would think they are seeing a layer of water along the direction of these rays.

may be of the order of the ocean depth itself. The intensity of the sound decreases as $1/R$ when the distance from the source increases. This decrease is less rapid than, $1/R^2$, that of the sound in the air (see Fig. 2), still leaving some doubts that the propagation of a sound can reach Bermuda starting from Australia. However, if these islands should fall in a region where the sound rays are focused (Fig. 7), then the height h can still weaken it. Even if one should think that the horizontal variations of temperature and salt amount can contribute to the vertical reflecting walls, still it is amazing that the sound can reach Bermuda, considering it must overcome obstacles such as the Cape of Good Hope, and that in the ocean the plankton and air bubbles should cause a certain amount of absorption.

The sound propagation along natural submarine channels is not the only example of waveguides realized by nature. Electromagnetic waves offer other examples. The most spectacular is that of a mirage: this is the consequence of a non-rectilinear propagation of light in an atmosphere warmed in an irregular way (Fig. 8). One could also recall the short radio waves that can travel long distances, thanks to the reflection onto the ionosphere in between 60 and 800 km altitude. Within certain conditions, a receiver can receive a signal emitted by a radio station located in another country.

Chapter 3

The Colours of the Sea and the Sky

The sky is blue in good weather and red at sunset. A few hours after dark, the sky sparkles with stars. During daytime, the clouds are white or more or less grey. When it rains, sometimes we see a rainbow... Which physical principles explain all those colours? In this chapter, we shall address these questions. In addition, we will study the flying habitants of the sky: the birds and the insects.

Seas and skies offer a variety of colours that have inspired many artists. The Russian painter Arkady Rylov has represented several of those colours in a painting exhibited at the gallery Tretyakov in Moscow (Fig. 1). White and somewhat dark clouds appear in the sky with different tones of blue. The surface of the sea is a darker blue. The body of the waves is almost black, while their tops are often white, thus forming "white horses".

The Colour of the Sea and the Strength of the Wind

The number and the amplitude of the small waves looking like white horses depend on the speed of the wind. This information is important for sailors: to determine it, they have a table of empirical correspondences written by the British Admiral Sir Francis Beaufort (1174–1857). For instance, the presence of a small number of small white waves indicates a wind blowing between 12 and 19 km h^{-1}, namely 7 to 10 knots. This is defined as a gentle breeze, and it corresponds to force 3 according to the Beaufort scale (Table 1).

The luminosity of the sea's surface is also related to the angle of observation. In fact, a light ray hitting the surface of the sea is in part reflected and in part refracted (see Chapter 2). The fraction of the light reflected, among other things, depends on the refraction index of the water and the angle of incidence of

Figure 1. "In the Blue Space", the painting by the symbolist artist Arcady Rylov (1870–1939).

Table 1. Beaufort scale.

Force	Terminology	Speed of the wind at the height of 10 m in knots	Speed of the wind at the height of 10 m in km/h	Appearance of sea
0	Sea calm	<1	<1	The sea like a mirror
1	Very gentle breeze	1 to 3	1 to 5	A few waves without any foam
2	Gentle breeze	4 to 6	6 to 11	Very small waves that do not break
3	Small breeze	7 to 10	12 to 19	Small waves (<60 cm), some sheep-like
4	Sensible breeze	11 to 15	20 to 28	Small waves (<150 cm) many sheep-like
5	Good breeze	16 to 20	29 to 38	Moderate waves and weak sprinkling
6	Fresh wind	21 to 26	39 to 49	Billows (4 m), crests of foam and sprinkling
7	Strong winds	27 to 33	50 to 61	Trains of billows up to 5.5 m that initiate to break
8	Hits of wind	34 to 40	62 to 74	Tornado-like foam and sprinklings, waves as high as 7.5 m
9	Strong hits of wind	41 to 47	75 to 88	High waves break and the billows reduce the visibility. Waves can be as high as 10 m
10	Storm	48 to 55	89 to 102	Waves up to 12.5 m with crests covered by foam and sprinklings
11	Violent storm	56 to 64	103 to 117	The sea is covered by foam and sprinklings moved by the wind, waves can reach 16 m
12	Hurricane	>65	>118	The sea is white, the air is full of foam and sprinklings, and the visibility is practically zero

Note: An idea of the wind speed can be found by resorting not only to the Beaufort scale but also to the contrast between the lightening of the sea compared with the sky. They are the same when the sea is totally calm, in which case, the horizon is hardly distinguishable. Usually, even a gentle wind can move the surface of the sea, making a certain contrast observable: thus, the sky is lighter than the sea, and the horizon appears as a well-defined line. The contrast has been studied for many years by Russian scientists on board the research ship R/V *Dimitri Mendeleev*.

the light ray. The more oblique the angle of incidence, the more the reflection is increased. Therefore, the sea's surface appears brighter near the horizon than near the observer.

What about the colour? On the surface, it is hard to define since it depends on a number of factors including the depth, the position of the sun, the colour of the sky, the presence of particles or algae in suspension, etc. All these factors affect the reflection at the surface, the scattering of the light inside the sea, and its absorption. But most of the time, the sea is blue. The reason for this is because water absorbs less light between 400 and 500 nm (blue) than the rest of the visible spectrum (see Panel below on this page). Yes, water absorbs little blue! A glass of water appears perfectly transparent. Only after a thickness of some metres does the absorption of the water become noticeable.

Colours of the Sky in a Good Weather Day

While the colour of the sea can be difficult to predict, in good weather, the colour of the sky can be explained according to the principles developed by English physicist Lord Rayleigh (1842–1919). In the absence of clouds, the colour of the sky is determined by the interaction of sunlight with components of the Earth's atmosphere, namely with inhomogeneities (fluctuations) of the density of nitrogen and oxygen molecules.

How do these molecules react when irradiated by sunlight? Let us refer to a monochromatic incident light, namely light of a defined wavelength or in other words of definite frequency ν and hence defined colour.

The vision of the colours

The human eye is sensitive to electromagnetic radiation in the range of 400–700 nm (see figure). The colours of objects are related to the light that they emit. This emission occurs because they are somewhat hot (like a piece of iron that is red) or because they scatter part of the light received from other sources. The light arriving at our eyes is in general polychromatic, namely it is the superposition of different wavelengths in different proportions. This composition fixes the so-called "perceived colour". An object absorbing all radiation thus appears black: an object emitting all electromagnetic radiation of different wavelengths between 400 and 700 nm with comparable intensity appears white.

(*Continued*)

(*Continued*)

Inside the eye, the colour perception is granted by cells called "cones" that cover all of the retina. There are three types of cones (see figure), and they transmit to the brain the signal that makes it to yield the sensation of colour.

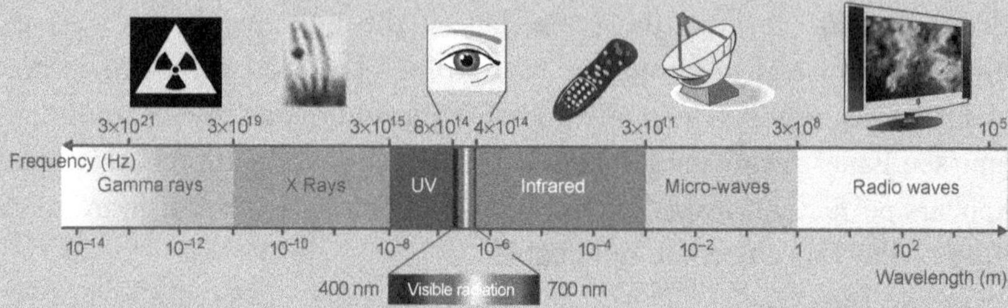

The different intervals of electromagnetic radiation and their applications. The narrow range, which extends between 400 and 800 nm (at frequencies between about 800 and 400 THz), corresponds to the visible range. Each radiation or "spectral colour" is identified by its wavelength λ, which is related to the frequency ν by the relation $\lambda = c/\nu$, where c is the speed of light in a vacuum.

The colours that are perceived are not only the ones of the rainbow, namely the "spectral colours" that one can obtain by decomposing the white light. The magenta colour, for instance, is obtained by combining red light around 680 nm with blue light around 480 nm. Furthermore, the same perceived colour can result from the composition of very different components. For example, an object can appear yellow because it emits monochromatic radiation of wavelength around 580 nm or by irradiation of white light that has been deprived of the short-wavelength radiations or as well by red light with superimposed green light.

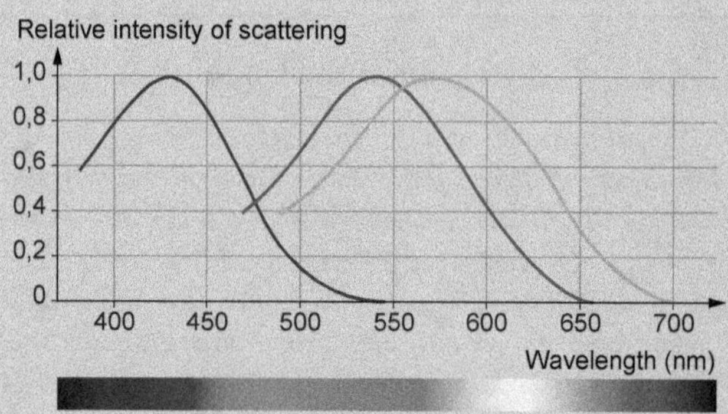

Sensitivity of the three types of cones versus the light wavelength.

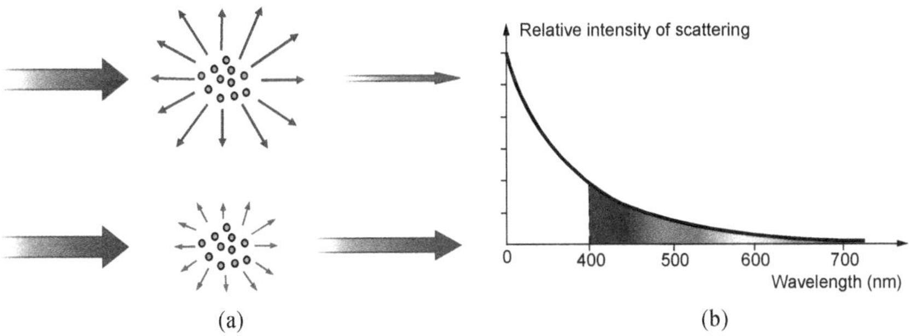

Figure 2. Relative intensity of Rayleigh scattering. For monochromatic incident light, (a) the molecules themselves emit radiation of the same frequency in all the directions. Blue light falls at a wavelength around 450 nm and red at about 650 nm. As a consequence of the Rayleigh law, (b) the fourth power of the ratio 650/450 is 4.3, and therefore the intensity of scattering in the blue range is about four times stronger than in the red.

The light is a combination of a magnetic and an electric field, oscillating at frequency ν, both of which are perpendicular to the direction of propagation. Due to the effect of the oscillating electric field, the electrons of the molecule also oscillate at the same frequency. Therefore, the molecule itself emits light at the same frequency ν in all directions (although not always with the same intensity in different directions). It is by a similar mechanism that TV antennas or radio transmitters emit radio waves. In the case of sunlight, the wavelength is much greater than the size of the molecules, and even variations in their density: we speak of Rayleigh scattering. The calculation shows that for a given intensity of incident radiation, the intensity of the scattered light is proportional to ν^4 (or to $1/\lambda^4$): this is the Rayleigh law (Fig. 2).

What is the relationship to the colour of the sky? As a consequence of this law, the intensity of the scattered electromagnetic radiation is much greater for the high-frequency components rather than for the low-frequency ones. Therefore, the molecules of the atmosphere scatter more the blue in comparison to the red, green or yellow components. Thus, it is the blue light rays which mainly arrive at our eyes. This is why the sky is blue! One could object that according to this argument, the sky should be violet since violet falls at a frequency even higher than blue. It is true that the eye receives more violet than blue, but there are also components of other colours, in particular, green: the final visual result is a matter of physiology (see Panel on page 27).

The Sky at Sunset and Twilight

At sunset, the sky close to the horizon turns a beautiful red colour (Fig. 3). Once again, this colour is related to the scattering of sunlight by the atmosphere.

(a) (b)

Figure 3. (a) During daytime, a sky free of clouds appears blue since the molecules of the atmosphere strongly scatter the blue components of sunlight. (b) At sunset, the sunlight that arrives at our eyes has passed through a large layer of the atmosphere.

The light is being scattered in all directions, and we receive a part of it, while another fraction is irradiated into space. This latter portion is small, but not negligible (Fig. 4).

In the range of visible radiation, the difference between the energy received by the upper layers of the atmosphere and that arriving at the ground is, in large part, due to scattering. We find that during daytime, the energy arriving at the ground is about 25% in the blue range while 10% is in the red. At sunset, these proportions are modified since the light has to pass through a very large part of the atmosphere (Fig. 5). The blue light is almost totally scattered, and an observer on the ground practically receives only red radiation. After the sun disappears beneath the line of the horizon, night progressively takes hold. The colour of the sky during nighttime is another matter (see Panel on page 32).

The Colour of the Clouds

In the Rylov painting, the clouds are white, grey or almost black, depending on their thickness and also from the position we are seeing them. In all cases, they are

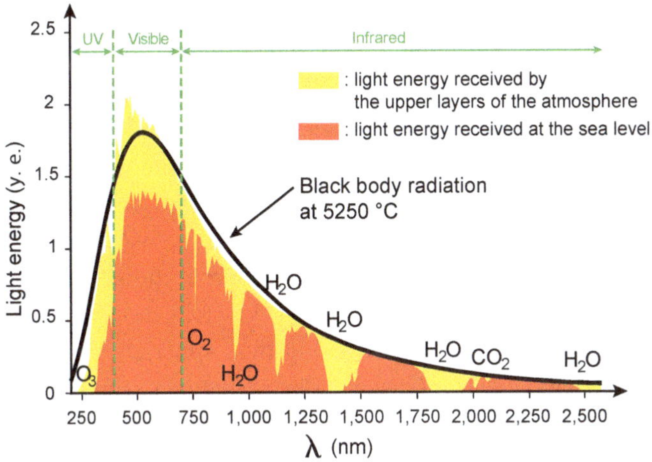

Figure 4. Light energy received by the upper layers of the atmosphere (in yellow) and at sea level (in red) during the day, considering scattering and absorption. The labels H_2O and O_2 report the absorption bands of the water and the oxygen, respectively. The energy on the *y* axis is reported in W m^{-2} of the surface and for nm of the wavelength.

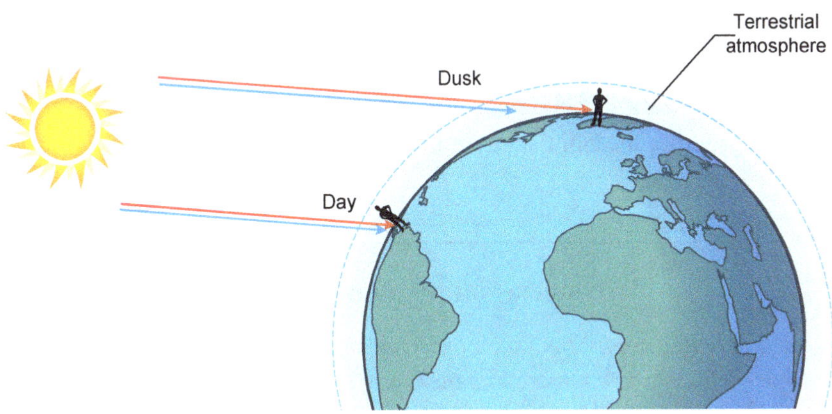

Figure 5. The radiation at long wavelengths emitted by the Sun in the red range is less scattered than the short-wavelength radiation in the blue range. Therefore, falling at dusk at a low angle, the blue ray is strongly scattered by the atmosphere and arrives at the Earth muffled unlike red. During the day hours, all the colours of solar light reach the Earth surface (proportions not real).

opaque: we do not see the Sun through a cloud, but we receive its light more or less strongly, depending on the cloud's thickness. This light is transmitted after scattering by the drops of water that form the cloud. This scattering is very strong, much more so than that by the variation in the density of the oxygen or nitrogen molecules discussed above. Why? The basic reason is that large objects scatter radiation more than small ones. For example, if a drop of water contains a million molecules (corresponding to a diameter of 0.04 micron), it scatters about a million times more strongly than would happen with a million isolated molecules. Should we then suppose that if the drop contains a billion molecules, then it would scatter a billion times more than if we were considering a billion isolated molecules? Not exactly! The diameter of this latter drop would be 0.4 μm, no longer negligible compared to the wavelength of visible light. In this case, the Rayleigh law no longer holds since the rays scattered by the different molecules are no longer in phase and cause destructive interference, as we shall describe in the following. The evaluation of the intensity of scattering by a spherical drop of radius R was performed by German physicist Gustav Mie in 1908. His result is an infinite series of terms that can be evaluated numerically. For a small drop ($R \ll \lambda$), we can limit the attention to the first term, corresponding to Rayleigh scattering. When the radius of the drop increases, we have to take into account more and more terms. When $R \gg \lambda$, the case is again rather simple: in this condition, geometrical (or ray) optics hold. What is found using geometrical optics? It predicts that the amount of energy intercepted by a sphere of radius R is proportional to the cross-section, namely to R^2. A large droplet intercepts a larger amount of energy than a small one. Furthermore, the calculation shows that the total intensity of the scattering by large droplets does not depend on the wavelength of the light. This explains why scattered light appears to become white when the incident light is white. Thus, sunlight being white, the light scattered by clouds is also white.

Mysteries on a moonless night

During a moonless night, the sky is black, apart from a few scattered stars. This can seem normal, but we remark that there are a huge number of stars, possibly an infinite number. An infinite number of stars should imply an infinite amount of light. Could the black sky suggest that the Universe is finite? This is what a German scientist, Johannes Kepler, thought at the beginning of the 17th century. During the 19th century, another German, Heinrich Olbers, argued that the closer stars could mask the ones farther away, so that, even in the case of infinite Universe, the intensity of light during the night would still not be infinite, even if it would be very great! The contemporary explanation is that the universe is not infinite in time. Since the Big Bang, a very dense and hot period about 13.8 billion years ago, the Universe has been in continuous expansion. As a consequence, the

(*Continued*)

(*Continued*)

light emitted by remote galaxies is shifted towards the red. When we view something from far away, we are looking back in the time since the speed of light is finite: thus we see the galaxies in the state they were in when their light was emitted, namely billions of years ago. From a certain distance, we can look back in time to almost the moment of the Big Bang, when the galaxies were not yet born and the Universe was dark: we reach the "cosmological horizon", the limits of the Observable Universe. We therefore cannot observe the entire Universe, whether finite or infinite, and the night sky appears black. In reality, it is not exactly black but full of electromagnetic radiation of wavelengths much greater than visible light (of the order of magnitude of millimetres instead of micrometres). This weak radiation is not visible to our eyes. It can be measured by a radiotelescope of sufficient sensitivity (see figure). Its fortuitous discovery by the USA scientists Arno Penzias and Robert Wilson in 1964 granted them the reward of the Nobel prize for Physics in 1978. It represents "a diffuse cosmological radiation" that is not emitted by the stars, nowadays, known as the Cosmic Microwave Background or CMB. It is the result of the expansion of the Universe following the Big Bang, and it is itself expanding. Thus, the wavelength of this radiation is increasing with increasing time.

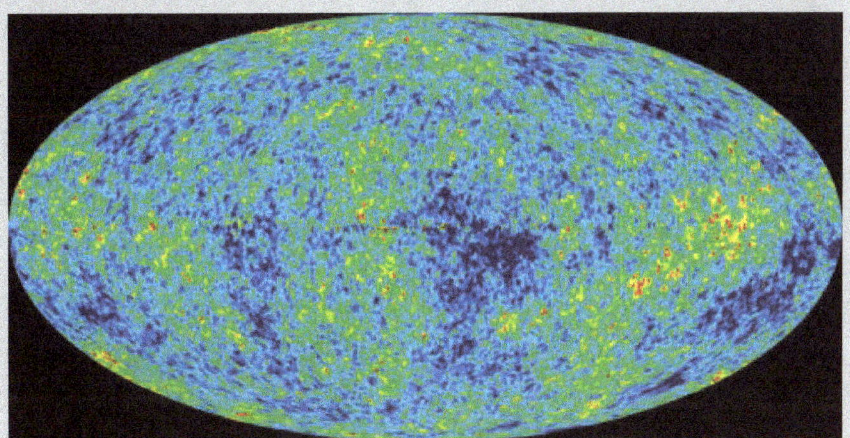

The first image of the Universe, namely the cosmological microwave background radiation, about 14 billion years old. The picture reflects the temperature fluctuations present in the Universe about 380,000 years after its birth, which correspond to the embryos of future galaxies. After the discovery in 1965, this relic from the very early universe has been studied by means of radiotelescopes on the Earth and by instruments carried by the satellites or balloons. This figure reports the first detailed image obtained by the satellite Wilkinson-Microwave Anisotropy Probe.

Interference and Coherence

The phenomenon of interference was demonstrated in the 19th century by means of a famous experiment performed by the English scientist Thomas Young. At that time, physicists were divided as to the nature of light: was it a wave-like phenomenon, as Young's experiment suggested or a flux of particles? In the fourth part of the book (see Chapter 22), we shall emphasise how both interpretations were correct.

The equipment used by Young (Fig. 6) included a monochromatic point-like light source placed behind an opaque plate with two small slits (of the order of 0.1 mm) spaced a little apart (a few mm). The light passing through the holes falls onto a screen. The image on the screen was a surprise: not a specific region of continuous light but a series of alternate light and dark. Why is this?

The intensity of the light at a given point M of the screen is the result of the superposition of waves passing through holes A and B. This phenomenon of the *algebraic* sum of the waves originating from different points is called interference.

The result of the sum can lead to a null or a very weak intensity. In this case, one speaks of destructive interference. It could also lead to a stronger intensity: the interference then is called constructive. The constructive or destructive character of the interference depends on the displacement of waves one from the other, namely the phase shift when they arrive at the screen (Fig. 7).

Along the axis SO, the waves due to holes A and B arrive in phase, and thus one observes a bright fringe. Once we move away from this axis, depending on the point we are referring to on the screen, the waves coming from the holes will have

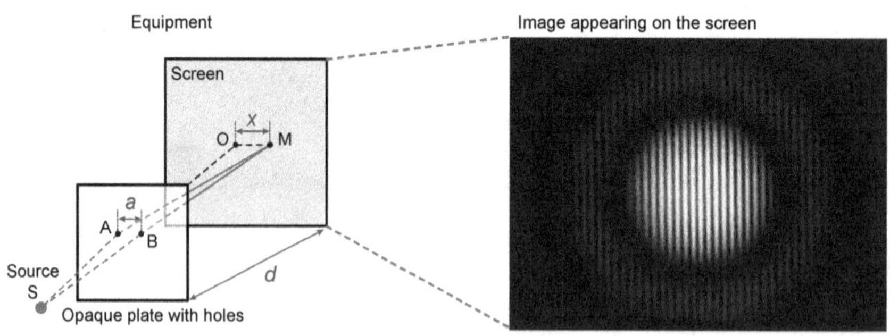

Figure 6. The Young experiment showing interference with the two holes. A source of coherent light illuminates the holes: on the screen, one observes alternating dark and light fringes. The rays arriving from the two holes A and B are said "to interfere". With a single hole, one would observe a single spot that cannot be a neat circle because of diffraction.

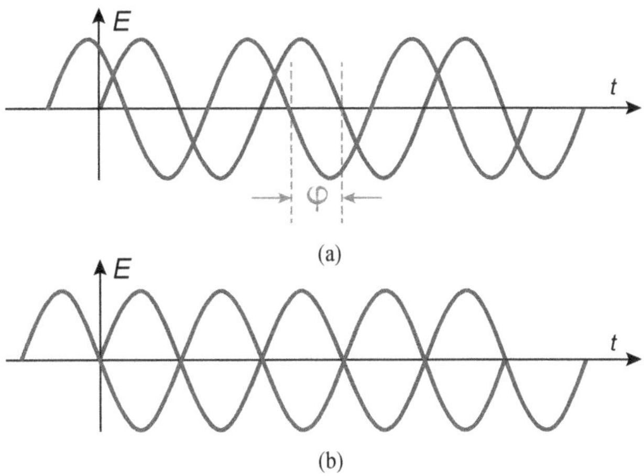

Figure 7. (a) Two waves having the difference φ in their phases. (b) Two waves of opposite phases interfere in a destructive way; the maximum amplitude of one corresponding to a minimum in the amplitude of the other.

travelled different distances. Their phases are becoming different, and a periodic sequence of bright and dark fringes is observed. We see destructive interference (dark fringes) when the difference in distance travelled is equal to a half-wavelength or to a multiple of that. We see constructive interference (bright fringes) when the difference in the paths is a multiple of the wavelength.

In the visible spectrum, the wavelength is about a micrometer, namely about one-tenth of the diameter of a human hair. However, the difference between two fringes on the screen is larger than that when the screen is at an appropriate distance d. The position $OM = x$ of the bright fringes can be obtained from the condition $AM - BM = n\lambda$, with n being an integer. If $a = AB$ is the distance between the holes, one finds that the distance between the fringes is $\lambda d/a$. For $\lambda = 0.5\ \mu$m, $d =$ 3m and $a = 0.5$ cm, then the fringes are 3 mm apart. Thus, we can distinguish the fringes with the naked eye, even though it is not easy. Experiments to show interference can be performed by students in a laboratory. We should therefore admire Young for his successful achievement in its original detection.

Instead of using two holes with the same bright source, we might try to use two different point-like sources emitting the same colour. However, in this case, the experiment fails! Interference is possible only when the sources are *coherent*, namely their phase difference maintains a constant value over time. Yet, unless we take specific precautions, two sources selected at random are not coherent. This difficulty in observing light interference could lead the reader to think it is an exotic phenomenon. This is not true! The colours of a soap bubble are an example (see Chapter 6). In that case, the interference is produced between the light directly reflected from the surface of the film before the entrance into the bubble and the ray reflected before its exit. Since the observation is ordinarily made by resorting to white light, the radiation that is in phase opposition cannot be seen, and so the light appears coloured. The particular colour depends on the position of the observer with respect to the ball and on the depth of the film. These *interferential colourations* seem to occur on the wings of butterflies such as the *Morpho*, and the plumage of the hummingbird as well as the cuticles of certain insects.

The Colours of the Rainbow

The rainbow is the result of the interaction of sunlight with water drops suspended in the atmosphere (Fig. 8). These drops of water have a size of the order of 0.1 mm, much larger than the wavelength of the light. As a consequence, the trajectories of the rays within the drop can be described by resorting to geometrical optics, namely one refraction at the entrance and another refraction at the exit, possibly separated by one or more reflections. The principal rainbow, usually the only observable one, corresponds to an internal intermediate reflection, while the secondary rainbow, sometimes observable, is due to rays that have been reflected twice inside the drop of water (Fig. 9).

Figure 8. Principal rainbow (the brighter, on the right) and the secondary rainbow. It is noted that their colours have a reverse order.

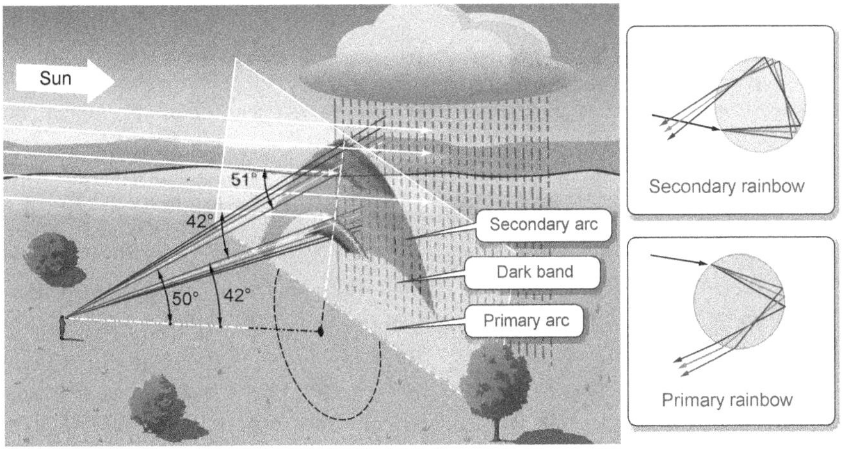

Figure 9. Trajectories of the light rays for a principal rainbow and a secondary rainbow. The average angles of deviation are 42° and 51°, respectively. Consequently, the rays that create the principal rainbow form a cone of revolution, its axis being given by the right direction sun-observer. From B. Valeur, *La couleur dans tous ses éclats*, Belin (2011).

For a given wavelength, the deviation of a ray from a water drop depends on the place where it hits the drop and is determined by the refraction law. Note the angles of exit and entrance cannot take arbitrary values. For the principal rainbow, the angle is between 0° and a value of the order of 42°, as the reader can confirm with a few calculations. Deviations by an angle less than 42° are permitted, but in the vicinity of 42°, the intensity is at a maximum. The same happens for the secondary rainbow in correspondence to the angle of roughly 51°. With the Sun behind us, we see two circular shining arcs. Between these two arcs, there will be a dark band: in fact, no ray can come out of the drop after one or two reflections within the range of the two angles given above. Thus, in the area between these two angles, the sky appears dimmer than elsewhere.

We have explained the occurrence of the two arcs but not yet their colours. In reality, the angle of deviation depends on the colour since the refractive index of the water increases with decreasing wavelength. Therefore, for a given angle of incidence, the angle of refraction increases with the wavelength, that is, moving from the blue towards the red. Thus, the deviation at the entrance and exit from the drop for the blue is larger than for the red. So, red is found at the exterior of the principal rainbow. The opposite happens for the secondary arc where the colours are interchanged: the red is at the interior. These peculiarities, related to the geometry and the law of refraction, are examples of the surprises that scientific calculations offer.

Snell, Descartes and Fermat

Let us return to the law that in some places is named Snell–Descartes or simply Descartes while elsewhere is the Snell law. It would seem that Descartes was the first to have published his treatise "Dioptrique" in 1637 although it was already known by the Dutch mathematician Willebrord Snell (1580–1626).

Snell was likely relying on experimental data, while Descartes wanted to demonstrate his law by assuming that the propagation of light rays was similar to the trajectories of a ball. Descartes' formulation, not so understandable, was criticised by Pierre de Fermat in a paper published in 1662, entitled "Synthése pour les refractions". The "Fermat principle" formulated in that manuscript stated that light follows the path that takes the shortest time to move from point A to point B (see figure). We leave for the reader the exercise to derive the Snell law on the basis of the Fermat principle. This should not be difficult if somewhat familiar with trigonometry and calculus. It is only necessary to search for the point C that minimises the time required by the light to travel along the trajectory ABC, which is equal to AC/c + BC/v, where c is the speed of the light in air and v = c/n the speed in water. The formulation by Descartes is mainly interesting from an epistemological point of view, while the Fermat principle also has a certain interest for modern Physics. In any case, it was Descartes who first explained the genesis of the two arcs in the rainbow, additionally providing an estimate of the corresponding angle of deviation.

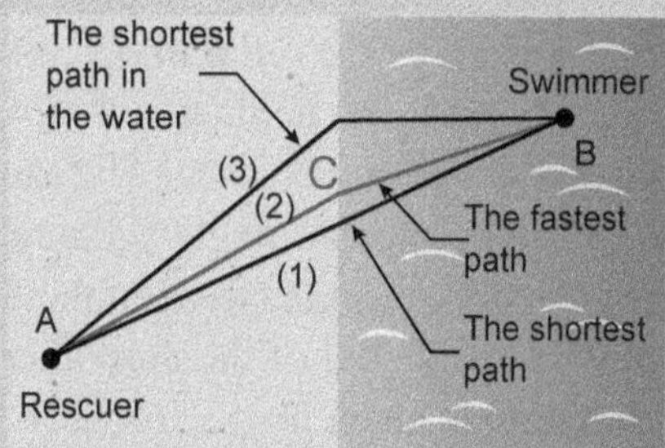

Analogy with the problem of a rescuer A who runs faster on the beach than they swim in the sea and must save a person B in danger, as quickly as possible. The shortest path, a straight line, is not the fastest since the rescuer has to spend too much time in the sea (1). If they limit the part of the path in the sea as much as possible (3), then they have to increase the path on the beach too much. The fastest path turns out to be the one (2) predicted by the Snell–Descartes.

What About the Birds?

In our analyses of the Rylov painting, until now, we have not taken into account the birds, which contribute to the charm of the sea. Now we will make up for that with the following riddle: knowing the mass of a bird, what is the frequency of the wing beat required for flying? Maybe the reader will not wish to search for the relation between the two quantities and will give up the attempt to understand?

Let m be the mass of the bird, S the total area of the wing, v the average velocity of a wing, t the time for a full beat, and ρ the specific mass of the air. During one beat of the wing, the bird forces a mass of air given by $M = \rho Svt$ to move, giving it a speed v, corresponding to an average acceleration v/t, and therefore to a force $F = Mv/t = \rho Sv^2$ that must be in equilibrium with the weight mg of the bird, g being the gravity acceleration. Thus,

$$v = \sqrt{\frac{mg}{\rho S}}.$$

The velocity of the wing is proportional to the number ν of beats of the wing per second, and at the length of the wing, i.e., to the square root of the surface $\sqrt{S}$. By assuming, with certain arbitrariness, that the proportionality factor is 2π, one derives

$$\nu = \frac{1}{2\pi S}\sqrt{\frac{mg}{\rho}}.$$

The mass of a heron (Fig. 10) is of the order of 1 kg. Its wingspan is about 2 m, and one can assume $S = 0.2\ \mathrm{m}^2$. With the approximate values $\rho = 1\ \mathrm{kg\ m}^{-3}$ and

Figure 10. The wings of the ash-coloured heron have a surface of the order of 0.2 m^2.

$g = 10$ m s^{-2}, the frequency of beats is about 3 bps, in good agreement with observation (from 2 to 3 bps during flight). Going further, let us assume the bird's body is roughly the same shape and with the same density. Then the area of the wing is proportional to $m^{2/3}$, and thus we can deduce that the frequency of the beats is proportional to $1/m^{1/6}$ from the equation above. In reality, we observe that the frequency of the beats decreases when the mass of the bird increases: the sparrow, whose mass is of the order of 20–30 g, flies with a frequency of 13 bps, the dove (having a mass of the order of 500 g) with one around 8–9 bps, while the buzzard (mass of the order of 1 kg) at 3 bps.

And what of insects? In the Rylov painting, we do not see any insects since they are too small. We would guess that the frequency of beats is very strongly increased for insects compared with that of birds, and this is in qualitative agreement with our estimate. An extreme situation is that of the mosquito which flies at about 400 bps. This is a pretty high frequency, meaning an audible sound is associated with the beats, and thus that we can hear when a mosquito attacks. Knowing that the mass of a mosquito is 2 mg and assuming that its wings have an area of the order of 10 mm^2, we would conclude that the frequency is about a factor 10 greater than the one evaluated by our formula. This should not be a great surprise, taking into account that such a small mass may be outside the threshold of validity of our description, being only approximately valid. Rather, it should be astonishing that it yields insights that are qualitatively correct even for quite different orders of magnitudes, from large birds to small insects. Might the painter Rylov have suspected, while painting his tableau, that he would evoke so many laws of Physics?

Chapter 4

Foucault's Pendulum and the Coriolis Force

By the beginning of the 19th century, practically everywhere in the world held the firm conviction that the Earth was spherical and was rotating on its axis. However, a real experimental proof was still missing. The demonstration was provided by a famous experiment suggested by Léon Foucault. The rotation of the Earth around its axis explains several meteorological and oceanographic phenomena, in particular. In order to understand these phenomena, we need to model them. For this, physicists appeal to a fictitious force that was introduced by Gaspard Coriolis.

In the year 1851, the Pantheon, in Paris, was the scene of an experiment carried out by physicist Léon Foucault (1819–1858). A ball weighing 28 kg was suspended at the top of the dome by means of a wire 67 m long, in this way, forming a pendulum somewhat similar to the one in an old-fashioned grandfather clock (Fig. 1). However, a pendulum can only swing in a given vertical plane: Foucault's pendulum could oscillate along any direction.

The experiment consisted of shifting the pendulum away from its equilibrium position (vertical) and then letting it oscillate. Friction being rather weak, the pendulum could oscillate a pretty long time without a noticeable decrease in the amplitude of the oscillations. What would the experimenters observe? In the beginning, it seemed that the pendulum remained swinging in the same vertical plane, the one defined by the axis of the pendulum and its initial position, as would be expected in light of the simple calculation learned in high school. However, after several minutes, it could be seen that the plane of oscillation of the pendulum was slowly rotating! And it always rotated in the same way, in a precise manner, that we are now going to describe.

Figure 1. A Foucault pendulum placed in the hall of the Pantheon in Paris, where the experiment designed by Foucault was carried out in 1851. The pendulum, once shifted from its equilibrium position, swings in a plane that is progressively rotating.

Foucault's Pendulum at the North Pole

Why is the plane of oscillation rotating? The Foucault experiment can easily be understood if we consider placing it at the North pole (or at the South pole). Let us imagine a pendulum whose equilibrium position coincides with the Earth's axis passing through the pole. We deviate it from the initial position, and allow to initiate the oscillations (Fig. 2(a)). For an observer whose position is fixed with respect to the Sun and the stars (the relative positions of the stars have to be considered fixed, namely, with a good approximation, time-independent), the pendulum oscillates along a fixed vertical plane (Fig. 2(b)). This is not the case for a terrestrial observer, since the Earth rotates around its axis, here assumed to be along the vertical at the point where the pendulum is suspended.

Then the Earth rotates with respect to the plane of oscillation of the pendulum, and the terrestrial observer, who does not realise he is rotating with the Earth, gets the impression that the oscillation plane of the pendulum is rotating; the ball appears to deviate towards the west. Remember, for us, the Sun appears to rise in the east and move westward across the sky, whereas in fact the globe rotates east while the Sun remains almost motionless. The reader can figure out that at the

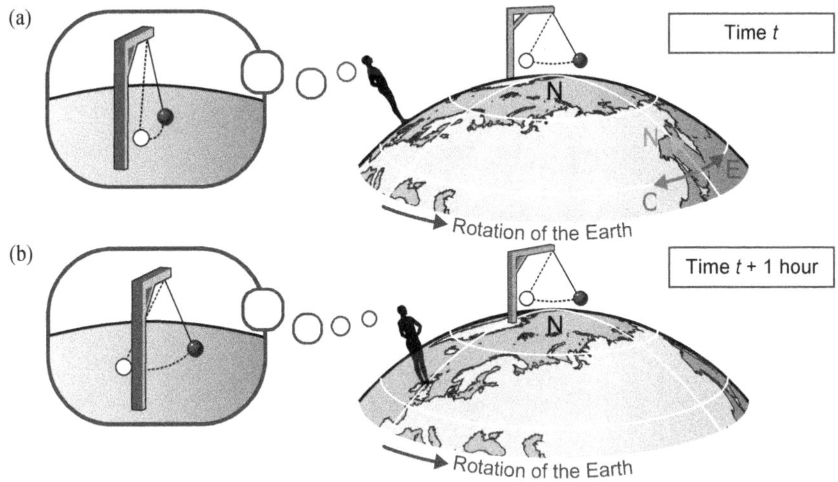

Figure 2. Evidence for the Earth's rotation by means of a Foucault pendulum. (a) Initial position of the pendulum at the North pole and of the observer located on the Earth. (b) After 1 hour, the Earth has rotated with respect to the stars (towards the east, as indicated by the arrow); the support is rotated, but the oscillation plane of the pendulum stays fixed. For the observer located on the Earth, the support has maintained its initial position, while the plane of oscillation of the pendulum yields the impression of having turned.

South pole, the deviation would be towards the east. For an object in vertical free-fall at the pole, there will not be any apparent deviation related to the Earth's rotation. On the contrary, at the equator, the apparent deviation of a falling object seen by an observer on the Earth would be at a maximum and towards the east (see the following). Returning to the pendulum oscillating at the pole, the pendulum will complete a full rotation in 24 hours. In contrast, in London, it would take more than a day to see the plane of oscillation complete a full rotation. Again, we shall discuss these aspects later on.

The Force Devised by Coriolis

It may be annoying having the requirement of an observer immobile with respect to the fixed stars in order to figure out what is seen by an observer on the Earth, who is rotating with the Earth itself. It would obviously be simpler for this terrestrial observer to discuss the motion of the pendulum by just considering that a force is acting on the ball which pushes it towards the west. We can indeed make this choice! Then we can analyse the motion of the Foucault pendulum in a frame of reference tied to the Earth, namely a "terrestrial frame of reference". In order

Figure 3. Gaspard-Gustave Coriolis (1792–1843). This portrait, made by Belliard from a painting by Roller, is one of the few known nowadays. The name of Coriolis, together with another 71 scholars, is celebrated in an engraving on the first floor of the Eiffel tower.

to account for the Earth's rotation, we have to assume that another force, besides gravity, is acting on the ball. This pseudo-force is known as "Coriolis force", after the French mathematician Gaspard-Gustave Coriolis (Fig. 3).

In order to go into more detail, let us now forget about the pendulum, where the oscillations complicated the understanding, and rather let us consider a bullet of mass m fired by a rifle. For further simplicity, let us assume that the bullet moves forward along a straight line, and we shall be concerned only with the effect of the Earth's rotation on its trajectory. This assumption is indeed unrealistic, as we shall see, just being a device to simplify the analysis. Let us imagine that the shooter is located at the North pole, in N (Fig. 4).

A bullet P having initial velocity v is fired towards the target C. The Earth rotates around its axis at angular velocity Ω, corresponding to one revolution a day.

After a time t, the Earth has rotated by angle Ωt and the target has been shifted with it. Therefore, from the point of view of the shooter at the North pole and looking towards the target, everything appears as though the target is fixed and the bullet has been deviated from the desired direction NC. The distance of P from the line NC, at a time t and when t is rather small, is approximately given by the product of the angle Ωt and the distance covered by the bullet, namely $\Omega v t^2$, having measured the angle in radians (the most convenient unit).

Then, *with respect to the Earth*, and for a small time, the bullet is moving linearly along the direction NC and in the meantime is being accelerated towards

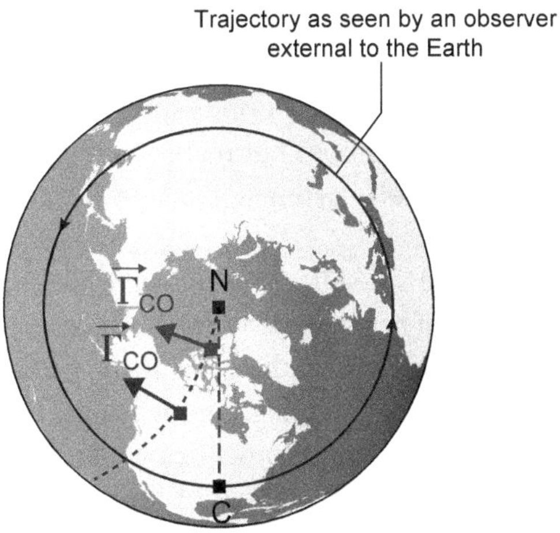

Figure 4. Deviation of a bullet due to the Coriolis force, in a frame of reference tied to the Earth. The bullet, shot along the direction of the target C from the North pole, is deflected towards the West with respect to the target. For an observer outside the Earth, the trajectory of the bullet is a straight line. For an observer fixed on the Earth, the trajectory of the bullet is deflected (the curvature being exaggerated). The bullet is subject to an acceleration F_{CO}, the Coriolis acceleration, reported at two different times.

its right, along a direction perpendicular to NC (it would occur towards its left if this were taking place at the South pole). Along this direction, the acceleration, called Coriolis acceleration, is $2\,\Omega\nu$ (having differentiated twice, with respect to t, the distance $\Omega\nu t^2$ covered by the bullet). From Newton's second law of motion (see Panel on page 46), this means that the bullet is subjected to the force $2m\Omega\nu$, directed along the Coriolis acceleration. This is the Coriolis force.

In general terms, the Coriolis acceleration is perpendicular to the instantaneous velocity (Fig. 4). Meanwhile, we have to remember that there is a vertical acceleration, directed downwards, that we have disregarded (but is not ignored when actually firing munitions!).

Let us clarify that the Coriolis force is a "fictitious" force, or, as often they say the force of inertia, because it is not caused by the physical impact of one body on another. When describing the motion of a body in a rotating coordinate system, the Coriolis force must always be taken into account when the body's velocity is not directed along the axis of rotation.

Could a shooter disregarding the Coriolis force fail to hit the target? Let us assume that the target is at a distance d of 100 m, and the velocity ν of the bullet is 1,000 m s^{-1}. The time required for the bullet to reach the target is $t = d/v = 0.1$s. The angular velocity Ω of the rotating Earth is 2π radians per day, namely 0.7×10^{-4} rad s^{-1}. The deviation y due to the Coriolis force is $y = \Omega \nu t^2 = 0.7$mm. Then the shooter does hit the target and does not realise that it has moved with respect to the line of the shot. We can also estimate the Coriolis acceleration that results in $2\Omega\nu$, namely 0.14 m s^{-2}.

In meteorology, where the velocities are smaller while the distances are much greater, the Coriolis force must be taken into account in order to describe the movements of masses of air or water. In fact, the deviation $y = \Omega \nu t^2$ due to the Coriolis force can also be written $y = \Omega x^2/\nu$ having introduced the distance $x = \nu t$ covered by the body. When the distance x is of the order of a hundred kilometres, for a velocity around 20 km h^{-1}, the deviation y turns out of the same order of the distance x. For instance, for a sea current with a velocity of 6 m s^{-1}, namely 22 km h^{-1}, the deviation due to the Coriolis force after 100 km of travel is 10 km! We shall return to the consequences of these effects in the subsequent part of this chapter.

Newtonian mechanics

At the beginning of the 17th century, the Italian scientist Galileo Galilei (1564–1642) made a first step towards the understanding of dynamics and gravitation. He showed that a free-falling body is being uniformly accelerated as it moves. At the same time, Galileo formulated the **inertia principle** *that "a body which is not subjected to any force moves along a straight line at constant velocity, if in motion, while it remains immobile if this is its current state". Nowadays, this statement appears self-evident. However, in the Aristotelean age, it was thought that a body was necessarily immobile when no force was acting on it. A short time after Galileo, the German astronomer Johann Kepler (1571–1630) discovered the laws controlling the motion of the planets: they describe ellipses with the Sun at one of the foci. The English scientist Isaac Newton (1643–1727) set out the basis for the mechanics we now call "classical", showing that those properties could be derived from a few rather simple hypotheses.*

As the first, he adopted Galileo's principle of inertia. Second guess, or second Newton law, which is often called the basic law of dynamics, says that $a = dv/dt$ of a point-like object, multiplied by the mass m, is equal to the sum of the external forces F applied to it

$$\Sigma F_{ext} = ma.$$

(Continued)

(*Continued*)

This law gives us the basic equation of motion and allows us to predict the position and speed of an object at any given time, provided that we know where it was and what speed it had at time $t = 0$. It is noted that the acceleration is defined as the derivative of the velocity with respect to the time, and it can correspond to a slowing down, at variance with normal language (in the case, when the vector **a** *is opposite to the sense of the trajectory)!*

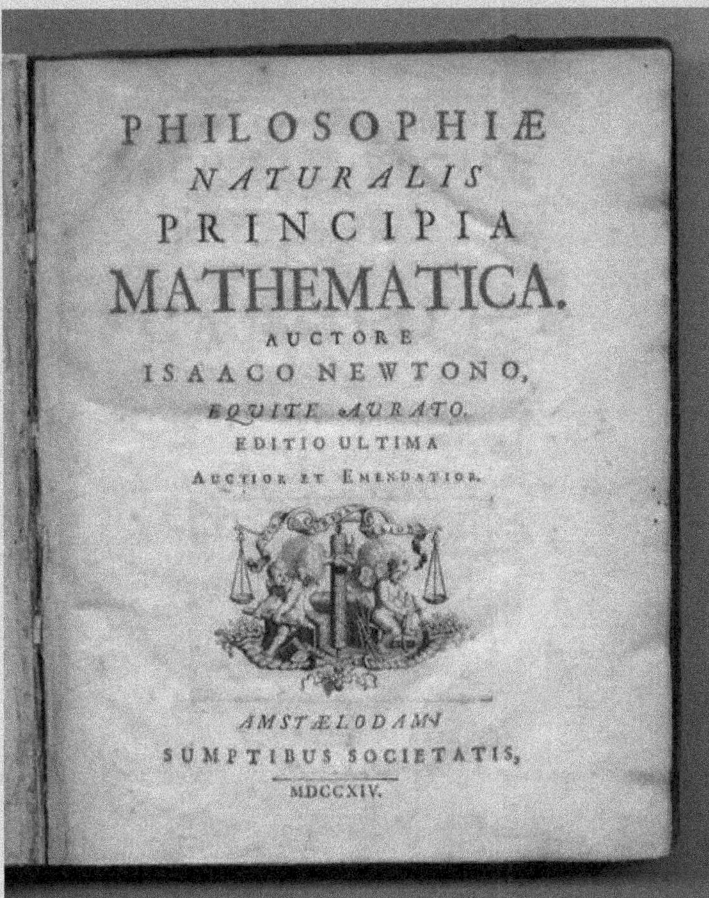

Page of the title of *Philosophiae Naturalis Principia Mathematica* (*Mathematical Principles of Natural Philosophy*) published in the year 1687, where Newton presented the fundamental results for mechanics.

(*Continued*)

(*Continued*)

It would seem that Newton's first law follows from his second law, by setting the sum of applied forces to zero. However, the ***principle of inertia*** *has its own deep meaning which gives it the status of a law. The modern formulation goes something like this: "there are certain frames of reference, called* ***Galilean****, for which a moving object maintains a constant speed when no forces act on it, or when the sum of the external forces is zero". In other words, this law allows one to select such frames of reference, in which the second and the third laws are valid.*

Newton's third law is the ***principle of action and reaction*** *and states that "all the bodies that create a force acting on any other undergo an opposite force, of the same strength and same direction of the one created by those bodies".*

Finally, Newton wrote the form of the force that the gravitational attraction creates on an object and formulated the ***law of the universal gravitation****. This states that two massive bodies attract each other by a force proportional to the inverse of the square of the distance between them.*

Newton's laws of motion and the law of universal gravitation could explain with a great precision all the problems of mechanics that were known at that time, for instance, the fall of an apple (see figure) or the motions of the planets! They have been used with a great deal of success in astronomy in order to evaluate the trajectories of satellites around the planets as well as to predict the return of Halley's comet in the year 1759. Unfortunately, real problems are often of such severe mathematical complexity that only approximate solutions are possible. This is the case, for instance, of the three-body problem (Sun, Earth and Moon, for example) under the reciprocal gravitational attractions, for which no strictly exact solution is possible.

Far from the poles

Until now, we have assumed our experiments are carried out at one pole. What happens if we move far from the poles?

With a few lines of equations, the texts of classical mechanics derive the following: for a bullet of mass m moving at the velocity v, the Coriolis force referred to the Earth is a vector perpendicular to the rotation axis of the Earth itself and to the velocity v having modulus

$$F_{\mathrm{CO}} = 2m\Omega v \sin\phi,$$

ϕ being the angle between $\boldsymbol{v}$ and the axis of rotation. In order to find the orientation of this force, one can apply the "left-hand rule" (see Panel on page 49).

The direction of the Coriolis force and the vector product

The Coriolis force acting on an object of mass m can easily be expressed on the basis of the vector product of the velocity $\vec{v}$ and the rotation vector $\vec{\Omega}$:

$$\vec{F}_{CO} = 2m\vec{v} \times \vec{\Omega}.$$

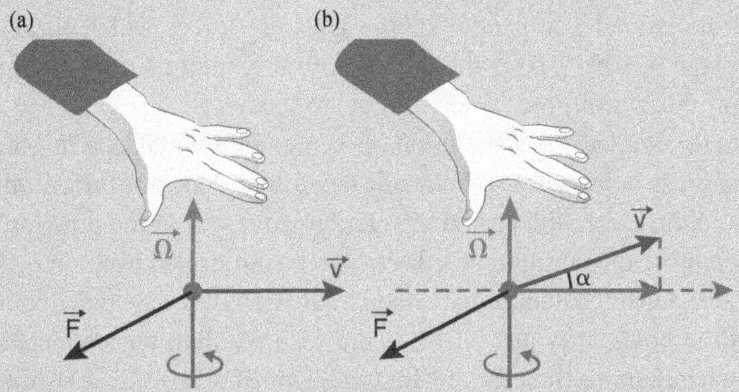

The left-hand rule in order to define the direction of the vector product $\vec{v} \times \vec{\Omega}$. The left hand is placed perpendicular to the vector $\vec{\Omega}$ so that this vector passes through the palm while the forefinger is directed as the vector $\vec{v}$. By opening the thumb, this indicates the direction of $\vec{F}$. (a) Case when $\vec{v}$ is perpendicular to the rotation axis; (b) general case.

We remind readers that $\vec{\Omega}$ is a vector associated with any rotating object (here the Earth), parallel to its rotation axis, having modulus given by the angular velocity Ω. What is the result of a vector product? The vector product of two vectors $\vec{v}$ and $\vec{\Omega}$ with angle α between them is a third vector, perpendicular to both of them and having modulus Ωv sin α. Its direction is given by the "rule of the left hand" (see figure). The vector product is a very useful formalism to describe several other physical phenomena, in particular when studying electromagnetism.

The Coriolis force has a horizontal component that causes the rotation of the plane of oscillation of Foucault's pendulum and that, in more general terms, causes the deviation of the object in motion towards the right in the Northern hemisphere and toward the left in the Southern hemisphere. However, the Coriolis force also has a vertical component, namely parallel to the vector of the force of weight of the object. This component can often be disregarded since, for the usual velocities, it is negligible in comparison to the weight.

By returning to the case of the bullet travelling at the velocity of 1,000 m s^{-1} the vertical component of the Coriolis acceleration can take a maximum value of 0.14 m s^{-2}, which is weaker than the acceleration due to gravity g by a factor of 70. For a tennis ball, which travels much more slowly, the ratio is even larger than 1,000.

Now let us go back to Foucault's pendulum and in particular to the rotation period of the plane of oscillation. From the equation $F_{CO} = 2m\Omega v \sin\phi$, we can deduce the horizontal component of the Coriolis force, which turns out to be $F_{CO} = 2m\Omega v \sin\alpha$, where α is the latitude of our experiment. This component is maximum at the North pole, where sin α = 1, and, in this case, we obtain the previous result: the plane of oscillation of the pendulum makes a full turn in 24 h. At the equator, α being zero, a Foucault pendulum does not turn at all. In other places around the Earth, the period of rotation is the one at the pole multiplied by sin α and therefore the duration of a day has to be multiplied by 1/sin α. The period of rotation of Foucault's pendulum is then a function of the latitude, being 24 h/sin α, while at the poles, it is just 24 h. At the 45th parallel, approximately close to much of Switzerland or the city of Montreal, sin α being $1/\sqrt{2}$, a Foucault pendulum will rotate slower than at the poles by a factor $\sqrt{2}$: instead of completing a revolution in 24 h, it will take 24 h times $\sqrt{2}$, namely about 34 h.

Foucault, Galileo and Aristotele

Among the many Foucault pendulums that have been functioning in the world, a special mention should be reserved for the one at the Saint-Isaac Cathedral in Leningrad, before this town became Saint Petersburg again in the year 1991. That Cathedral had been modified into a museum of atheism. What is the relationship between Foucault's pendulum and religion?

The presence of a Foucault pendulum in an anti-religious building recalls the conviction of Galileo by the Saint Office in the year 1633 (see figure). Galileo was claiming that the Earth is rotating around the Sun for over a year, while it is also rotating around its own axis once a day. The Cardinals instead were convinced that according to the Bible the Sun was rotating around the Earth with no rotation of the Earth around itself. Moreover, the following statement was claimed: "We judge and proclaim that you, Galileo Galilei, are strongly suspected of heresy with respect to the Saint Office for having believed and declared a doctrine untrue and contrary to the saint and divine Scripta, by claiming that the Sun is at the centre of the universe,

(Continued)

(Continued)

that is not moving from East to West and that the Earth is moving while is not at the centre of the world."

Galileo facing the Inquisition.

As a consequence of this judgement, Galileo was forced to renounce his theory. However, in claiming that the Earth was rotating around its axis, he did not have real experimental evidence. He had only extrapolated the observations by Copernicus and Kepler who had claimed how astronomical motions were better described in a consistent way in the assumption that the Earth was rotating! The experiment by Foucault in the year 1851 provided direct evidence of the rotation of the Earth around itself and finally ended a controversy that had begun nearly 2200 years before, when Aristotle had described a universe perfect and eternal, with the Earth at its centre.

In fact, Foucault was not the first to provide an experimental proof of the Earth's rotation. In the year 1833, the German scientist Ferdinand Reich had let some stones fall inside a deep mineshaft, detecting a deviation from the strictly

(Continued)

(*Continued*)

vertical (about 28 mm for a fall of 158 m) that was in agreement with the hypothesis of Earth's rotation, as it could be derived on the basis of the Coriolis force (it is noted that for a body in freefall, the deviation from the vertical due to the Coriolis force is towards the East, as can be calculated from the rather complicated solution of the equations of motion; the deviation would be zero at a pole and a maximum at the equator). Furthermore, in the year 1661, a few years after Galileo's death, the outstanding scholar Vincenzo Viviani had carried out an experiment very similar to the one of Foucault. Unfortunately, he could not provide a complete explanation, in part because he was unable, but also perhaps being afraid of the Holy Inquisition.

Where to see a Foucault pendulum?

A Foucault pendulum is such a fascinating object that the reader should certainly visit a museum where it is in operation. In Paris, the pendulum was returned to the Pantheon in 1995 and can be visited when is not in repair. Another one can be seen at the Museum of Arts and Metiers. In London, there is one at the Science Museum and many in America, although the one at the Smithsonian Institute has sadly been permanently removed. Why not make your own Foucault pendulum? Unfortunately, the difficulties are rather serious. First, the pendulum should be sufficiently big that the effects of the friction can be limited. The one in Leningrad is 98 m long. Thus you would need a support sufficiently tall from which to suspend the pendulum. Once the pendulum is properly suspended and then set in motion, the pendulum has the tendency to take the privileges of freedom that are not allowed to its counterparts in common watches: it could develop some twisting, and this would increase the friction. Even worse, instead of remaining in a plane of oscillation that slightly rotates, the pendulum will have a tendency to describe a cone around that original plane. The difficulties related to the phenomena recalled above can be appreciated by the students who visit the pendulum at the University of Grenoble, for instance.

Another Fictitious Force: The Centrifugal Force

Another example of force being "fictitious" is provided by the *centrifugal force*, where the consequences are more easily realised than those due to the Coriolis force. That force appears in a rotating frame of reference with respect to the fixed stars, and it tends to force the objects outward from the centre of rotation of this frame. In everyday life, for example, that force allows us to dry household linen in a washing machine: the water is forced out through the holes of the rotating drum. We also experience the centrifugal force when travelling in a motorcycle (Fig. 5) or in a bus when going along a narrow curve, and we are projected towards the outside of the curve and often finish up on a neighbouring traveller.

What is the value of this centrifugal force? Let us refer to a child on a merry-go-round of radius R rotating at angular velocity ω radians per second. The child lets go of a teddy bear. By neglecting the weight, the bear moves along a straight line at the speed $v = R\omega$ with respect to the ground (by definition of angular

Figure 5. The motorcyclist is leaning towards the interior of the curve in order to cancel the effect of centrifugal force. In a narrow curve, the speed of rotation is increased, and therefore the centrifugal force also increases and becomes relevant.

velocity). For a short time, for the child, the bear travels by uniform accelerated motion, with acceleration given by $v^2/R = \omega^2 R$. This acceleration is related to the fact that one describes the motion of the bear with respect to the rotating frame of reference. From the fundamental principles of dynamics in the rotating frame, the bear of mass m would experience a force $\boldsymbol{F}_{ce}$ arising from the centre of the merry-go-round and having modulus $m\omega^2 R$. This is the centrifugal force!

As with the child and their bear, we are subjected to a centrifugal force due to the rotation of the Earth. It is about 300 times weaker than the gravitational force experienced due to our weight, and we are hardly aware of it. It has also the effect of causing a slight swelling of the Earth, which at the equator is about 43 km larger than the distance from pole to pole, and this difference is about 0.3% and therefore without any spectacular consequences.

On the other hand, the centrifugal force due to the Earth's rotation has no influence on the motion of objects with respect to it, for instance, on the motion of the wind, that is the movement of air with respect to the ground. In fact, the centrifugal force is acting at the same time on the air and on the ground, and it only depends on the distance from the centre of rotation. In contrast, the Coriolis force is zero for an object connected to the ground in the terrestrial frame of reference, but it acts on the air in motion. As a consequence, it plays a significant role in meteorology, as we are going to address.

Meteorological Manifestations of the Coriolis Force

An important effect of the Coriolis force is that vortices form in the atmosphere around high-pressure or low-pressure zones and rotate in a well-defined sense, though in different directions in the Northern and Southern hemispheres (Fig. 6).

We might think that masses of air flow directly into an area of low pressure (also defined as de-pressure, often written D in meteorological maps). In reality, the winds deviate due to the Coriolis force. In the Northern hemisphere, the winds rotate around the de-pressure area in an anticlockwise direction. In the case of an anticyclone, centred around an area of high pressure, the winds rotate in a clockwise direction. In practise, they do not create a complete vortex as represented in Fig. 6. However, we can formulate the rule that in the Northern hemisphere the winds have the low-pressure area to their left and the high-pressure area to their right (Fig. 7). The opposite happens in the Southern hemisphere (see Panel on page 56).

Another manifestation of the Coriolis force is the direction of the trade winds that systematically, between the 30th parallel South and the 30th parallel North, flow towards the West. The major cause of these winds is convection (see Chapter 7): the warm equatorial air rises and leaves room for colder air coming

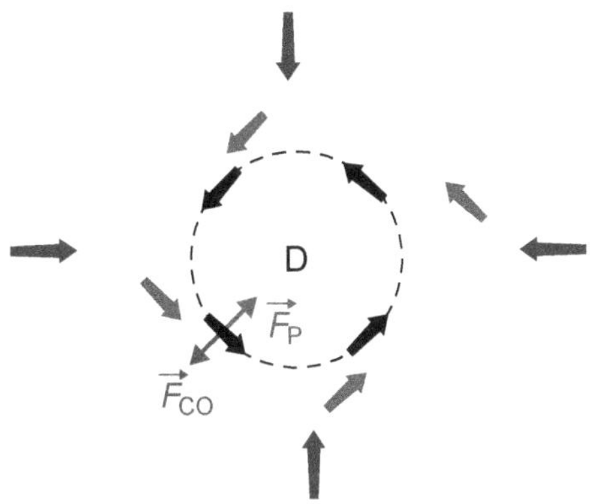

Figure 6. The winds around the centre D of a low-pressure region in the Northern hemisphere. The pressure difference causes a flow of air towards the centre of low pressure, as indicated by four entering external arrows. This wind deviates because of the Coriolis force and moves towards the centre with an anticlockwise rotation (grey arrows). The black arrows indicate the approximate direction of the wind resulting from the equilibrium of the Coriolis force and the force due to the pressure.

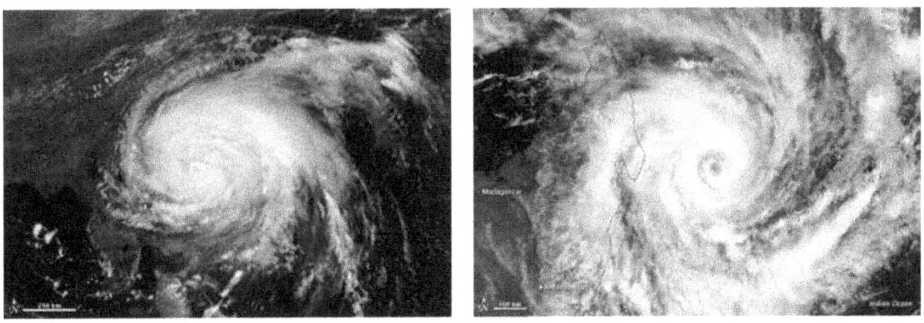

(a) (b)

Figure 7. The Irene tornado approaching the Bahamas (Northern hemisphere) in August 2011, as seen from a satellite. (a) The masses of air rotate in an anticlockwise direction in the Northern hemisphere; the opposite would happen in the Southern hemisphere. (b) The tempest Bingiza over Madagascar in February of the same year.

from higher latitudes. As a consequence, winds arise and are directed from North to South in the Northern hemisphere and from South to North in the Southern hemisphere. The Coriolis force pushes both these winds towards the West (Fig. 8).

The Coriolis force does not act only on the winds but also on the ocean currents and the tides (see Chapter 5). The effect of the Earth's rotation on the displacements of water masses was already addressed by Laplace in the 18th century, well before Coriolis' observation. However, Laplace did not introduce a fictitious force to take it into account. The idea of Coriolis was not accepted without reservations. A member of the Academie des Sciences in the year 1859 wrote,

In the bathrooms of different hemispheres

Anybody can see that when a sink is emptying, a vortex is formed. What is the direction of rotation? A common error is the belief that in the Northern hemisphere, the rotation is always anticlockwise (Fig. 7), while in the Southern hemisphere, it is the opposite. Evidently, a certain effect of the Coriolis force is acting when a vessel is emptying, but it is necessary for the container to be very large for the effect to dominate.

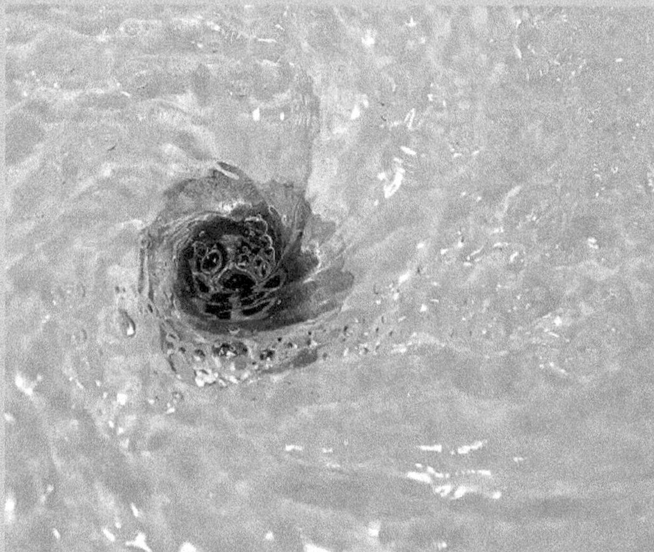

Vortices can arise during the emptying of the container: at variance with the usual belief the direction of rotation depends on other circumstances and not the hemisphere in which the sink is located.

The experiment was carried out at the Massachusetts Institute of Technology by Ascher Shapiro in 1962. His container had a diameter of 2 m and a depth of around 15 cm. It took about 20 min for the container to empty, and a vortex (evidenced by a floating object) rotating anticlockwise was indeed formed after about 15 min.

For our own washbasins of much smaller size and faster emptying, the direction of rotation of the vortex is more closely related to the geometry of the vessel (like asperities or the impulsion initially given to the water, etc.). Thus, in practise, rotation in both directions is possible.

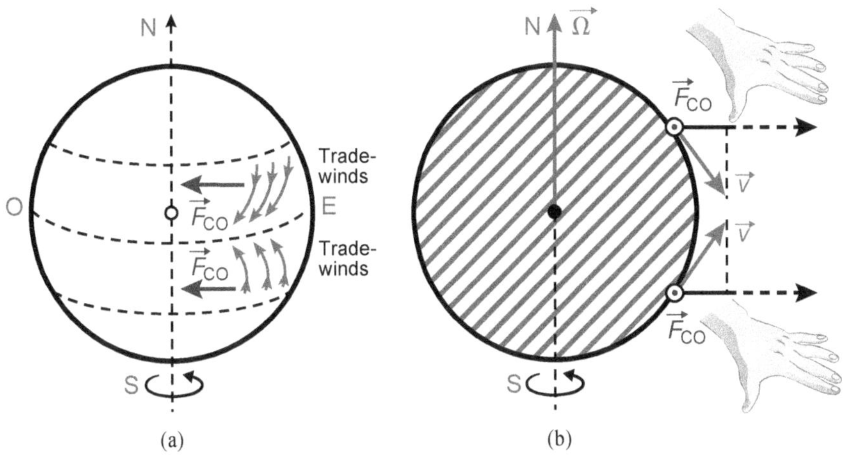

Figure 8. Trade winds are directed towards the West for the Earth's rotation because of the Coriolis force (a), as can be understood by applying the left-hand rule (b).

"These fictitious forces lead to exact results; however, just because they are fictitious, they are not suited to achieve a good understanding of the mechanisms of the phenomena by providing analysis of the real causes." This remark lets us realise the innovative character of the concept developed by Coriolis, its role and usefulness being nowadays undisputed.

Returning to the Baer Law

According to the Baer law, the right bank of the rivers in the Northern hemisphere is more rugged than the left one, while the opposite effect occurs in the Southern hemisphere. This observation has been made for Siberian rivers, the Danube and the Nile. A possible explanation of this phenomenon rests on the Coriolis force that implies a deviation of the water in the rivers towards the right-hand bank (Fig. 9).

Due to friction at the banks, the current is faster at the surface than at the bottom of the river, and therefore the Coriolis force is stronger at the surface. As a consequence, we see a vertical circulation of the water that favours the erosion of the right-hand bank and the depositing of materials on the left-hand bank. The mechanism has a certain analogy to the formation of meanders (see Chapter 1). However, if one tries to estimate the effect of the Coriolis force, it turns out to be very small. For this reason, it is difficult to believe that the problem which was the aim of the article by Einstein and of the related controversy for many weeks at the French Academie des Sciences has been fully solved.

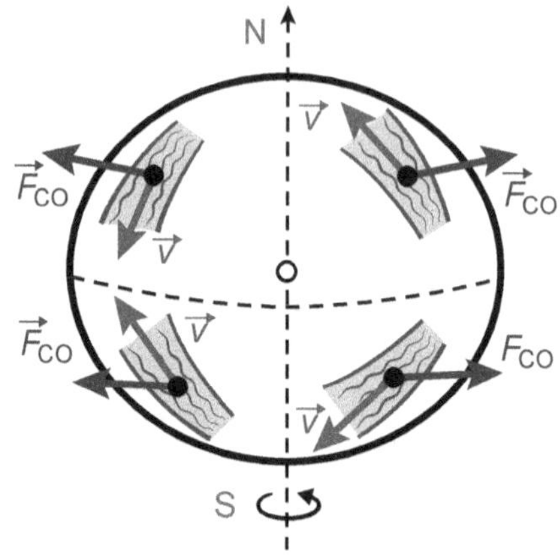

Figure 9. The Coriolis force F_{CO} diverts the currents of the rivers towards the right-hand bank in the Northern hemisphere and towards the left-hand bank in the Southern hemisphere.

Figure 10. The Coriolis platform in Grenoble. It was deconstructed in 2011 and rebuilt in 2017.

Conclusion

In the attempt to solve a problem of mechanics, researchers have the choice between two methods. In general, astronomers opt for the system of non-rotating coordinates, referring to the faraway stars, being conventionally fixed. The other method is to choose the frame of reference fixed to the Earth, and this requires the introduction of fictitious forces such as the Coriolis force or the centrifugal force. This is done in meteorology. It isn't always simple to take the Coriolis force into account in the calculations, and thus it appears convenient to look for experimental confirmation. This is the purpose of the "Coriolis platform" installed in Grenoble, a basin 13 m diameter, rotating up to six revolutions a minute (Fig. 10).

Chapter 5

The Ebb and Flow of the Tides

It is claimed that the sailor Pytheas, from Marseille, in the 4th century BC, already suspected that the Moon had some role in driving the tides, having noted that the rhythm of the tides corresponded to the rotation of the Moon around the Earth. Nowadays, it is known that the tides are indeed due to the action of the gravitational forces by the Moon and the Sun on the great masses of water. In this chapter, we are going to describe in detail their behaviour.

The tides are a spectacular phenomenon in particular areas, for example, on the sides of the English Channel (Fig. 1), and they can even be dangerous for walkers who like to venture on the beaches.

From Newton (see Chapter 4), we know the law of physics that is at the origin of the tides. Just after the formulation of the law of the universal gravitation, it was argued that the tides were an application of gravity. The Moon attracts the water in the ocean, and a kind of large bulge is formed. This bulge remains directed towards the Moon while the Earth is rotating around its axis.

When this mass of water meets the coast, it ascends, and the tide rises as the water flows in. Then it ebbs away, the tide receding when the Moon has rotated. This description raises some questions. First, the Earth rotates with a period of 24 hours, while the Moon does not effectively change its position during a day (it takes about 27 days to rotate around the Earth). Thus, we should expect just one tide a day. Instead, there are two tides in 24 hours! Second, why is the role of the Moon so relevant for the phenomenon of the tides despite the fact that the Earth is much more attracted by the Sun?

The answers to these questions can be found in the phenomenon of gravitation, the laws of which were discovered by Newton.

Figure 1. Reefs at Étretat during the high and the low tides. The difference in the height of the water between the two states in a given place can be around 10 m. In a sea almost entirely closed, like the Mediterranean, that difference is only of the order of 12 cm.

Figure 2. Newton, a fraction of seconds before his discovery of the law of universal gravitation. The same fruit that one day had caused Eve's misfortune made the fortune of Newton.

Newton, the Founder of the Modern Physics

The story goes that Newton figured out the law of universal gravitation when an apple fell down from the tree while he was relaxing beneath it (Fig. 2). This was considered proof that the Earth acted on the apple with an attractive force.

Obviously, this force was not acting only on apples, but, in the same way, on all the objects that are close to the Earth. Why only the Earth? Newton had an ingenious intuition: the attraction had to be universal and thus also acting between the Sun and the planets and more generally involving all massive objects!

The fall of the apple and the motions of the planets can be explained with the assumption that two objects of mass M and m, at the distance D, act on each other by an attractive force given by

$$F = \frac{GMm}{D^2},$$

where G is a constant, called the gravitational constant (Fig. 3).

From the fundamental principle of dynamics (see Chapter 4), by considering the force that the Earth exerts on the apple, assuming we can neglect friction, then the apple moves towards the ground with uniformly accelerated motion, in agreement with Galileo's discovery a few years before (see Chapter 4). Then, the acceleration is $g = GM_T/R_T^2$ where R_T is the radius of the Earth and M_T its mass. The radius (about 6,400 km) had been known for a long time, and the gravitational acceleration (around 10 m s^{-2}) was experimentally measurable. By assuming that the density of the Earth is approximately uniform, Newton could estimate the order of magnitude of the mass of the Earth and, at the same time, estimate the gravitational constant G.

The precise value of G, as it is known nowadays, is $G = 6.674 \times 10^{-11}$ m^3 kg^{-1} s^{-2}. It is not that large! For a proton and an electron, gravitational attraction is negligible in comparison to the electrostatic attraction that also involves the inverse of the square of the distance. For a large object such as the Earth, it is just sufficient to keep us fixed to the ground or to cause serious problems if we fall from a certain height!

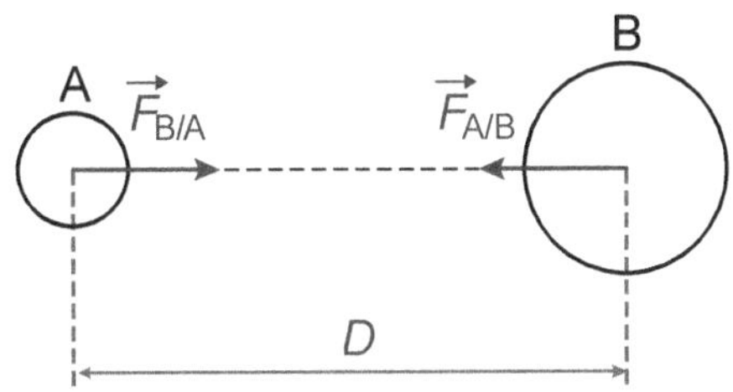

Figure 3. The law of gravity. Two objects A and B at the distance D exert on each other an attractive force proportional to $1/D^2$.

The Sky Falls on Us

As opposed to the apple, the Moon does not fall to the Earth, and the Earth does not fall into the Sun. Why, in spite of the fact that gravitational attraction applies to any object? A simple calculation can explain this, but we can understand that, in Newton's time, this apparent paradox was astonishing to many people. And in a certain way, it still remains a little surprising for us. Thus we are going to provide the explanation.

If the Earth were suddenly to lose the gravitational attraction due to the Sun, it would continue its path by uniform motion because of the inertial principle (see Chapter 4). Therefore, the Sun's attraction does not allow the Earth to fly away, but it is insufficient to cause the Earth to fall into it. In order to understand this, we must recall the concept of centrifugal force (see Chapter 4) that occurs when the fundamental principle of the dynamics is applied to a rotating frame of reference. If the system rotates with an angular velocity Ω around a fixed center O, then the centrifugal force acting on a body of mass m at a distance D from the center is

$$F = m\Omega^2 D.$$

The direction of this force is outwards with respect to the circular trajectory. For the Earth, rotating around the Sun, the centrifugal force F_2 exactly balances the gravitational force F_1 due to the Sun, and thus the planet does not fall into the Sun itself. This equilibrium occurs when the Earth rotates around the Sun along a circular path of radius D_s, with the angular velocity

$$\Omega = \sqrt{\frac{GM_s}{D_s^3}},$$

where M_s is the mass of the Sun.

In reality, the orbit of the Earth around the Sun is not a circle but an ellipse.[1] That can be deduced from the law of universal gravitation, but, in this case, the calculations are more complex.

As remarked by Newton, the law of gravity explains both the motion of the Earth around the Sun and also of the Moon around the Earth. We can show that it explains the tides as well.

---

[1] The eccentricity of the terrestrial orbit (ratio of the distance between the foci and the major axis) is 0.017, not far from the value zero pertaining to a circle. The one of the ellipse that the Moon describes around the Earth is 0.055.

The Origin of the Tides

In analysing the tides, if we take into account the interactions involving the Earth, Moon and Sun, the calculations are too complex. Thus we shall start by assuming that the Moon does not exist, and the Earth and the Sun are isolated bodies. This assumption is obviously unrealistic in dealing with the tides, but it simplifies a preliminary calculation.

We already argued that the centrifugal and gravitational forces acting on the Earth balance each other. This is true at the Earth's centre, but it is incorrect if we consider a point on the surface of the Earth. At a point closer to the Sun (such as the point A in Fig. 4), the distance D to the Sun is smaller, and the attraction by the Sun (proportional to $1/D^2$) is therefore stronger, while the centrifugal force (proportional to D) is weaker. Therefore, the resulting force is directed towards the Sun. The water at A is attracted towards the Sun, and a high tide occurs!

At a point more distant from the Sun (such as point B) where the distance R is larger, the attraction towards the Sun is weaker, and the centrifugal force can push the mass of water farther away from the Sun. Thus, another high tide is produced. If the Sun were the only celestial body causing the tides, we would observe exactly two high tides a day, one at noon when the gravitational attraction dominates and one at midnight when the centrifugal force is strongest. As a consequence, assuming the Earth is a purely water world with no land masses, then it would take the shape of an ellipse (a little bit stretched out, as we will see) or rather the shape of a rugby ball (see Fig. 4). The stretching out of the points of the ellipse is limited by the gravitational attraction of the Earth on the masses of water, obviously much stronger than that due to the Sun or the Moon.

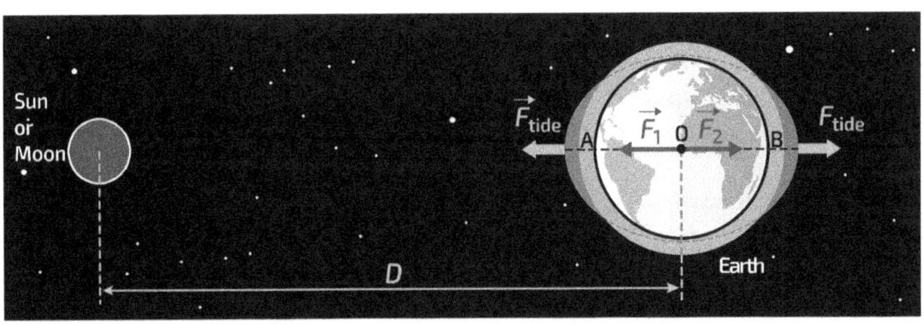

Figure 4. Attraction due to a celestial body (Sun or Moon) at the ground of the Earth. At the centre of the Earth (at O), the gravitational force is balanced by the centrifugal force. At A, the gravitational force is stronger than the centrifugal force. At B, the gravitational force is weaker than the centrifugal force. Therefore, the ocean water masses are attracted by the celestial body at A, while they are repelled at B.

The analysis made in regard to the action by the Sun is equally valid for the Moon: the bodies of water closer to the Moon are attracted, while the ones farther away are repelled. Which is the more effective action, that due to the Sun or the one due to the Moon? To find out, a simple calculation is required (see Panel on page 77). It transpires that the force acting at the point A closer to the attracting celestial body (Moon or Sun) is proportional to M/D^3, where M is the mass of the attractor and D the distance from the Earth. The quantity M_S/D_S^3 is about half of M_M/D_M^3. Thus, the influence of the Moon on the tides is roughly a factor two stronger than that due to the Sun! Since the Moon does not move quickly around the Earth, two tides do occur around every 24 h. However, because of the rotation of the Moon around the Earth, there is a little more than 12 h between the two tides (12 h and 25 min).

Thus the effect of the Sun on the tides is far from being negligible. When the three celestial bodies lie approximately in a straight line, which happens a little more than once a month (Fig. 5), the effect of the Moon combines with that of the Sun, and therefore the tides are particularly marked. These "spring tides" correspond to the full Moon or the new Moon, while the weak tides, also known as "neap tides", occur in the first and the last quarters (Fig. 6). On the other hand, the tides are particularly strong at the equinoxes since, at that time, the Sun is in the equatorial plane of the Earth.

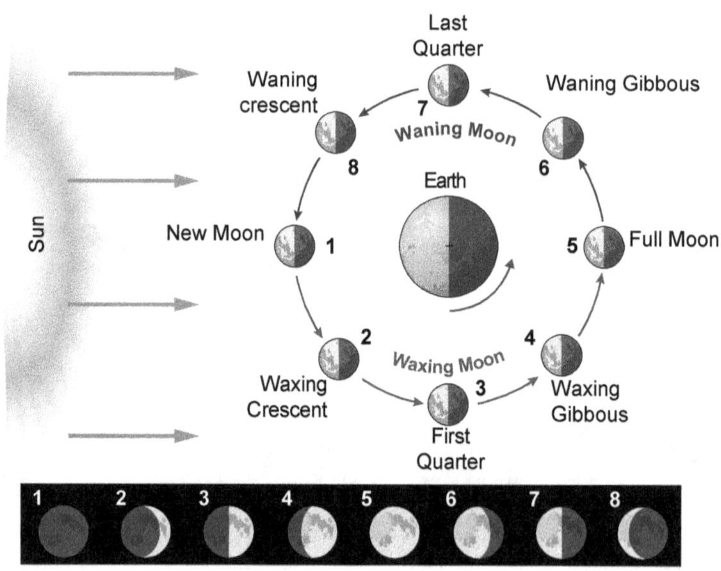

Figure 5. Phases of the Moon. For a terrestrial observer, the fraction of the Moon that is in bright light is related to its position along the orbit with respect to the Sun. About 15 days separate the new Moon from the subsequent full Moon.

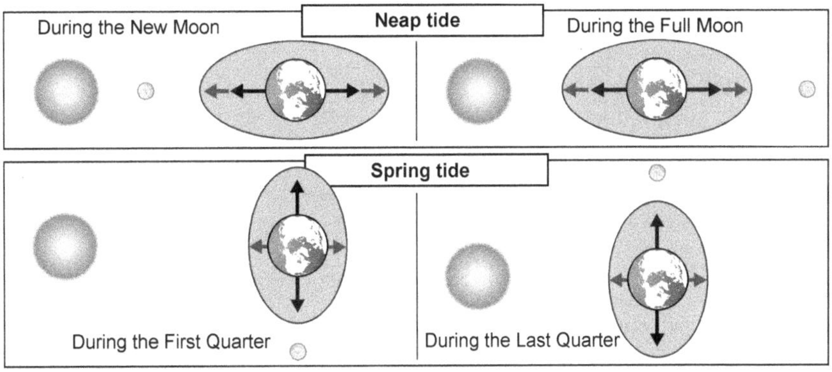

Figure 6. At the full Moon and at the new Moon, the effects of the Moon and the Sun sum together, and the tides are particularly strong. Conversely, the tides are attenuated in the first and the last quarters.

The fight between the Moon and the Sun for greater influence

Does the Moon or the Sun have the most influence on the tides? To estimate that, let us consider a small quantity of matter (for instance, water) having mass δm at a point M on the Earth's surface at a distance $\mathbf{r}$ *from the centre O. This matter is attracted towards the Sun by the force* $\boldsymbol{f}_S = -GM_S\delta m\mathbf{r}'/(\mathbf{r}')^3$, *where* M_S *is the mass of the Sun, G is the Newton gravitational cstant and* $\mathbf{r}' = \mathbf{D}_S + \mathbf{r}$ *is the vector joining the centre of the Sun to the point M, while* D_S *is the vector connecting the centre of the Sun to O. At the point O, the force is* $\boldsymbol{f}_0 = -GM_S\delta m\mathbf{D}_S/D_S^3$. *At any point M, one can write* $\boldsymbol{f}_S = \boldsymbol{f}_0 + \delta\boldsymbol{f}_S$, *where* $\delta\boldsymbol{f}_S$ *is related to r as shown in Fig. 4. This is the driving force causing the tides. From a simple calculation, we derive, at the surface of the Earth,*

$$\delta \boldsymbol{f}_S = 3GM_S\delta m\frac{\boldsymbol{x}_S}{D_S^3}$$

where $\boldsymbol{x}_S = (\mathbf{D}_S \cdot \mathbf{r})/D_S^3$. *For the Moon, we obtain an equivalent equation where the term* M_S/D_S^3 *is substituted by* M_M/D_M^3. *Thus in order to compare the effects of the Sun and the Moon, we must compare the terms* M_S/D_S^3 *and* M_M/D_M^3. *From the* Table *below, we immediately see that the force related to the Moon is a little more than twice the one related to the Sun.*

Taking into account that $M_{M,S} = 4\pi\rho_{M,S}r_{M,S}^3/3$ ($r_{M,S}$ *is the radius of the Moon or the Sun, respectively), we can conclude that the tidal force is proportional to the*

(*Continued*)

(*Continued*)

product of the density of the celestial body $\rho_{M,S}$ *and the cube of the angular size* $(r_{M,S}/D_{M,S})^3$ *of the latter. Coincidentally, the Sun and the Moon appear the same size in our skies, meaning their angular sizes are the same, while the average density of the Moon (3.34 gcm*$^{-3}$*) is approximately twice that of the Sun (1.41 gcm*$^{-3}$*), which again justifies our above estimation of their relative contributions to the tidal forces.*

	Mass (kg)	**Distance from the Earth (km)**	M/D^2	M/D^3
Sun	1.99×10^{30}	149,598,000	0.89×10^{14}	0.59×10^{6}
Moon	7.35×10^{22}	384,400	0.50×10^{12}	1.30×10^{6}
Earth	5.98×10^{24}	–	–	–

Data for the Sun, the Moon and the Earth. The ratios in columns 4 and 5 allow us to compare the relative effectiveness of the gravitational attractions that the Sun and the Moon produce on the Earth and their effects on the tides.

The Moon during the night in a clear sky has inspired a number of beliefs and superstitions. While the Moon to a large extent has an effect on the tides, the effects on the sap flow of the trees or on the mood of human beings, as well as on werewolves, are still awaiting a scientific demonstration.

(*Continued*)

(*Continued*)

However, the gravitational force due to the Moon on the Earth is much weaker than that due to the Sun. This is because the force due to a mass M at the distance D is proportional to M/D^2 *and the quantity* M_S/D_S^2 *is about 200 times the value of* M_M/D_{MP}^2 *as reported in the table. And what about the influence of the Sun on the motion of the Moon? We must compare* M_S/D_S^2 *to* M_T/D_T^2, *where* M_T *is the mass of the Earth. These two quantities are of the same order of magnitude (see table). Therefore, we cannot neglect the effect of the Sun on the motion of the Moon.* I. Martin, C. Gutzwiller, *Rev. Mod. Phys.* 70, 589–639 (1998).

Tidal Heights and Predictions

The ingenious and simple theory by Newton cannot predict the height of the tides. In fact, it would predict a tidal range (the difference between the high and the low tides) of some tens of centimetres. In reality, the tidal range can reach tens of meters at the coast. Furthermore, it varies considerably from site to site, and this can hardly be explained on the basis of what has been addressed above. Newton had to assume that the surface of the ocean was always in equilibrium and could strictly obey the forces acting on it. About a century later, it was a French mathematician and physicist, Pierre-Simon Laplace (1749–1827), who emphasised that a correct theory had to be dynamical. Resonance phenomena indeed play a great role in the mechanisms of the tides: the tides on a certain day are affected by those on previous days and influence subsequent ones. The Coriolis force also plays a role by shifting the currents within bodies of water: this is the reason why the tides at the coast of the English Channel vary by some tens of meters on the French side, while on the English coast are limited to several meters.

The height of the tides is related not only to the motions of the Moon and the Sun with respect to the Earth but also to the structure of the coasts and underwater relief. It would require a very complex and difficult calculation to take these factors into account. Fortunately, it is possible to predict the tides with high confidence by considering the sea level, at a given point, as a sum of sinusoidal functions of the time, $\sum a_i \sin(\omega_i t - \alpha_i)$. About 12 terms are sufficient. The frequencies ω_i are well known, while the coefficients a_i and the de-phasing factors α_i are experimentally estimated, for each point of the coast.

Days Lasting Longer… and a Moon Getting Ever Farther Away

Observations reveal the tides occur about 12 min later compared with the real motion of the Moon. By assuming the Earth is an elliptical shape, this delay means that its major axis is not exactly aligned towards the Moon. In fact, it forms an angle ϕ of the order of 3° with the Earth–Moon axis (Fig. 7). Indeed, the water, due to the inhibition of its movement by friction of the bottom of the ocean and the coast, does not have time at each moment to take the most energetically favourable position. This friction leads to the conversion of some of the kinetic energy of the Earth's rotation into heat. The tides slow down the rotation of the Earth! As a consequence, the length of the day is increasing, as was claimed in the 19th century by English physicist Lord Kelvin. This increase has been estimated with rather good precision, by looking at fossil corals present in the Indian Ocean for 400 million years. These corals exhibit a succession of rings, related to the alternation of day and night (*daily rings*). By studying these rings, scientists have found that the solar year, namely the time for the Earth to complete one full revolution around the Sun (which is the same today as it was 400 million years ago), took 395 days in that earlier time. This means that the length of each day was only 22 h!

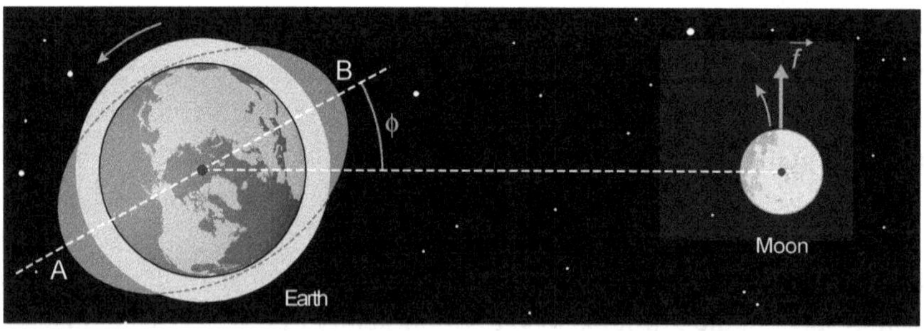

Figure 7. The consequence of the tidal delays for the movement of the Moon. The Earth and the Moon are seen by an observer above the North pole. The Moon rotates in the same direction as the Earth but with a much slower angular velocity (about one revolution a month). The tides on the Earth are delayed by an angle ϕ (not to scale) with respect to the motion of the Moon. Because of this delay, the attraction force implies a small component $\vec{f}$, perpendicular to the axis Earth–Moon, that pushes the Moon farther away from the Earth.

The delay of the tides has another consequence: the Moon is moving away from the Earth! In fact, one can refer to a rugby ball with its axis not oriented along the line Earth–Moon. Then, the attraction force by the Earth on the Moon is not directed exactly towards the centre of the Earth: a component perpendicular to the Earth–Moon axis arises. This force can push the Moon away from the Earth at a rate of 3.8 cm a year … we are moving away more and more from our neighbour, the Moon.

Energy from the tides

The tides involve a great amount of energy that we could try to benefit from. This is the purpose of tidal power stations *(see figure) that take advantage of the water currents flowing due to the difference in the heights of the tides, to produce electric energy. Is tidal energy really inexhaustible and cost-free in the same way as solar or wind power? We have seen how the tides cause a certain slowing down of the rotational motion of the Earth. In principle, the use of tidal energy should accelerate this slowing down.*

Let us estimate the maximum energy that we could obtain from tides. We will give an order of magnitude rather roughly by assuming that each day the tide raises the height of the water by $h = 1$ m for a surface given by R_T^2. We obtain energy of the order of $g\rho R_T^2 h^2$, where ρ is the specific mass of the water, which is 10^{18} J day^{-1}. Over the course of a year, this value corresponds to about the total energy budget of the entire world in 2008 (about 5×10^{20} J). Now let us compare this value to the rotational kinetic energy of the Earth. This energy turns out to be of the order of $M_T R_T^2 \omega^2/5$, the mass of the Earth being $M_T = 6 \times 10^{24}$ kg, its radius $R_T = 6.37 \times 10^6$ m and ω the angular velocity of one revolution per day, namely 7×10^{-5} rad s^{-1}. We obtain about 2.4×10^{29} J. Dividing this result by 5×10^{20} J, we can deduce that, even if human beings extracted all the energy from the tides (which sounds practically impossible), the Earth could still continue to rotate for half a billion years.

(Continued)

(*Continued*)

In Northern France (the Bretagne), this tidal barrage on the river Rance has been operating since 1967. The barrage, which also serves as a bridge, functions in two senses: the turbines are rotating twice a day, both in the ebb and in the flow. The installation uses the variation in the level of the tides in order to produce electric energy.

Chapter 6

Bubbles and Droplets

Bubbles and droplets are fascinating objects for children, for grownup children, and for scientists. "Blow a soap bubble and observe it. You may study it all your life and draw one lesson after another in physics from it." This was written by English physicist Lord Kelvin. Therefore, we shall devote our attention to the subject in this chapter. We shall see why droplets, as well as bubbles, like to take a spherical shape, how to obtain bubbles with a cylindrical shape or a saddle horse and also how to create a microphone by means of water flowing from a tap!

Why the Rain Falls Drop by Drop

Often water takes the shape of droplets, with a size of the order of millimetres. This can easily be seen just by looking at the falling rain. Why does this happen? Why does water from a dropper come out while light pressure is applied and always in the form of droplets with a rather well-defined diameter?

Minimising potential energy and surface tension

Objects have a tendency to minimise their potential energy, namely the energy that they have acquired as a consequence of their position in space and interactions with neighbours. It is a consequence of this principle that a billiard ball will fall down a hole, as sometimes happens to human beings: they minimise their potential energy due to the gravitational attraction of the Earth.

On the other hand, droplets have a tendency to acquire the geometrical form that minimises their surface energy. In fact, in order to increase the surface of a liquid, one has to provide energy. This is what we call "surface energy" (or "interfacial energy"). What is the reason for this behaviour?

Molecules near the geometrical surface of a liquid are in a specific situation: instead of being surrounded by molecules of the same type, as happens for molecules in the interior of the liquid, they have identical neighbouring molecules on only one side, with a few molecules of air as neighbours too. We note that, in general, the molecules of a liquid attract each other and thus the molecules at the surface, being more isolated, are in a less favourable energetic state. This is why energy is required in order to increase the surface area of a liquid (this also explains why egg whites have to be whisked in order to form peaks). This energy, for a unit area of surface, is called the coefficient of surface tension. It keeps the surface of a liquid as a taut layer and prevents wrinkles. The coefficient of surface tension of most liquids has been evaluated (see Table 1), and it is usually expressed in joules per square meter.

Table 1. Surface tension coefficients for different liquids.

Liquid	**Surface tension (in mJ m^{-2} or mN m^{-1})**
Water (25°C)	72
Water (100°C)	59
Ethanol (25°C)	22
Mercury (25°C)	485

One can consider the coefficient of surface tension as a force per unit length, expressed in newtons per meter. This way of considering the surface tension suggests a simple experiment that allows us to visualise this phenomenon (Fig. 1).

Using metal rods, we shall construct a rectangular frame, with one of the sides of length being a movable rod. The frame is immersed in water with soap powder diluted in it. Then we extract the frame so that a rectangular film is obtained. Under the action of the surface tension, the film begins to contract by attracting the mobile rod.

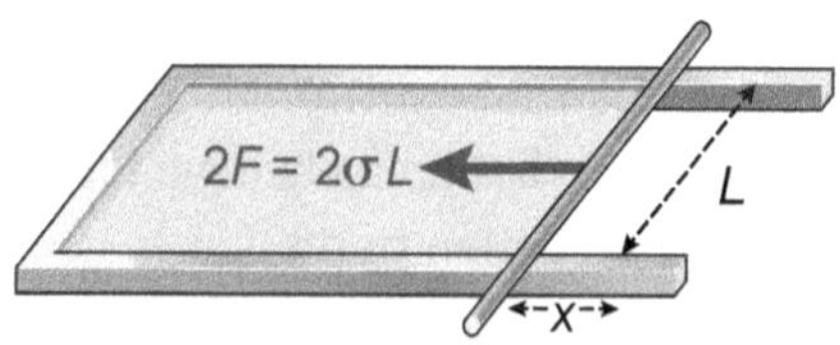

Figure 1. A film of soapy water creates a force $2F$ (since the film has two faces) acting on a mobile rod. When the rod has been shifted by a length x, the work $2Fx$ is equal to the decrease of the potential energy $2\sigma Lx$ of the film. Therefore, the surface tension σ is given by the force per unit length F/L.

Similar to surface tension, we can define an *interfacial tension* between two immiscible liquids (two liquids that do not mix each other), as well as between a solid and a liquid.

The equilibrium shape of a single drop is the one minimising the surface energy. This energy is given by the product of the surface area with the coefficient of surface tension. Therefore, the equilibrium shape is the one that for a given amount of liquid minimises the surface area. This condition defines a sphere! Thus water droplets or those of other liquids have to take a spherical shape (Fig. 2). That said, this shape may be perturbed by several factors, which we shall address as we progress through this chapter.

Size of the drops

The equilibrium form of a droplet in air is to be spherical. What about the size of this sphere? The growth of a drop is usually affected by its weight. As an example, let us consider the formation of a drop in a typical medical dropper (Fig. 3). When a certain pressure is applied to the ring nut, the liquid comes out from the dropper. The drop is not exactly spherical, but still has a dimension roughly equal to R in all directions (Fig. 3(a)). Therefore, its mass is of the order of ρR^3 (ρ being the

Figure 2. Water droplets on a spider's web. The shape is spherical, with the exclusion of the bigger drops which are more affected by gravity due to their greater weight.

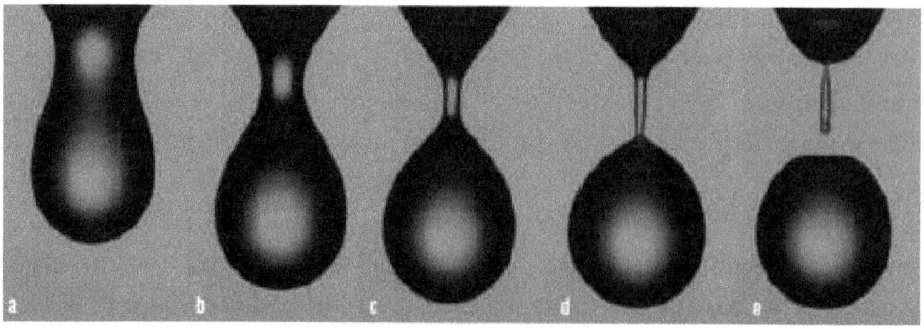

Figure 3. Detachment of a drop. The drop, its shape being a portion of sphere, if affected only by its surface tension, in turn is stretched out by its weight.

specific mass of the liquid) and its weight is of the order of $g\rho R^3$ (where g is the acceleration due to gravity). Its surface energy is about σR^2, σ being the surface tension of the liquid. Because of its weight, the drop begins to detach (Fig. 3(b)). Meanwhile, it has moved by a length of the order of R, and its gravitational potential energy has decreased by a term of the order of $g\rho R^4$ (the variation in the potential energy being given by the product of the weight times the difference in the height). During this process, the surface energy of the drop is increased by a term of the order of σR^2.

Due to its weight, the fall of the drop can only happen when the decrease in potential energy is larger than the increase in the surface energy; therefore, at the condition, when the radius R is larger than an amount R_1 of the order of

$$R_1 = \sqrt{\frac{\sigma}{g\rho}}, \tag{1}$$

a drop of smaller radius would still remain attached to the dropper.

More generally, the value R_1, called the "capillary length", is the length above which the role of the weight becomes more relevant than the surface energy. For example, if we place a small amount of liquid on a plane surface, then a drop is formed having a free surface of roughly spherical shape. However, when the radius of the drop is larger than R_1, a flat coating is obtained. In conclusion, the radius of the drops is not larger than R_1 (see Panel on page 67). In the case of water, by taking $g = 9.8$ m s^{-2} and $\rho = 1{,}000$ kg m^{-3}, the capillary length is of the order of 3 mm.

Soapy Bubbles

We can create drops of water in air as well as bubbles of air inside water, and also bubbles of air within air itself using soapy water. In fact, bubbles created in

Why raindrops do not get big

Why we do not see big raindrops? In Paris, a group of physicists decided to study the problem by letting drops of water of different radii fall from a height of 8 to 12 m, while photographing them (see figure). Small drops were found to retain their spherical shape. The bigger drops were instead observed to flatten, and, above a certain critical size, they became like sacks. The researchers noticed that air enters this "sack" during the fall of the drop, which eventually tears it. Thus, a droplet larger than a certain critical size could not reach the ground whole. Therefore, we could thank our good luck that allows us to avoid drops as big as centimetres when raining!

The critical size is of the order of the capillary length. This could be surprising since the role of air resistance is clearly relevant in the experiments, while in Eq. (1) it does not appear. However, when a drop is falling from the height of 12 m, after some time, it reaches a constant (terminal) velocity of some metres per second (about 9 m s^{-1} for the big drops and some m s^{-1} for the smallest drops), due to the equilibrium between its weight and the air resistance. Thus the force due to the air resistance is equal, in absolute value, to the weight, even though its distribution on the surface is different. While we are fortunate not to experience large raindrops, unfortunately large hailstones can reach the ground. This happens when the initial drops are becoming large because of several upwards currents. Some attempts have been made in the hope of preventing that process; the idea has been to crush the clouds by means of particles that should induce the formation of usual-sized drops but has not led to any real successes.

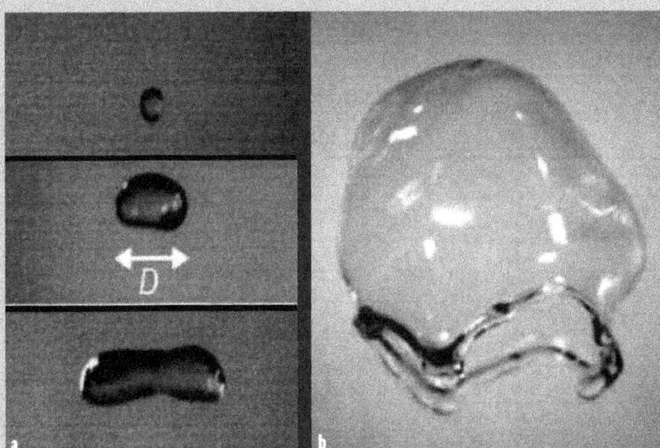

(a) Shape acquired by a drop of water falling in air as a function of its size. From the top, drops having diameter *D* less than the capillary length, about equal to it and finally larger than it [from Reyssat *et al*., *Europhys. Lett.* 80, 34005 (2007)].

(b) Shape of the sack acquired by an initially spherical droplet having radius around 18 mm, during its fall. The droplet was progressively filled with air, and finally had to break into pieces (*Ibidem*).

this way are relatively stable, unlike those made just by using pure water (Fig. 4).

The molecular structure of the film in soap water offers enough material for a lecture on physical chemistry. The soap provides particular molecules, defined as *surfactant*, characterised by a part, the head, known as a hydrophile (namely something that likes water) and a tail that does not like water, called a hydrophobe. In order to keep the head in the water and the tail outside it, these molecules favour the formation of a surface by aligning themselves perpendicular to it (Fig. 5).

The presence of these molecules decreases the surface tension. The equilibrium shape of a soap bubble is the one that minimises the surface energy: a sphere. The little bubbles, just as with the little drops, are therefore spherical! But the large bubbles are also spherical since they are hardly affected by gravity: the film is very thin and therefore light. Thus, they are suitable objects for the study of surface tension and its effects. If the shape of the bubble depended only on surface tension, then it would continue to reduce its surface: the bubble would become smaller and smaller and in the end would disappear. Since the bubble contains air, a decrease in the diameter implies an increase in pressure: when the internal pressure becomes larger than atmospheric pressure, then the walls of the bubbles are repelled, and an equilibrium sets in.

Figure 4. Soap bubbles. The beautiful colours are due to interference processes (see Chapter 3).

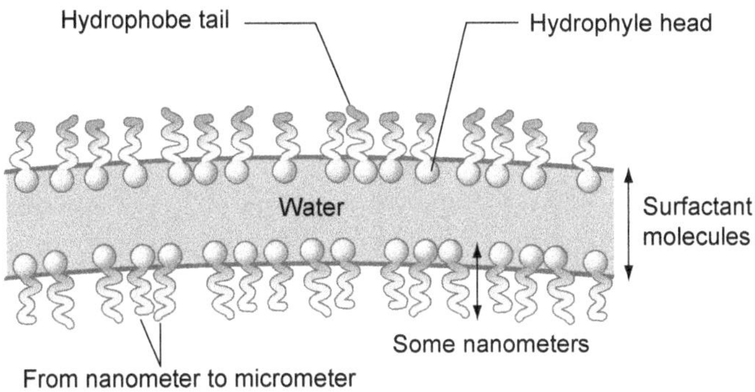

Figure 5. Scheme of the film of soap water. The surfactant molecules decrease the surface tension and make it difficult for the bubbles to dissolve. The hydrophile head is in general electrically charged and strongly interacts with the water molecules. Thus a kind of electric dipole is created (see Chapter 16).

The Laplace formula

What is the extra pressure that stabilises a soap bubble? For a spherical bubble of radius R, the calculation is simple. The surface energy is given by the product of the surface tension multiplied by the surface area, namely $S\sigma' = 4\pi R^2\sigma'$, where $\sigma' = 2\sigma$ is twice the surface tension of the soapy liquid since the film has two faces. A small increase in the radius δR implies a variation in the surface area given by $8\pi R\delta R$ and therefore a variation in the surface energy equal to $8\pi R\delta R\sigma'$ (Fig. 6).

This variation of the surface energy must be compensated by the work done by the pressure forces on the walls when the increase in the radius of the bubble occurs (the work done by a force is the energy transferred to the system when this force is moved). This work is equal to the overpressure ΔP multiplied by the variation in the volume of bubble, namely $4\pi R^2\delta R\Delta P$.

Therefore, a soap bubble having radius R is in equilibrium when the pressure of the air inside is larger than the pressure outside by the amount

$$\Delta P = 2\sigma'/R. \tag{2}$$

This equation is known as the "Laplace formula", in honour of the physicist who derived it in the year 1806 (see Chapter 5). The overpressure increases as the radius of the bubble decreases. You can easily test its validity by connecting two bubbles of different sizes with a thin tube: the smaller bubble will immediately grow, and the larger bubble will shrink!

For a bubble of the order of one millimetre in size, the overpressure is of the order of one-thousandth atmospheric pressure. For a gas bubble inside water,

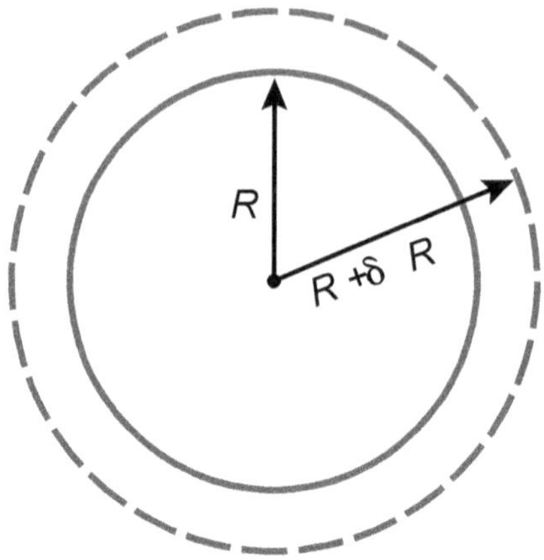

Figure 6. For an infinitesimal variation in the radius of the bubble in its equilibrium state, the variation of the energy must be zero.

$\sigma' = \sigma$, and the excess pressure turns out to be half that of a soap bubble of the same radius.

Bubbles in contact and foams

Thanks to the Laplace formula, we can predict the form of a system of many bubbles as it occurs in a foam. Let us consider two bubbles of radii R_1 and R_2 (Fig. 7).

The overpressure inside each bubble is $\Delta P_1 = 2\sigma'/R_1$ and $\Delta P_2 = 2\sigma'/R_2$, respectively. The film separating the two bubbles is the upper portion of a sphere whose radius R_3 obeys Eq. (2) with $R = R_3$ and $\Delta P = \Delta P_2 - \Delta P_1$, and therefore

$$\frac{1}{R_1} = \frac{1}{R_2} - \frac{1}{R_3}.$$

This relation allows us to establish the geometry of the two bubbles and the interface. Another geometric property is obtained by writing that the surface tension forces act at any point A of the circle Γ that limits the equilibrium of the interfaces (namely that the sum of their vector forces is zero). There are three of these forces, each being tangent to one of the spheres 1, 2 or 3, and they tend to restrict the corresponding top portion of the sphere. Now these forces per unit length are given by σ'. To achieve equilibrium, they must define, two by two, an angle of 120° (Fig. 8).

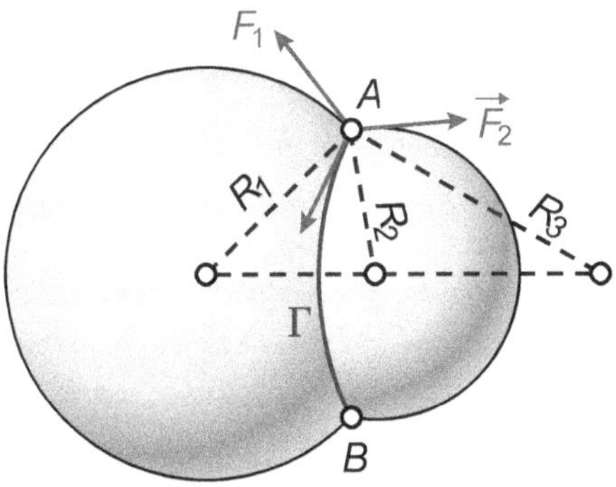

Figure 7. Two bubbles in contact. The planes tangent to the spheres and their interface (line Γ) must define an angle of 120°, and the radii of the bubbles follow the relation $1/R_1 = 1/R_2 - 1/R_3$, R_2 being the radius of the smaller bubble. An equilibrium sets in between the forces of surface tension $\boldsymbol{F}_1$ and $\boldsymbol{F}_2$, acting at the interface that tends to reshape the two spheres, and the pressure inside the bubbles, which is higher than atmospheric pressure.

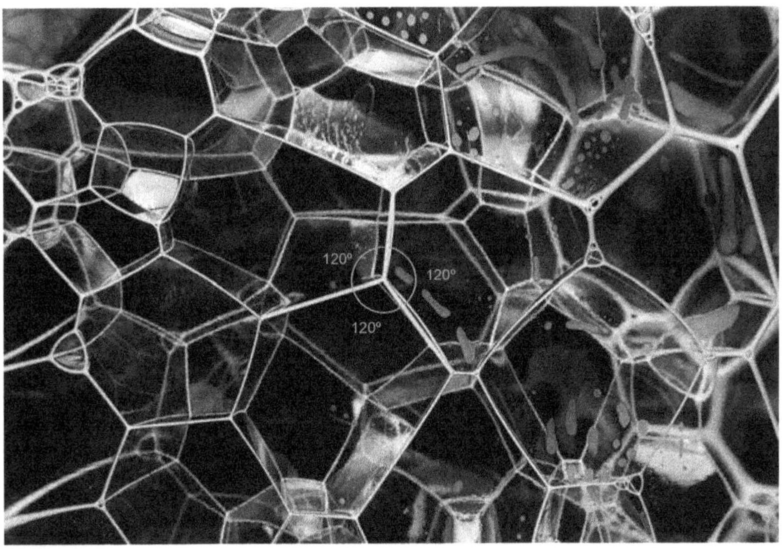

Figure 8. Layer of soap bubbles on a flat surface. The angles which join three bubbles together are all 120°. In a thick foam, the six walls in contact with four neighbouring bubbles take on a tetrahedral symmetry: their corners forming an angle of 109.5°.

Similar considerations allow us to derive the form of a drop lying on a plane (see Panel on page 82). When the foam contains a large number of bubbles, the resulting structure must obey conditions that generalise the ones we have derived for two bubbles.

A droplet on a flat board

What is the shape taken by a droplet when deposited on a board? In contrast with the case of contact between two bubbles where only the surface tension plays a role, in this case, we have here different interfacial tensions: σ_{lg}, σ_{ls}, *and* σ_{sg} *respectively corresponding to the interfaces between liquid and gas, between liquid and the supporting board, and finally between the board and gas. Depending on the values of these parameters, the drop takes a certain form on the board. The corresponding distribution can be defined according to the value of the angle* α *of contact of the drop (see figure).*

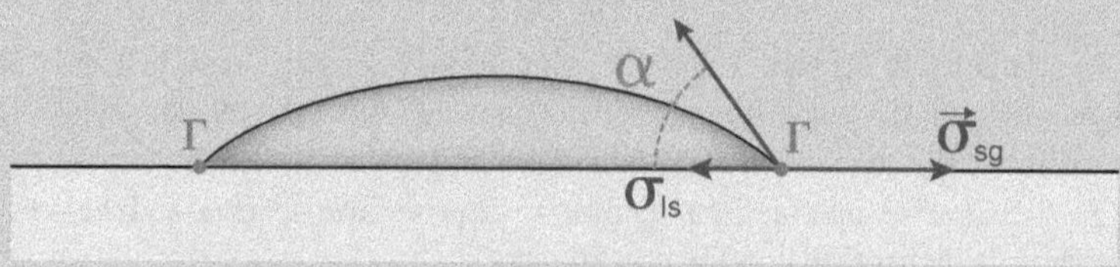

A liquid drop partially wetting the solid board.

On the line Γ *common to all the three phases, two forces per unit length are active:* σ_{ls} *and* σ_{sg} *parallel to the board, and a third force of value* σ_{lg}, *which is tangent to the drop. The line* Γ *cannot be detached from the board, that is fixed. In order to have equilibrium, it is just necessary that the component of net force parallel to the board is zero, namely*

$$\sigma_{sg} = \sigma_{lg} \cos\alpha + \sigma_{ls}.$$

Mercury drops that do not wet their supporting board. The smallest drops are spherical: others are flat because of their relatively large weight.

(*Continued*)

(*Continued*)

This is the Young–Dupré equation. The cosine of the angle α *is between 1 and −1: this requires* $-\sigma_{lg} < \sigma_{sg} - \sigma_{ls} < \sigma_{lg}$. *If this condition is met, we can say that* ***partial wetting*** *is occurring. Then, the drop forms a spherical top portion.*

If $\sigma_{sg} - \sigma_{ls} > \sigma_{lg}$, *then the drops spread out as much as possible, forming a very thin layer. So-called total wetting occurs. If* $\sigma_{sg} - \sigma_{ls} < -\sigma_{lg}$, *the drop is no longer attached to the board, and no wetting occurs.*

The reader has perhaps had the chance to see little drops of mercury (for instance, when an old-type thermometer was broken) running across a board (see figure) or water drops running down the feathers of a duck: these are examples of the latter case.

Strange soap bubbles

The list of soap bubble shapes does not only include the sphere. When a film of the solution is not free but is constrained on a particular framework, it can take very strange forms that we might not expect! Let us begin by immersing two rings of about the same size in soap water. By careful manipulation, we can obtain a cylindrically shaped bubble with spherical ends (Fig. 9).

The pressure difference ΔP between the interior of the bubble and the outside is related to the radius of the cylinder by an equation analogous to the Laplace formula, without the factor 2:

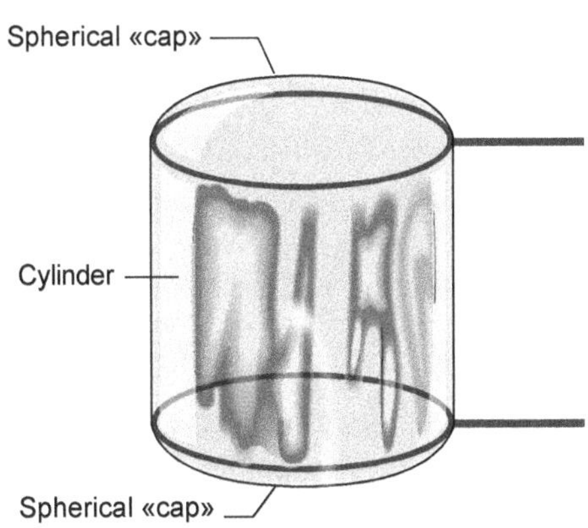

Figure 9. Cylindrical soap bubble made using two rings. Two spherical ends complete the bubble.

$$\Delta P = \sigma'/R. \tag{3}$$

The two ends are spherical, and the radius of each corresponding sphere is $2R$ as follows from Eqs. (2) and (3).

What happens if we break the ends? Then the pressure difference between the interior and the exterior of the film falls to zero. The film held by the two rings cannot remain cylindrical but deforms in order to reduce its surface energy and therefore its surface (Fig. 10).

The surface created in this way, with the shape of a saddle, is called a catenoid. From a mathematical point of view, this surface is obtained by rotation around an

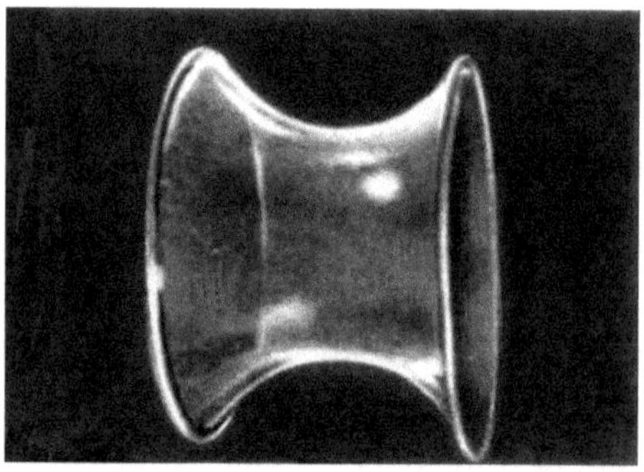

Figure 10. Surface formed by a soap bubble between two parallel rings is called catenoid. Any of its longitudinal section is concave, while any transverse section (circle) is convex.

Bending, average bending, chain and catenoid

A plane curve (under certain conditions of continuity, derivability, etc.) at each point possesses a curvature of radius R that can be defined as the radius of the circle that best approximates the curve around that point. We can also define the curvature (or bending) $\rho = 1/R$ *at each point. In turn, the surface at any point A (see figure) is characterised by two principal radii of curvature:* R_1 *and* R_2*. They correspond to the minimum and maximum curvature radii at this point when the surface is cut by a plane passing through the normal at point A. The radius of curvature is considered positive when the section is convex, and negative if it is concave (in figure,* $R_2 < 0$ *while* $R_1 > 0$*). Then the average curvature* γ *is defined:* $2\gamma = 1/R_1 + 1/R_1$*. A necessary and sufficient condition to have a minimal surface is that the average curvature*

(Continued)

(*Continued*)

is zero everywhere. In other words, the two principal radii of curvature are equal in absolute value and of opposite sign. A large variety of minimal surfaces exist. However, in the group of revolution surfaces, only the catenoids have that property. The reader having some familiarity with differential calculus will easily derive the equation for the hanging chain: $y = \alpha \cosh(kx)$. *Then a rotation around the x-axis generates a catenoid with zero average curvature. At the same time, we can demonstrate that this surface is the equilibrium form of a film of soap water bounded by two parallel rings when they are not too far apart (see Fig. 10). If the rings are separated by too much, then we end up with two discs in the interiors of the rings while the catenoid disappears. When the pressure on the two sides of the film of soap water is not the same, the surfaces have uniform average curvature, but, in this case, it is not zero. For example, this happens when the soap bubbles are formed on a wireframe. The cylindrical bubble capped by two spherical ends (Fig. 9) provides a particularly simple example of such a surface: the average curvature* γ *is everywhere equal to* $1/(2R)$, *where R is the radius of the rings.*

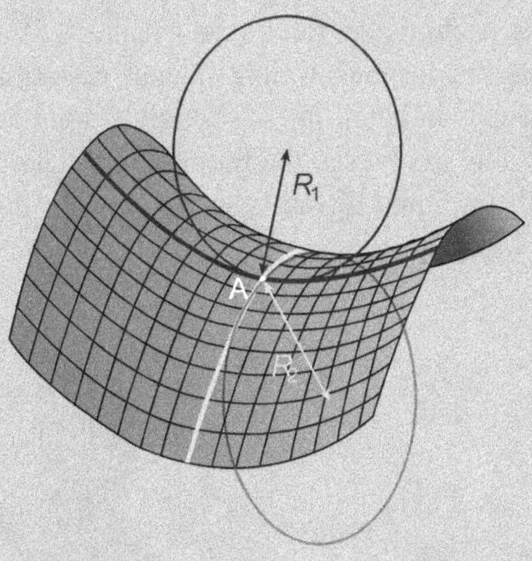

Geometry of a minimal surface (form of a film of soap water when the pressure on both sides is equal). At all the points A on this surface, the curves drawn on it have curvatures turned upwards (dark curve) or turned downwards (light curve). Geometry of a minimal surface (form of a film of soap water when the pressure on both sides is equal). At all the points A on this surface, the curves drawn on it have curvatures turned upwards (dark curve) or turned downwards (light curve). The dark curve minimises the curvature upwards, while the light curve minimises the curvature downwards. For a minimal surface, the two curvatures $1/R_1$ and $1/R_2$ have the same absolute value.

axis of a hanging chain or cable, suspended at its two ends and curved by its weight. If we change the support, for any given geometry, the shape acquired by the film (when the pressure at the ends is the same) will always be the one minimising: a *minimal surface* (see Panel on page 85).

At the Exit from a Tap

Let us leave bubbles behind and instead return to droplets, specifically to the familiar situation of a poorly fitted kitchen tap, which lets drops fall at a regular rate (Fig. 11).

Their fall is very fast, and often we are unable to distinguish the details that instead a suitable high-speed camera can detect. However, even without such a high-speed camera, Belgian physicist Joseph Plateau (1801–1883) examined the form of the drops in detail. A fine experimentalist, he decided to compensate for the weight so that the drops could fall at a more moderate speed, allowing the human eye to observe the detail. Instead of having the drops fall in air, he had them fall in an immiscible liquid, with density close to one (the value for water) — see Panel on page 88. The Archimedes force (see Chapter 15) acting on the drops almost exactly compensates for their weight. Things appear as if the drops have lost almost all their weight, and their fall is strongly slowed down.

Thus, Plateau was able to observe the formation of the drops at the outlet of the tap. It turned out that a liquid thread is set up between the forming drop and

Figure 11. A badly fitted tap lets drops fall. The dynamics of the formation of the drops is complex and was studied in detail during the 1990s.

Two experiments along the Plateau tracks

In the following, we describe two experiments that, without having the accuracy of the one carried out by Plateau, can be performed in our kitchens.

After some centimetres, the cylindrical flow of the water breaks up into drops.

(*Continued*)

(*Continued*)

Experiment 1
Avoiding the weight of the drops
Fill a glass at least 12 cm deep with oil and by means of a pipette add a solution of alcohol in water (70% by volume) having previously slightly coloured this. The small drops take several seconds to reach the bottom, while larger ones may take a second or so, so you will have the chance to make a suitable observation. As Plateau observed, you will be able to see that pretty large drops (of the order of one centimetre across), have a near-spherical shape. On the contrary, drops of this same size would be strongly deformed if falling through air, due to their weight. This is not too surprising, because if we introduce an analogy of "capillary length", similar to the term in Eq. (1), we need to substitute the density with the difference between the densities of the two liquids: the capillary length becomes very much longer. We can estimate the speed of the falling drops, by comparing it to the Stokes formula, which we will deal with in Chapter 15.

Experiment 2
To demonstrate the Rayleigh–Plateau instability
Let us gently open a tap. We see that the drops fall one after another. Then open the tap a little more so that a thin and continuous flow of liquid occurs. In order to show the fragmentation of the flow related to the Rayleigh–Plateau instability, take one of the many plastic cards that nowadays have invaded our life, and place it well below the tap, in the water flow. You will hear the noise of falling drops, and your fingers holding the card will begin to shake. This phenomenon will disappear if the card is placed in the upper part of the flow, which has a well-defined cylindrical shape (see figure).

the tap, which gradually becomes thinner and thinner until the drops separate. Then the filament forms a second "satellite" drop as can be seen in the last part of Fig. 3 in this chapter. This satellite drop, which is systematically present in drop formation, is the discovery of Plateau.

Now we shall describe another of the experiments he carried out. When a tap is open in such a way that we have a very thin flow, we can see that it is cylindrical and lasts only in the upper part. A little lower down the flow loses its regular shape, without it being possible for our eyes to see what is really happening (see Panel above on this page). But we can at least imagine it. The cylindrical shape requires a large surface energy. Thus the filament can reduce its energy by breaking up into drops (see Fig. 11). A cylindrical filament is unstable! This instability has been called Rayleigh–Plateau instability, the theory having been previously developed by Lord Rayleigh (see Chapter 3).

The fragmentation of the flow is anticipated by the appearance of swelling and shrinking which occurs while the drop is formed. High-speed photography reveals that a small drop is created in between two drops of normal size, an analogue of the Plateau satellite. In reality, before their separation, the two drops remain connected by a thin filament that, after some time, changes its shape to a small drop. The dynamical behaviour of the flow from the tap is relatively complex. The drops oscillate between two shapes, namely long and flat ellipses, before turning into a spherical shape. These oscillations have been revealed by modern photographic techniques as well as by using the old-fashioned stroboscope invented by Plateau that can illuminate the object being observed in rapid steps.

Physicist–Musicians

The first observations of the fragmentation of liquid jet were made by French physicist Felix Savart (1791–1841), who gave his name to the unit measuring the pitch of musical notes. Being an expert in acoustics, he wanted to study the effect of sound waves on liquid jets. He observed that the generation of a musical sound of the appropriate frequency in the vicinity of a jet could cause its fragmentation: the cylindrical part of the jet disappeared, and the jet was divided into drops from the very top. According to Savart, the drops started to form as they exited the tap. At the beginning, these are simple bumps that become more and more pronounced as the liquid falls to the point where they completely separate. These swollen drops, near to each other, cause a faint sound at a precise frequency (Fig. 12). A musical note, emitted in unison, can interact with the jet and transform it into a sequence of drops!

British physicist John Tyndall (1820–1893) repeated Savart's experiment on one jet 27 m high. Due to this height, he could obtain a uniform cylindrical and transparent jet: in fact, the sound emitted by one organ can induce turbulence and split the jet into a number of small drops. Tyndall allowed the jet to fall on targets placed at different heights, above and below the "critical point" where the jet would break up. Here is what he observed: "When a water jet falls on a liquid surface placed above the critical point and the pressure is not too great, then the water enters in the liquid without noise, but when the surface of the liquid is below the critical point, then one can hear the noise, and a large number of small droplets is observed." That comment gave American inventor Alexander Bell the idea of the water microphone (Fig. 13). His jet was much smaller than 27 m and high and was

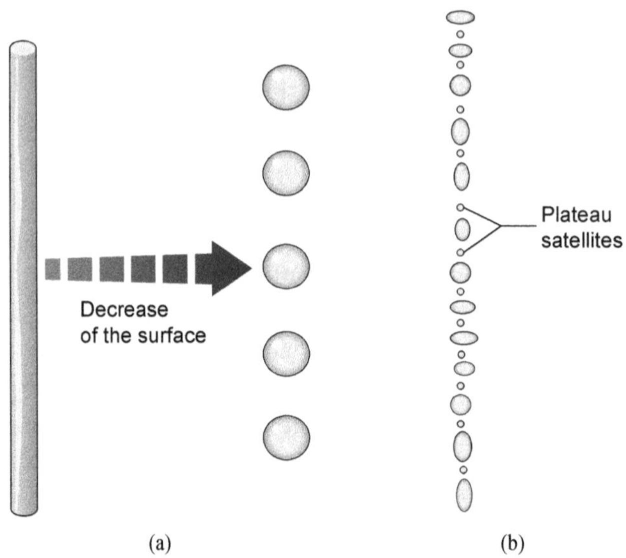

Figure 12. A cylindrical water jet tends to fragment into drops (a) in order to decrease the surface energy. In fact, under some conditions, a small drop (the Plateau satellite) appears in between two large drops. Because of the deformation when the drop leaves the tap, oscillations occur, and this explains the nonspherical shape which is observed at a precise moment (b).

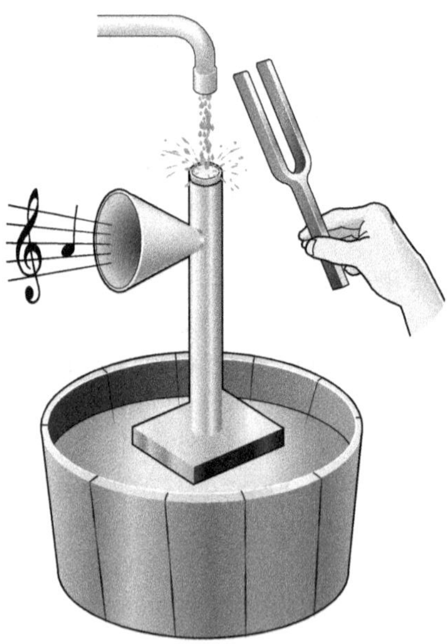

Figure 13. The water microphone of Alexander Bell.

falling on a rubber membrane instead of the liquid surface. The membrane was set at the top of a tube in which another funnel-shaped tube was embedded. In agreement with the experiment by Tyndall, the lower part of the water jet was broken into drops upon reaching the basin. Due to the resonator represented by the tube and the funnel, the tapping of the drops was amplified. When a vibrating tuning fork was placed in the vicinity of the water flow, it could induce a splitting choir of drops: the tick-tock of a clock could be heard in a large hall.

A man named Donat claimed in a popular text that at the end of the 19th century he tried to use such a device to transfer his voice. In fact, the jet had projected his speech but in a way so unpleasant that all those present had left!

The water microphone is not the most important invention by Alexander Bell, better known for the achievement of the telephone. It should be added that the paternity of the telephone seems to be jointly assigned to Elisha Gray (1835–1901) and to Antonio Meucci (1808–1889). These three brilliant inventors were American, but the first was born in Edinburgh and the third near Florence. After a conflict over the patent, nowadays the paternity is generally assigned to Bell.

Chapter 7

The Climate: Why Summers Become Hotter?

Over time, a climate became established on the Earth that was favourable for the origin and continued existence of life. This is due to the Sun warming our planet with its rays, and the natural greenhouse effect, plus the complex dynamic balance between the oceans and the atmosphere. In this chapter, we shall look at the basic physical mechanisms for maintaining the temperature of the Earth's surface, making it comfortable for the human body. However, human activities and industries now take place on such a scale that they nowadays have a measurable and significant impact on the natural world.

The weather is mainly controlled by the laws of physics, although chemistry and biology also play a part. These laws are well known, and meteorologists apply them in an attempt to predict the weather for the coming days. As everybody knows, sometimes they fail. They use deterministic laws that in principle should allow them to predict the future so long as the present state is known. However, the present is never exactly known, and furthermore the equations can be solved only numerically, to a certain level of precision. Their solutions may also be extremely sensitive to even small errors in the data. It is now common to claim (by exaggerating!) that the simple beat of a butterfly wing, by slight modification of the atmosphere, could induce radically different meteorological conditions, for instance, causing a storm in another part of the Earth. This extreme sensitivity to a small lack of precision is defined as "deterministic chaos".

The task of the climatologist seems, in a sense, simpler than that of the meteorologist. If he makes predictions, then the latter should be applicable over fairly large regions and over long periods of time. So, in climatology, climate predictions are made, which will be established in 50 or 100 years. Does this mean that deterministic chaos has no effect on the climate? Hardly. However, chaos is not the

Figure 1. Climate history is revealed by ice cores drilled in the Antarctic. Over the last 450,000 years, five glacial–interglacial cycles have occurred. These cycles are strictly connected with the orbital parameters of the Earth, the interval in between two glaciations being around 100,000 years (Courtesy of © Thibaut Vergoz/CNRS Images).

main enemy of a climatologist — first of all, he must take into account the complex, diverse and interdependent physical phenomena that form the climate. These phenomena occur on different, up to cosmic, scales, in very different time intervals (they can last for days, months, centuries, millennia — see Fig. 1). Certain processes essential for the formation of the climate can occur at any altitude of the atmosphere, at any depth of the seas, etc.

Terrestrial Radiation Balance

One of the easier physical quantities to predict is the temperature. The Earth's temperature depends in large part on the heat received from the Sun, which drives all climate mechanisms. This heat is arriving to us in the form of electromagnetic radiation, part of it being in the visible range (see the figure in Chapter 3). The Earth returns part of this energy by itself emitting electromagnetic radiation back into space. This irradiation, being in the range of relatively low frequencies, cannot be seen by our eyes but has the crucial role of moderating the temperature at the surface, thus allowing human life to exist.

Before analysing the climate further, let us pause to examine some characteristics of electromagnetic radiation.

From atomic to black body spectra

All bodies, brought to a certain temperature, emit light or more generally electromagnetic radiation. The radiation emitted by a body at a given temperature is characterised by its spectrum, namely by the plot of the power emitted as a function of the wavelength λ or the frequency $\nu = c/\lambda$. A simple method to examine the light emitted by a body is to place a prism in the trajectory of the beam and analyse what happens on a screen: the different spectral colours that form the light are separated (see Chapter 3).

The radiation spectrum of a body depends on its chemical nature, temperature and aggregation state: whether it is a gas, a liquid or a solid. An atomic gas (for example, mercury vapour) at low pressure emits radiation of well-defined frequencies; its spectrum consists of clear lines (Fig. 2). This fact expresses a fundamental property of atoms: the portions of energy that the latter can emit or absorb are discrete (it is explained by quantum mechanics, see Chapter 22).

Each line of the emission spectrum corresponds to the transition of an atom from a state with energy E_2 to another state with a lower energy E_1. This transition is accompanied by emission of radiation of frequency $\nu = (E_2 - E_1)/h$, where $h = 6.67 \cdot 10^{-34}$ J·s is Planck's constant.

This single act of irradiation corresponds to a "quantum of light", a *photon*. On the contrary, if a monochromatic ray of frequency ν is sent on the atom, it can

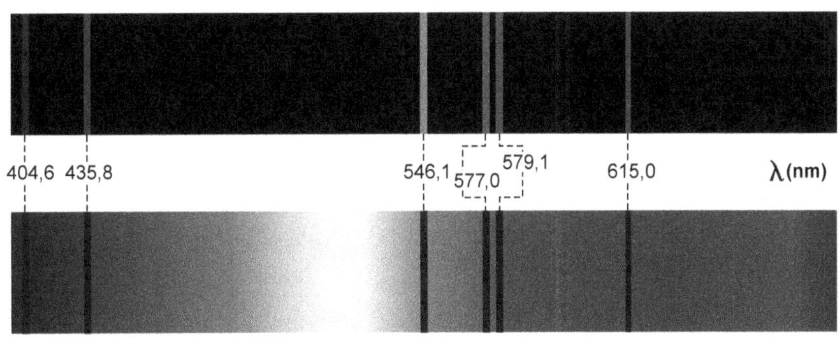

Figure 2. Emission (top) and absorption (bottom) spectra of mercury vapour obtained by letting the light beam pass through a prism. Similar to the generation of the rainbow by water droplets, the prism causes the dispersion: the refraction index depends on the wavelength (see Chapter 2). The spectrum emitted by a monoatomic gas excited by the temperature is composed of monochromatic radiation (spectral rays) that is characteristic of its chemical nature.

absorb the photon only when the frequency corresponds to the transition between two energy states differing by $\Delta E = h\nu$. Therefore, the atoms can absorb only the radiation that they can emit. Thus an emission spectrum is somewhat similar to the absorption one: the emission rays appear as dark lines in the absorption spectrum. This holds for a gas composed of atoms. What about a gas given by an assembly of oxygen molecules O_2 or nitrogen molecules N_2, as with the air around us? The principle is similar to that of atoms: a generic frequency cannot be absorbed or emitted. However, for certain frequency ranges, the energy levels in the molecules are very close to each other so that the molecular emission spectra are quite different from the ones in atoms, as we shall address later on.

What about solids or liquids? Their spectra do exhibit ranges of frequency where they cannot absorb or emit radiation. On the other hand, the emission spectrum (and the absorption as well) is practically a continuum in other frequency ranges, with some alternation of maxima or minima so that the spectrum is usually a complex curve.

Among 19th century physicists, particularly in Germany, there was some controversy regarding that difference. They referred to a particularly simple emitter of radiation. For instance, a body capable of absorbing all the radiation hitting its surface. This conceptual object was introduced by German physicist Gustav Kirchhoff (1824–1887). It was defined as *black body*: in fact, if a body should absorb all the visible radiation, it would appear black. We shall see that the Sun, and to a certain extent the planets, can be approximately a black body.

The Planck formula

In reality, a black body appears black only at low temperatures. When heated, it emits light. What is the colour of this light? In order to answer this question, we must derive the distribution of radiation by a black body, for a given temperature T, as a function of the wavelength λ. This was the problem addressed by Kirchhoff in the year 1859.

It is not obvious that there is a solution: we might think that the problem is absurd, in the sense that there could be a body that absorbed light with no emissions. Such a body, placed in front of a black body emitting radiation, would represent a refrigerator without any energy cost: the black body would become colder by transferring heat to the other body just by means of irradiation. This refrigerator with no cost would contradict the principle formulated by Sadi Carnot in the year 1824 (see Fig. 3). According to this principle, in order to be able to transfer heat from a cold object to a warm one, namely to achieve refrigeration, energy, whether chemical, electrical or mechanical, must be expended. From the

Figure 3. Physicist Nicolas Léonard Sadi Carnot (1796–1832) with the dress of polytechnic student. The Carnot family includes a number of eminent persons: Lazare Carnot (1753–1823), war Minister during the First Republic and mathematician; his nephew Marie-Francois Sadi Carnot, President of the Republic from 1887 to 1894; also the chemist Marie-Adolphe Carnot (1839–1920), brother of the President, gave his name to a mineral, carnotite.

Carnot principle, which became the second law of thermodynamics after reformulation by physicist Rudolf Clausius (1822–1888), Kirchhoff was able to deduce that all black bodies emit light in the very same way (see Panel below on this page). More precisely, the amount of energy emitted in the wavelength interval between λ and $\lambda + d\lambda$ by a given surface element dS of an absolutely black body with a temperature T is equal to the value of $Q(T, \lambda)\ dSd\lambda$, whatever it may be.

Proof of the Kirchhoff principle and the properties of a black body

Imagine two black bodies A and B of the same temperature. Let's place them in a thermally insulated closed chamber with reflective walls. According to Carnot's principle, in the absence of work produced by external forces, heat transfer always goes from a hotter body to a colder one. Since both bodies have the same temperature, there is no net heat transfer from one to the other, which means that the energy received by body A from body B is equal to the energy emitted by it in the

(*Continued*)

(*Continued*)

direction of body B. Therefore, if we replace one of the perfect black bodies, for example, body A with a perfect black body C of the same shape, but of a different chemical nature, then it will have to emit the same amount of energy as body A since it receives this same amount. Thus the amount of energy emitted from a unit surface per second at a given temperature (that is, the radiation power) for any perfect black body C will be the same as for a perfect black body A. Thus, the radiation power from a unit surface for all black bodies is determined only by temperature.

Now we place a colour filter between two perfect black bodies, which allows only light with wavelengths close to a certain value of λ to pass through. The same reasoning as with the absence of a filter shows that the radiation power at a given temperature within a given range of wavelengths (those the filter transmits) is the same for all perfect black bodies. Furthermore, it will be understood that the total radiated power of the radiation is proportional to the surface area of the body. Therefore, we can conclude that the power $Q(T, \lambda)\,d\lambda$ emitted by a unit area of a perfect black body in the wavelength interval $d\lambda$ depends only on the temperature T and the selected wavelength λ (the function $Q(T, \lambda)$ is called the spectral density radiation power). With equal temperatures, this power is approximately the same for both a coal block and a tungsten wire since both of these objects, with some assumptions, can be considered to be perfect black bodies.

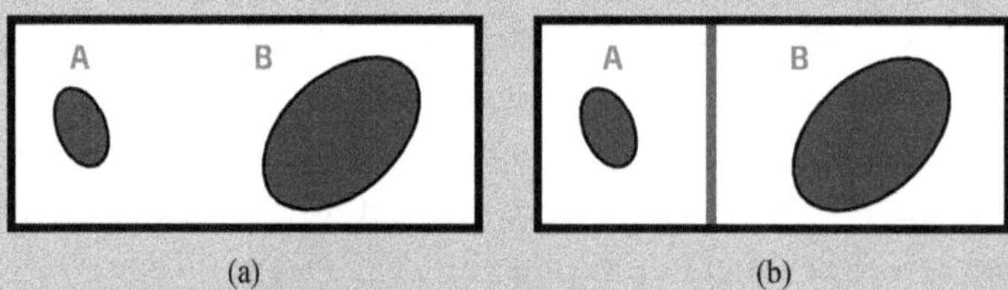

(a) Two black bodies in equilibrium have the same temperature, which means that each receives as much energy as it emits. If one of the perfect black bodies is replaced by another perfect black body of the same shape but of a different chemical nature, it will absorb the same energy, and therefore must emit the same energy. Thus the energy emitted by any perfect black body of a given shape at a given temperature does not depend on its nature.
(b) If you add a colour filter, the reasoning remains valid. Thus the energy emitted in a given frequency band by a perfect black body of a given shape at a given temperature does not depend on its composition.

Therefore, in order to derive the function $Q(T, \lambda)$, we can consider a system where the calculations are the the simplest, specifically a cubic cavity where the electromagnetic radiation (or if one prefers, the photons) is embedded. From the small hole made in a face of the cavity, the photons can escape. From this model situation, it is possible to derive the expression for $Q(T, \lambda)$:

$$Q(T,\lambda)=\frac{2\pi hc^2}{\lambda^5}\frac{1}{e^{\frac{hc}{k_BT\lambda}}-1}$$

where c is the speed of light in a vacuum, h is the Planck constant and $k_B = 1.38 \times 10^{-23}$ J K^{-1} the Boltzmann constant. It should be remarked that to derive the above expression, the quantum mechanical properties of the oscillator (the light wave inside the cavity) have to be taken into account. Thus, this famous derivation given by Max Planck (1858–1947) in 1900 not only provided the solution to a difficult problem that many experimentalists as well as theoretical physicists had been involved with. In addition, the expression for $Q(T, \lambda)$ given above marked the beginning of the era of quantum mechanics (see Chapter 22). In fact, at the heart of Planck's solution was the hypothesis that radiation energy takes only discrete values which are multiples of a certain *quantum* of energy. This hypothesis inspired Einstein to postulate the existence of the photon and allowed him to explain the photoelectric effect. For that, he received the Nobel reward in Physics in the year 1921. The historical importance of the black body is clearly evident.

From black bodies to stars

In accordance with the Planck formula, the emission spectrum of a solid is continuous, at variance with that of atoms which are given by discrete lines. The power emitted by a black body is maximum for a wavelength $\lambda_{\max}$ that is inversely proportional to the absolute temperature:

$$\lambda_{\max} T=\frac{hc}{4.965k_B}=0.0029\,K\cdot m.$$

This is the Wien law. The maximum of the wavelength shifts towards ever smaller values when the temperature of the black body is increased (Fig. 4). Thus on increasing temperature, the black body progressively changes colour from red to white according to the progressive shift of its spectrum towards the blue.

The emission spectrum of solid metals corresponds pretty well to that of the black body. This is the case for the filaments of the incandescent bulbs that were used in our homes during the 20th century (Fig. 5). The metal normally used is tungsten since its melting temperature is rather high (3422°C), thus

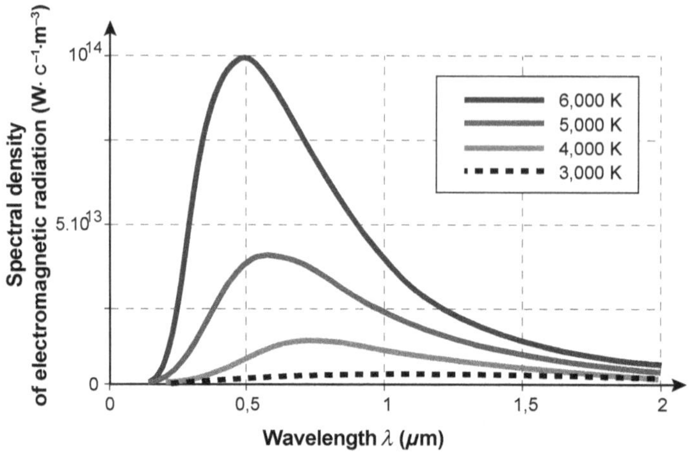

Figure 4. Emission spectra of a black body at different temperatures as a function of the wavelength. It should be noted that absolute temperature T is related to the Celsius scale C by $T\,(\mathrm{K}) = T\,(\mathrm{C}) + 273.15$. The temperature reported in the figures can be compared to that of the surface of the Sun (5,800 K) or that of the tungsten filament inside old-fashioned incandescent lightbulbs (about 3,000 K).

Figure 5. The filament of an incandescent bulb is not very different from a black body, and when it is warmed up to about 2,000°C, it emits a white light. The filament is heated due to the Joule effect (Chapter 16). At the same time, a large portion of the emission spectrum lies beyond the visible range. This represents a significant energy waste, leading to the production of incandescent bulbs being halted at the end of the 20th century.

being able to sustain temperatures that lead to the emission of a practically white light.

Thanks to the Wien law, it is possible to estimate the temperature of an incandescent body just by looking at the maximum in its emission spectrum. That law has also an important application in astronomy, letting us estimate the temperature of the surface of stars. Like other stars, the Sun displays an emission spectrum similar to that of a black body (we can compare the curve in Fig. 4 corresponding to 6,000 K to the one in Chapter 3), with the addition of the absorption rays characteristic of those elements present in its atmosphere. What about the radiation emitted by planets such as the Earth?

The Temperature of the Earth

We have seen that the Earth receives energy from the Sun and loses a large part of it by emitting electromagnetic radiation back towards space, after having made use of that energy. This "use" is of particular importance since solar energy enables life on Earth. The minimal use that occurs on all planets is to keep them cool. Radiation from the Sun maintains a more or less constant surface temperature, which would otherwise inexorably decrease. The electromagnetic power emitted by planets towards space is a function $F(T, \lambda)$ of the temperature T of the surface. On the other hand, the power $P(\lambda)$ received by the planet from the Sun is well known. Since the two powers are approximately equal,[1] we can write $F(T, \lambda) = P(\lambda)$, and this equation in principle determines the temperature T of the surface. We still have to find the function $F(T, \lambda)$. Are we allowed to assume that this function is close to that of a black body? This hypothesis is approximately correct when the planet does not have an appreciable atmosphere. Fortunately for us, the Earth does have an atmosphere (Fig. 6). Without this protecting layer, the average temperature of the Earth would be –16°C, not so favourable for life. Instead, it is 15°C.

The greenhouse effect

What is the role of the atmosphere in causing such a difference of about 30°C? The atmosphere collects a good fraction of the radiation emitted by the ground,

[1] This description is simplified. In fact, an appreciable part (on average 30%) of the energy received by the Earth is reflected (by clouds or by sand) or scattered (by the air and clouds). Thus in the equation $F(T) = P$, P must be considered to be the power which is absorbed by the Earth not the one globally received.

Figure 6. Picture of the atmosphere surrounding the Earth, taken from space. Its density decreases with increasing height. Its depth is much smaller than the radius of the Earth: 90 percent of its mass lies within a layer only 16 km wide around the Earth.

through a mechanism that was explained by Joseph Fourier (1768–1830), at the beginning of the 19th century. He wrote, "the temperature is increased by means of the interposition of the atmosphere because the heat does not find a great obstacle in penetrating the atmosphere when it arrives from the Sun in form of visible light, rather than the case when that heat is in non-visible form". Nowadays, the non-visible heat is known as infrared radiation (with wavelengths approximately in the range of 0.7–500 microns); disregarding the detail of the vocabulary, the analysis by Fourier is correct. The atmosphere allows solar radiation to pass through in large part unimpeded, the maximum intensity being in the visible range which can "penetrate the air", while the infrared radiation emitted by the ground back towards space is, to a large extent, absorbed by the atmosphere and only reaches space after a long pathway (Fig. 7): instead, it is absorbed and then re-emitted a number of times. This mechanism is called "greenhouse effect" since in a garden greenhouse, we want to limit the energy lost through the use of glasses or plastic sheets, which are opaque to infrared radiation.

Thermal heating of the terrestrial atmosphere by convection

As described above, the temperature at the surface of the Earth is determined by the equilibrium of the heat received from the Sun and the energy that is irradiated towards space. This energy is in part emitted by the ground but to a large extent by the atmosphere, at a height that depends on the wavelength. Therefore, we have

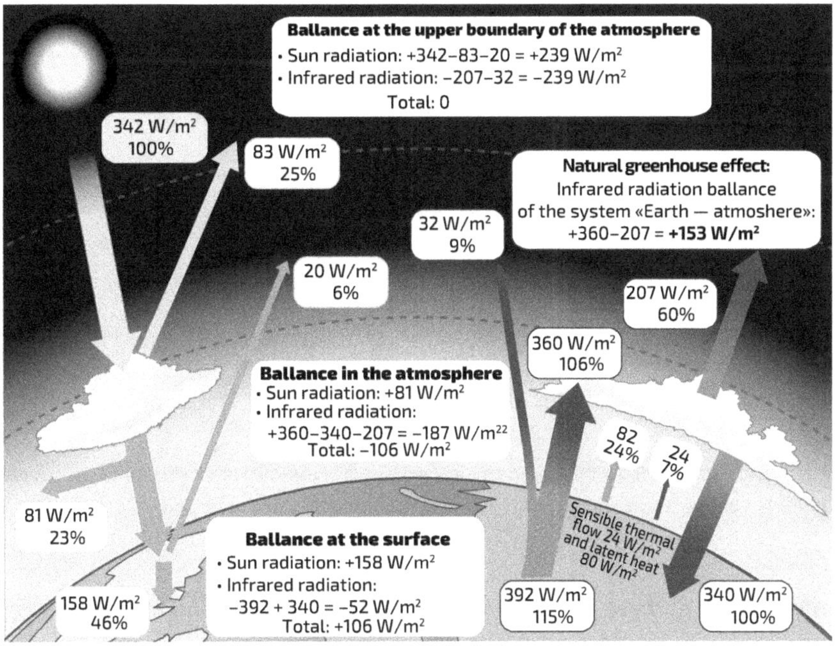

Figure 7. Radiation balance in heat exchanges between the surface of the Earth and the terrestrial atmosphere. Heat arriving from the Sun is in part reflected, in part scattered and in part absorbed. From R. Delmas *et al.*, *Atmosphere, Oceans and Climate*, Belin (2012).

to study how the atmosphere is organised. The pressure P of the atmosphere is a decreasing function of the height z since the force acting on a surface is due to the weight of the column towering above. The mechanical equilibrium implies that $dP/dz = -g\rho$, where g is the acceleration due to gravity, while ρ is the density of the air. This latter value is proportional to P/T, according to the Gay-Lussac law, T being the absolute temperature. Therefore, the decrease in pressure with increasing altitude is described by an exponential function. Very approximately, the pressure decreases by a factor of 2 when the height is increased by 6,000 m. However, the atmospheric pressure at altitude can vary by several percent during the day, while at sea level, it varies from one point to another with respect to the "normal pressure" of 101.3 kPa even at the same time. It might be surprising to learn that the low pressure we may detect in a given place is not suddenly compensated for by air coming from regions where the pressure is higher. Often it is the Earth's rotation that hampers a quick compensation and the Coriolis force (see Chapter 4) and can stabilise the low or the high-pressure regions for several weeks.

In low layers of the atmosphere, the temperature generally decreases with increasing altitude. The reason is that the heat coming from the Sun is mainly

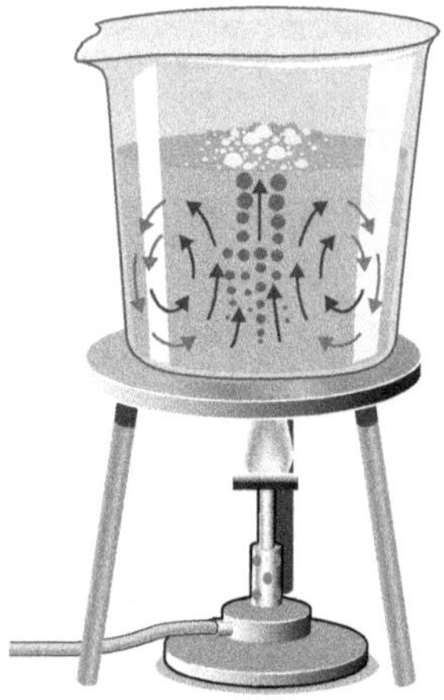

Figure 8. Evidence of convection currents by heating a glass of water. The liquid in contact with the bottom is getting warm and goes up, inducing a circulation in the vessel.

absorbed by the surface of the Earth. Then it is distributed along the atmosphere. This redistribution in part occurs through irradiation, but mainly because of the convection (Fig. 8). In other words, the warm air near the surface of the Earth rises, while the cold air from the layers at higher altitudes tends to come down.

In the atmosphere, large masses of air are in motion, sometimes on the scale of the entire globe (Fig. 9). These masses of warm air rise into regions of lower pressure and therefore their volume expands. According to the laws of thermodynamics, the expansion implies a reduction in the temperature (air is a poor conductor of heat, so the expansion is adiabatic, see Chapter 13; expansion of a gas is just one of the classical procedures for the cooling process).

For dry air, the extent of the cooling can easily be estimated: it is found that the temperature decreases by about 6.5°C for an increase in altitude of about 1 km. However, the decrease of temperature on increasing altitude is not an absolute law: fluctuations can change the local temperature. Furthermore, it is important to note that the temperature only decreases up to a certain altitude. In fact, at about 11 km, the average temperature is around –56°C, but on ascending further, the temperature increases! (see Fig. 10).

Indeed, convection only involves the part of the atmosphere called the "troposphere", that portion below a certain altitude. Above it, we enter the "stratosphere" where heating is no longer provided by the irradiation from the surface of the Earth but by the Sun. Ultraviolet rays can maintain a temperature of around 0°C at an altitude of roughly 50 km. These rays are absorbed by molecular oxygen O_2 which reacts to form ozone O_3 and also some heat.

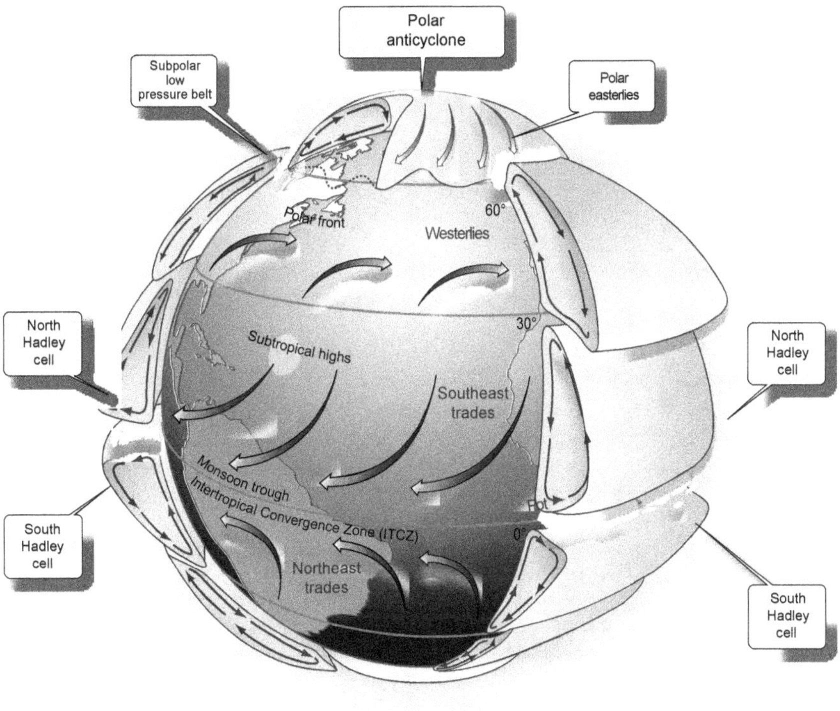

Figure 9. The large convective motions of the air in the atmosphere and the circulation of the winds. The motions near the Equator have characteristics, with convection channels known as Hadley cells. They are responsible, through the Coriolis force, for the trade winds which are systematically directed from East to West (see Chapter 4).

The water cycle plays an important role in heat exchanges at a global level. Air near the surface of the oceans acquires water vapour originated by evaporation. The cooling of this related to its upwards motion causes it to condense and produce water vapour which, in turn, provides some heat (see Chapter 15). Thus clouds are formed and possibly the small droplets that will create rain and thus return to the ground.

Other heat exchanges occur through ocean currents. In the climate, they play a role that's less evident than the atmosphere, but it is well known that just ocean circulation can transport heat to regions that are at the same latitude but at different longitudes. For instance, the Gulf Stream provides considerable warming to the European coasts so that the climate is more gentle than that of Canada. Those currents have worldwide amplitude and great stability.

Even if convection is the main mechanism for transporting heat within the troposphere, we should not disregard the importance of radiation. This is entirely

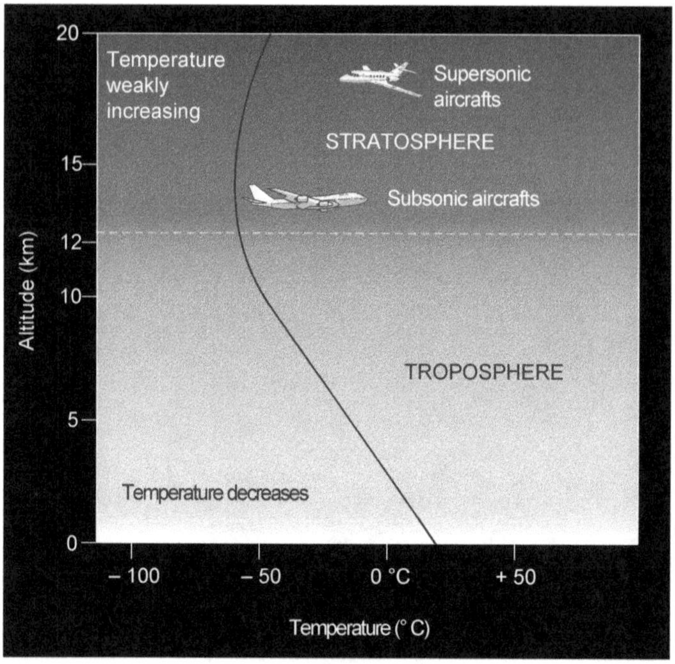

Figure 10. Variation of average temperature with altitude (dark line), in the lower layers of the atmosphere. The troposphere, which measures approximately 12 km is the place where most atmospheric phenomena occur.

responsible for the loss of energy towards space and at the same time for the transport of the energy stored in the ground or in the oceans towards the lower layers of the atmosphere.

Responsibility of the Various Molecules in the Greenhouse Effect

The atmosphere is composed mainly of nitrogen (N_2) by a fraction of about 78% and oxygen (O_2) by 21%. At variance with what Fourier wrote, it is not these gases that absorb the infrared radiation. The absorption is mainly due to water vapour (H_2O), carbon dioxide (CO_2) and methane (CH_4) (these so-called greenhouse molecules make up just over 1% of air by mass), as well as clouds. It was John Tyndall in the middle of the 19th century who realised that nitrogen and oxygen molecules do not absorb infrared radiation.

The emission (or absorption) of electromagnetic rays by molecules is due to the oscillatory motion of the negative electric charge with respect to the positive charge. In the visible range, these oscillations involve transitions among the electronic levels. In the infrared range, it is the chemical bonds between atoms that are vibrating: the lengths of the bonds and the angles between them are involved in the vibrations around their average values. Thus, in a water molecule, we have oscillations of the positively charged hydrogen atoms with respect to the negative oxygen atoms (these charges are related to the different values of the electronegativity for the hydrogen and oxygen atoms, see Chapter 16). In an analogous way, in the carbon dioxide molecule, it is the carbon atom (positively charged) that oscillates with respect to the oxygen atoms, negatively charged. These two molecules lose their symmetry during the vibration, and this implies a displacement of the negative charges with respect to the positive electric charges (Fig. 11).

On the contrary, when molecules of oxygen or nitrogen vibrate, their centre of symmetry (also the centre of gravity of the positive and negative charges) is conserved. This inhibits the absorption and the re-emission of infrared radiation in these two molecules.

The absorption of infrared radiation by the greenhouse molecules (H_2O, CO_2 and CH_4) depends on the wavelength in a complex way. The difference between the absorption spectrum of these molecules and the atomic line spectra (Fig. 2) or the continuous blackbody spectrum (Fig. 4) is striking. In certain ranges of wavelength, the "forbidden bands", absorption and emission, are not possible.

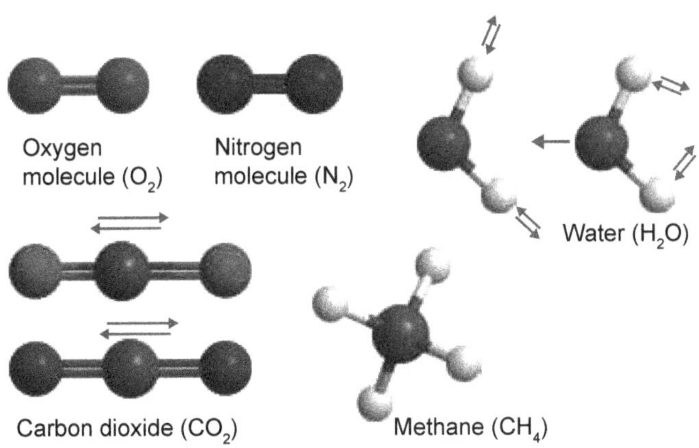

Figure 11. Some of the molecules present in the atmosphere. The dioxide molecule O_2 is symmetric, and it keeps this symmetry during vibrations. It cannot absorb or emit infrared radiation. The same occurs for the nitrogen molecules N_2. The water molecule is asymmetric, and when it vibrates, the positions of the negative electric charges change with respect to the positive ones, which allows the molecule to absorb or emit infrared radiation. The molecules of carbon dioxide and methane are on average symmetric, but when they vibrate they lose their symmetry, and this allows them to absorb or emit infrared radiation.

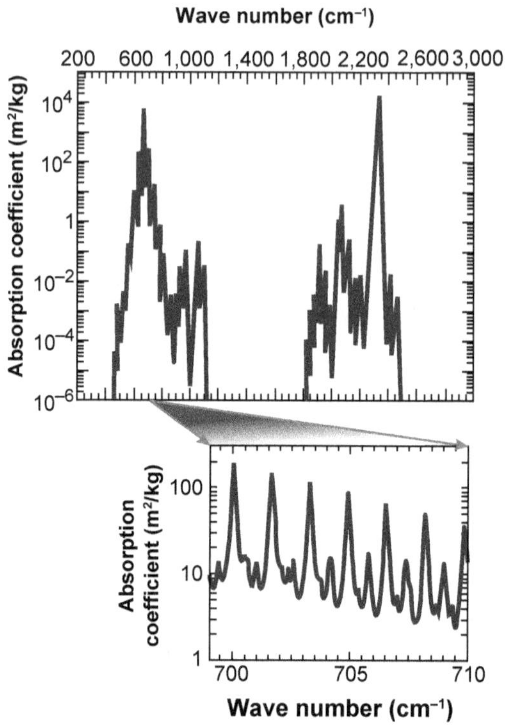

Figure 12. Absorption coefficient of carbon dioxide at 20°C and at pressure of 10^5 Pa, at two different scales. The wavenumber k is given along the x-axis, namely the inverse of the wavelength multiplied by 2π (i.e., $k = 2\pi/\lambda$). The absorption coefficient is defined by the relation $k\rho$ (ρ being the density of the absorption medium) equal to the probability of absorption of a photon per unit length. The sizeable variations of the absorption coefficient (by a factor of 10) imply considerable fluctuations in the contribution to the greenhouse effect corresponding to small variations of the wavelength. From R. T. Pierrehumbert, *Principle of Planetary Climate*, Cambridge University Press (2010).

In contrast, in the ranges defined as "permitted bands", the intensity can change in a regular way by a factor of 10 (Fig. 12).

Independent of their wavelengths, infrared photons emitted from the Earth's surface can take very different paths. They can travel through the terrestrial atmosphere without any absorption when the wavelength corresponds to a forbidden band (for water vapour as well as for carbon dioxide). When the wavelength corresponds to a maximum for absorption by a greenhouse gas molecule, the photons are rapidly captured, within every few metres. The photons are immediately re-emitted by the molecules at the same altitude, and so on. The photons that are finally directed towards space are the result of processes at high altitudes, where the temperature is much less than that at the ground. Thus, in the infrared range, there are fewer photons directed towards space than are absorbed, and the Earth warms, for a greenhouse effect that can be defined as "natural".

The Influence of Human Activity

During the 19th century, an additional greenhouse effect was initiated that has been added to the "natural" one related to the water vapour and carbon dioxide naturally present in the atmosphere. Humans have burnt combustibles defined as "fossil fuels" that have taken nature millions of years to produce: coal, oil and gases. Their combustion generates carbon dioxide, and, in just a short time, there is nowhere it can be stored. Through rain, it could reach the oceans, but the ability of the sea to store carbon dioxide is limited (fortunately for the fish). The concentration of this gas has increased pretty much from the industrial age (Fig. 13). If no changes regarding energy use are taken at political levels, there is the risk that levels could increase even more. It is likely this would raise the temperature in the lower atmospheric layers above dangerous levels.

Additionally, human activities are going to increase the amount of other gases that normally would be negligible. In fact, current intensive agriculture practises produce significant amounts of methane. This is a "new" gas in the atmosphere. It has long been known, for instance, that methane produces explosions in coal mines, often known as "firedamp" shots. Its concentration in the atmosphere has jumped from the middle of the 20th century onwards. This increase in the concentration of methane, due to the intensive growth of ruminants, has created novel barriers for the passage of infrared radiation towards space. Methane has become a further aggravating factor for global warming.

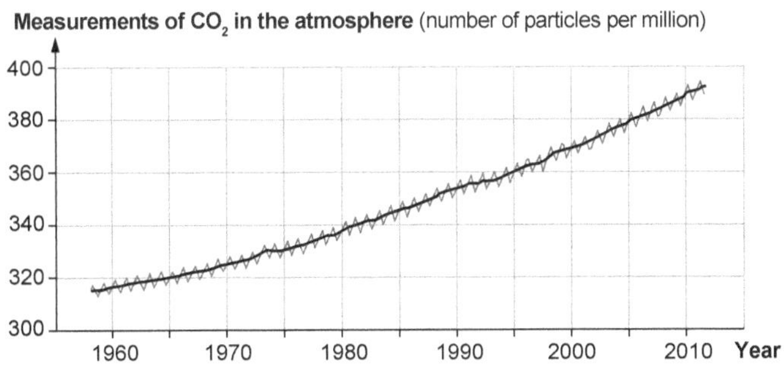

Figure 13. Evolution of the CO_2 concentration in the atmosphere, in parts per million (ppm) as measured at the observatory of Mauna Loa, in the Hawaii archipelago.

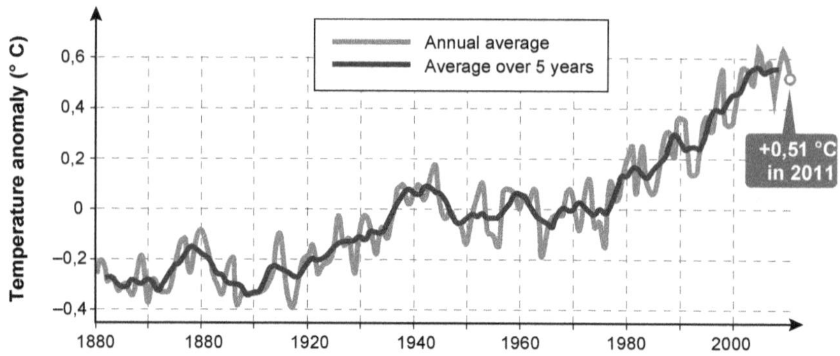

Figure 14. Evolution of the temperature from 1880 to 2010. The dark line tracks the values averaged over five years, smoothing fluctuations from year to year.

This warming is absolutely proven (Fig. 14): practically, it is half a degree from 1970 to the present. Can it be definitively linked with the increase of carbon dioxide and methane, due to human activities? Very likely yes, according to the experts of the IPCC (*Intergovernmental Panel on Climate Change*) although the precise estimate of its impact is not easy to determine. The direct effect of carbon dioxide is relatively weak, but a number of indirect factors can amplify or even compensate for it.

One amplifying factor involves the concentration in the atmosphere of water vapour: the excess carbon dioxide causes a certain increase in temperature, raising the pressure of the water vapour. This implies a sizeable increase in the water vapour in the atmosphere which aggravates the greenhouse effect!

Conclusion

In the future, the climate will certainly depend on human activities. Even if the warming throughout the 20th century has only been half a degree, the calculations indicate that the 21st century could experience warming between 1 and 4 degrees, depending on the different scenarios assumed. The calculations are not entirely reliable, but given the estimates of the consequences of human activities, it is extremely unlikely that there will not be some consequences for the climate. Global warming would certainly lead to sea-level rise (it rose 3 mm between 1990 and 2010), because of the thermal dilatation and extra melting of the glaciers and other consequences more difficult to estimate and possibly even worse.

According to the IPCC, since 1970, there has been a significant increase in the frequency of so-called "extreme events", namely heat waves or heavy rains.

An analogous tendency is present in several regions in terms of droughts or, by contrast, tropical cyclones. Another qualitative issue, involving droughts or heavy rains, is the following: as a consequence of the increase in average temperatures, we should also expect temperature gradients to increase, causing stronger convective currents. This would be the case, for example, with the Hadley cells in Fig. 9 and their related trade winds. The effects of these convective currents would become more marked: humid areas would become more humid and dry areas drier.

Will humans be able to control this increase in temperature? The reduction of greenhouse gases conflicts with economical and sociological difficulties that are beyond the scope of this book. However, we note the increased awareness of the problems, as emphasised by the creation of the IPCC mentioned above. This organisation, collecting scientists from all over the world, advises governments and is a topical example. The IPCC, in conjunction with the former American Vice President Al Gore, was awarded the Nobel Peace Prize in 2007.

Chapter 8

Footprints in the Sand

What is more commonplace than a footprint in the sand? On the other hand, grains of sand display counterintuitive properties. Laboratory materials are not required to demonstrate this: simple kitchen tools will allow you to perform experiments, and for those lucky enough, to try on the beach!

British physicist Osborne Reynolds (1842–1912) was a distinguished specialist in the field of hydrodynamics. Perhaps with the aim of studying the motion of the waves, he walked on sandy beaches. He made the following observation, presented in 1885 at the meeting of the *British Science Association*: "When a foot is put on the sand left compact by a descendent tide, the area nearby the foot becomes immediately dry" (Fig. 1). According to Reynolds, the pressure made by the foot causes the sand to spread out, and, as a result, the water is pulled through the larger gaps between the surrounding grains of sand, causing the sand to become dry.

This loosening phenomenon seems contrary to common sense. Why does the pressure increase the space between the grains of sand, thereby forcing the water to leave? The answer has to do with the structure of this mixture of water and sand. For simplicity, we will assume that the grains of sand are spherical and the same, and this will lead us to the problem of packing hard balls. And later in this chapter, we will ask ourselves: how to stack atoms?

Stacking the Spheres

Could we fill all a space by stacking hard spheres? Obviously not, as some room is always left in between the spheres. The fraction of the space filled defines the density of stacking. The less room left free in between the spheres, the more compact the stacking and the greater the density. How do we achieve the maximum

Figure 1. Dry area formed around a foot placed on the sand.

density in stacking identical hard spheres? The response to this question will make clear why sand dries out when a foot is placed on it.

If the World Were Flat...

To begin, let us consider the problem in two dimensions instead of three: in this case, we are placing identical discs onto a plane. First, we take three discs. The most compact arrangement is when each disc is in contact with the other two, therefore when the centres of the three discs form an equilateral triangle. If we have many discs, the most compact arrangement will be achieved by placing together multiple such triangles of three discs (Fig. 2(a)). As can be easily determined, the fraction of the plane covered by the discs is $\pi/(2\sqrt{3})$, namely 90.7%. It is interesting to compare this value with the one obtained with the arrangement of discs in Fig. 2(b), where they cover only the fraction $\pi/4$, namely 78.5%.

Our intuition suggests that the maximum packing is the one in Fig. 2(a). Each disc is in contact with six others, and it is not possible to place a larger number of discs in contact with any given one. However, the fact that the number of discs in contact is the maximum does not guarantee that the fraction of the plane covered by the discs is really the maximum possible. The exact proof was only found in the 20th century.

Stacking and Tiling a Plane

It is noted that for the most compact stacking, the discs are situated within hexagons that totally cover the plane, as with paving tiles (Fig. 2(a)): one says that the

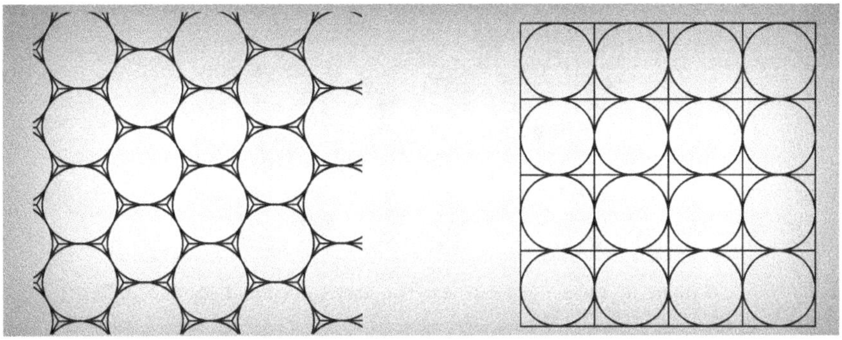

(a) (b)

Figure 2. Two arrangements of identical discs on to a plane. (a) The arrangement is the most compact; (b) a less compact disposition, in which the discs are situated within squares.

Figure 3. Bees achieve the tiling of the plane by hexagons when they are building up their nest. It is likely that this arrangement is even comfortable for their larva. Furthermore, it allows them to economize on beeswax. In fact, the shortest lattice of walls to mark the limits of a given planar area is just the tiling by regular hexagons.

hexagons tile the plane. This hexagonal disposition is the one created by bees to stack the alveolus of their nest (Fig. 3). There are only two other ways to tile the plane by means of identical regular polygons: by means of squares (Fig. 2(b)) or triangles (Fig. 4).

This result can easily be proved. In a regular polygon having n sides, each angle is given by $\alpha = 180°–360°/n$. On the other hand, if each vertex of the

Figure 4. Tiling a plane by means of triangles. The centres of the discs in Fig. 2(a) form such a "triangular lattice".

The software assisting the proof

At odds with classical mathematical demonstrations written in the human languages, the proof provided by Hales was in part based on computational language. Using computer software to achieve some proofs is something that's developed over recent decades. The software helps mathematicians provide formal verification.

The complete proof given by Hales takes the form of a series of articles covering more than 250 pages, and the software requires three gigabits of memory. Did anyone completely read and understand these articles by Hales? The Referees having to check the articles (published in the Annals of Mathematics *in 2005) were not confident that the proof did not have any errors, given the reliance on computer assistance. Only in 2014, did the team led by Hales provide a proof that was considered conclusive! Regardless, everybody was convinced that the stacking originally devised by Kepler was the most compact possible. If not, during the course of three centuries, somebody would surely have found a better arrangement, wouldn't they?*

polygon is common to $(m - 1)$ other polygons, then $\alpha = 360°/m$. The integer numbers n and m must satisfy the condition $2/m = 1 - 2/n$, which is possible (as can be easily verified) only by choosing $n = 3$, 4 or 6. In other words, it is not possible to tile a floor by regular pentagons!

If the compact stacking of discs does not have many practical applications, one can consider an equivalent problem in the real three-dimensional world: it is simply necessary to substitute the discs with regular cylinders. For instance, cylindrical electric wires are usually assembled in compact bundles. This is also the case for superconducting cables, as we shall see in Chapter 25, which are made by bundles of superconducting wires, each wrapped by a sheath of copper. In the beginning, the wires have the classical cylindrical shape, but, after a strong pressure, they take on the form of hexagonal prisms!

Spheres in the Real World

Now, let us step out of the plane and enter the three-dimensional world, the real space for both our spheres and of the world as well. How can we stack the spheres in the most compact fashion? Let us try by considering it reasonable to begin with a layer as suggested in Fig. 2. Then we place on top of it a second identical layer, taking care to ensure that the contact points are the most numerous possible. In this case, each sphere of the layer above is in contact with three in the layer below and vice versa (Fig. 5(a)). Then we place a third layer so that each of these spheres is in contact with three in the second layer, and so on. The result is that every sphere is in contact with 12 other spheres, six from the same layer, three in the layer above and three in the one below (Fig. 5(b)). Then we continue with the same procedure.

Have we solved the problem? Is the volume occupied by the spheres really the minimum possible? Yes, claimed Johann Kepler in the 17th century, better known for his discovery that the planets move on elliptical orbits around the Sun, which is at one of the two foci. However, Kepler did not have a proof for his claim. In contrast to the situation in the plane, rigorous proof in three dimensions is really very difficult! So difficult that the problem of stacking spheres

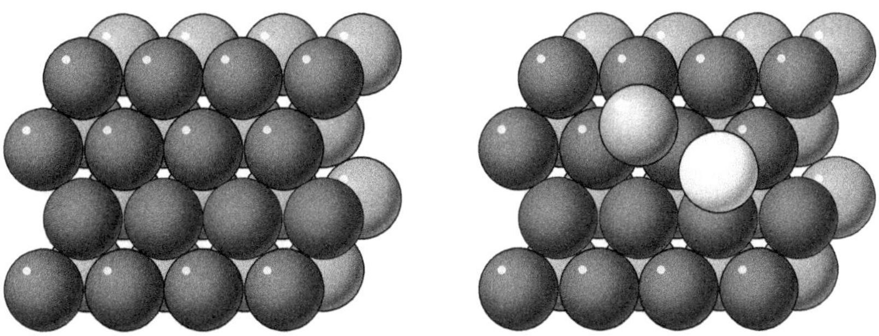

(a) (b)

Figure 5. Building for stacking spheres in the most compact fashion. (a) First, we place one layer compacted as much as possible (light spheres) and onto it an identical layer is deposited (dark spheres), in the way that each dark sphere is in contact with three light ones. (b) The third layer can be deposited in two different ways, six spheres being in vertical contact with the spheres in the first layer (as the striped sphere) or displaced as the spotted sphere.

Figure 6. Practical example of stacking of hard spheres, on a stall of oranges.

(sometimes called "the Kepler problem") is included in the famous list given by the German mathematician David Hilbert in an address at the beginning of the 20th century which outlined, in his opinion, the most important mathematical problems of the day. Only rather recently, in 1998, the American mathematician Thomas Hales claimed that he had solved the problem (see Panel on page 120). Unsurprisingly, Kepler was right, the stacking of compact layers as suggested is the most compact possible arrangement (Fig. 6).

The percentage of space filled by the spheres is about 74%. More precisely, a rather courageous reader could evaluate that it corresponds to $\pi/(3\sqrt{2}) \approx 0.74$.[1] We see that the fraction of space occupied is not that large.

A Problem with Multiple Solutions

Surprisingly, the stacking solution defined above is not unique. When the second layer is deposited on top of the first, there are two ways to do it, which are equivalent. In contrast, the third layer can be deposited in two ways that are not equivalent. It could be placed on the vertical of the first layer (Fig. 7(a)) or a little displaced (Fig. 7(b)). These two choices appear anytime a new layer is added.

[1] The main steps of the calculation are the following: in each layer, the N_1 centres of N_1 balls of radius R form a lattice of N_1 lozenges having area $2R^2\sqrt{3}$ (the fraction of the plane covered by the discs, as estimated above, being $\pi/(2\sqrt{3})$). The distance between two layers is $2R\sqrt{2/3}$. Having N balls, they occupy the volume $4NR^3\sqrt{2}$, being $4\pi NR^3/3$ the total volume of the balls.

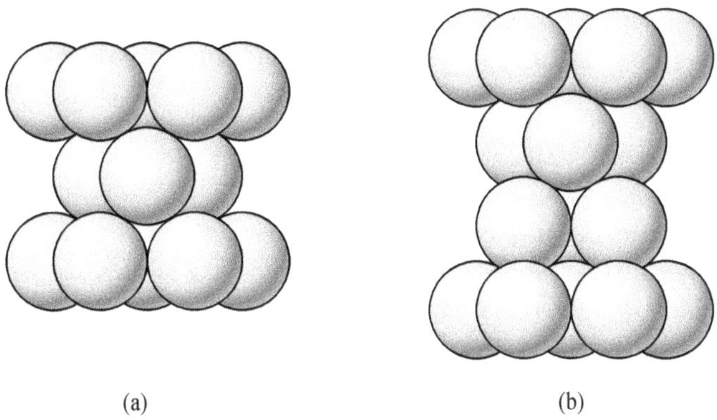

Figure 7. Compact packaging balls. (a) The case when the third layer is placed strictly above the first one (hexagonal packing). (b) The case when the third layer is displaced regarding the first. Wherein the fourth layer turns over the first (cubic face-centred lattice).

Thus the Kepler problem has infinite solutions. This nonsingular solution is at the core of the difficulty of the problem (see Panel on page 120). Two solutions are of particular interest:

(a) the one obtained by placing the *n*th layer vertically above the $(n - 2)$th one, this for each *n*;
(b) the one obtained by placing the *n*th layer shifted with respect to the $(n - 2)$th layer.

The natural world displays several examples of constructions that follow one of the rules mentioned above, as in certain crystals. The crystals are formed by packing atoms, molecules or ions, arranged periodically in the space (see Chapter 9). Rule (a) leads to the formation of the "hexagonal compact" crystal; Rule (b) gives rise to a "face-centred cubic" crystal. Several elements, for instance, cobalt, obey such an atomic arrangement.

Randomly Packed Balls

Let us return to the problem mentioned at the beginning of this chapter. What is the relation of the compact packing of balls with footprints in the sand?

When the reader finds themself on a sandy beach, they could perform the following simple experiment to clarify the observation made by Reynolds. First, take a jar made of somewhat flexible plastic material and fill it with sand. Then add water to just slightly above the level of the sand (Fig. 8).

From spheres kissing to the Kepler problem

In the stacking described above, each sphere is in contact with 12 neighbours. But if we consider only the ball in the centre, would it be possible to place more balls around it? If so, how many in total? This maximum number is known in English as the kissing number (the number of contacts between the spheres).

The corresponding problem in two dimensions has already been addressed a few pages above. It is easy to visualise, by reasoning as well as by experience, that the kissing number in the plane is 6. If we duplicate the arrangement, we can tile a plane.

In three dimensions, the solution is not so apparent. To develop an understanding of it, nothing works better than an experiment. First, get a sufficient number of billiard balls or ping-pong balls and fix them using some glue (not too rapid-acting): in order to not get dirty, you can put the balls in a transparent sack and try to obtain a compact pile. You will see that is not possible to place more than 12 balls around a central one. However, you will see some room in the pile and might think that another ball could be packed if someone had a particularly brilliant idea. In the 17th century, this was the subject of a dispute between Isaac Newton and the mathematician David Gregory. This thought that it should be possible to find a way to include an additional ball around the central one. Nowadays, we can demonstrate that Newton was right; however, the first proof wasn't given until 1953.

Could the discovery of the possible number of contacts in the space provide the response to the Kepler problem? No! There is an infinite number of ways to place the 12 balls around the central one, and this implies the difficulty of the problem. One of the ways corresponds to the face-centred cubic lattice (see Chapter 9). Then the centres of neighbouring balls form a polyhedron that is not regular, defined as a Kepler cubic-octahedron (see below). It has 14 faces, with eight equilateral triangles and six squares. On the triangular faces of the cubic-octahedron, the three balls have the maximum packing since their edges are equal to the diameter of the spheres, and then they cannot get any closer. It is not the same for the faces of the squares.

If we were to place the 12 peripheral balls without any particular rule, in general, their centres form icosahedrons (a polyhedron with 20 triangular faces) that can be regular or not. If we calculate the length of the edges, we find a value greater than the diameter of the ball. Therefore, we can modify the position of the peripheral balls, always with the condition that they touch the central one.[1]

[1] *Here are the main steps in the calculation for regular icosahedrons. The volume of the cubic-octahedron is* $5a^3\sqrt{2}/3$, *a being the diameter of the ball. The volume of the icosahedrons is*

(*Continued*)

(*Continued*)

Even when we accept that the packing of those polyhedrons is the most compact, there remains the question: cubic-octahedrons or icosahedrons? The regular icosahedrons have a volume a little smaller, but it is not possible to fill up the space by means of a stacking regular icosahedrons. This highlights the difficulty of the Kepler problem...

How many identical coins can be placed on a table in a way that they touch the central one? It is easy to demonstrate that the maximum number is six.

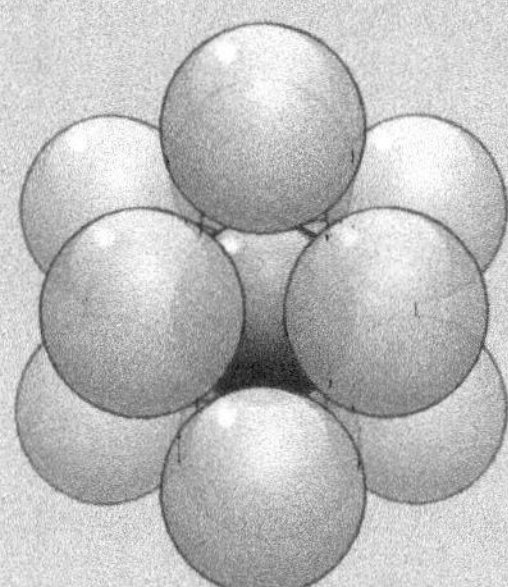

In three-dimensional space, we can place a maximum of 12 balls around a central ball. The external balls still leave some room in between them, and this might make us think that another ball could be placed.

$5(3 + \sqrt{5})b^3/\sqrt{12}$, *where b is the length of the edges. The square of the ratio b/a is* $8\ sin^2(\pi\ \pi/5)/[3 + sin^2(\pi/5) - cos(\pi/5)]$, *with the angles given in radians. The ratio* $b/a = 1.044$.

(*Continued*)

(Continued)

(a) (b)

(a) The figure obtained by an atom's nearest neighbours in a body-centred cubic crystal is the Kepler cubic-octahedron. (Remark: We need to shift this polyhedron in such a way that the vertical axis of Fig. 7(a) is not the one of Fig. 7(b).)
(b) Another polyhedron with 12 vertices, the regular icosahedron.

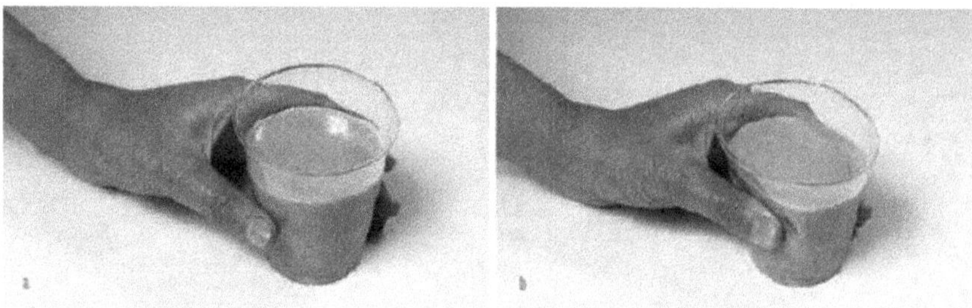

Figure 8. Evidence of the effect related to spreading out of the sand under pressure. (a) The plastic jar containing sand filled with water up to the level of the sand; (b) when the plastic jar is pressed in the middle, the water disappears from the surface of the sand.

Then squeeze the top of the jar, and the water will disappear, penetrating into the gaps between the grains of sand. Everything happens as if the sand "expands", and its upper boundary rises above the water level. This expansion effect only occurs if the initial arrangement of the grains of sand was compact enough.

Why should the departing tide leave the sand in a particularly compact configuration? In order to answer this question, we have to consider the mechanisms that can make the sand more or less compact. Let us begin with a simple experiment that can even be performed in a kitchen (maybe you have already done it!). We take a glass tankard, possibly with a graduated scale, and fill it with powder

Physics and fakirs

A counterintuitive property of granular media was already empirically known by the Indian fakirs. One of their tricks was to push a knife several times inside a vessel containing rice, filled up to the top. After some time, the knife was firmly fixed inside the rice, and, by lifting it, the entire vessel could be seen to lift. The craft of the fakir was to shake the vessel while adding the rice so that a reasonable packing was achieved. While pushing the knife, the grains of the rice were moved, and the room in between them was increased. The increase of the pressure and the friction produced on the knife could make it hard to remove the knife itself. During one of his conferences, physicist Pierre-Gilles de Gennes presented this very simple experiment: the knife was substituted with a piece of a broom handle and the rice with sand, while the fakir was substituted by the award of a Nobel prize.

The variant of the experiment of the Indian fakir.
(a) Fill the container tightly with sand with a stick held in the middle.
(b) Gradually compact the sand by tapping on the sides of the container.
(c) Grab the stick and pull it up: the container rises with it!

sugar. When we slightly shake this vessel, we will note that some room for a new amount of sugar will appear. In 1960, the Britain C. David Scott studied this effect quantitatively. He put identical small spheres in vessels of different sizes and evaluated the degree of filling. He found that the degree of filling is related to the way the spheres are added. If the spheres are allowed to fall into the vessel one by one, the degree of filling is about 60%, much smaller than the 74% in the case of a compact arrangement of the spheres. By shaking the vessel while it is being filled, then the degree is 64%, better but still less than the maximum value of the compact configuration.

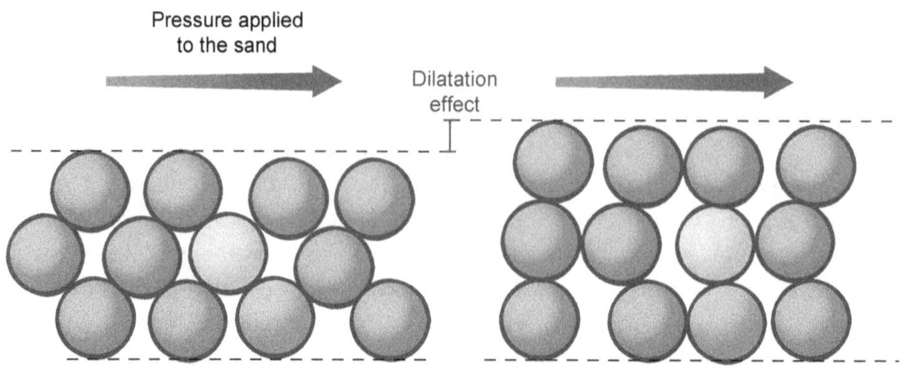

Figure 9. Explanation of the sand loosening. Due to the force applied by the foot, one of the grains (here the light one) is shifted under another grain (grey colour). From E. Guyon *et al.*, *What Fluids Say*, Belin (2011).

Therefore, the results obtained by Scott confirmed, in a quantitative way, our observation: the compactness of granular samples, such as the sand, is indeed increased when we shake the vessel. What is the role of this shaking? A ball at the bottom of a cavity is in stable equilibrium and will remain there if the shaking of the system is moderate. By contrast, a ball that is near the top of a pile is in a condition of unstable equilibrium. This is approximately what happens in our case: when we shake the balls in the vessel, some slip into an available space. The overall level of the filling material is then decreased, the centre of gravity of the whole system is lowered, and its gravitational potential energy is decreased.

We now have the explanation of the problem pointed out by Reynolds. As a consequence of the ebb and flow of the waves, the sea shakes the sand so that the grains attain the condition of compact configuration (although likely not the most compact). When the foot of Mr Reynolds is placed on the sandy beach, it induces a kind of deformation: the grains are somewhat shifted, and an increase of the free space in between them occurs (Fig. 9). Then the water is sucked into the cavities created in between the grains, and a dry area is formed around the foot.

Conclusion

In this chapter, a pleasant walk along the beach allowed us to study the mathematical problem of the optimal way to pack hard spheres in order to minimise the free space around them. Even though atoms are not balls, certain properties are similar to them. In Chapter 9, this correspondence will drive us to deal with the properties of matter.

Chapter 9

From Crystal Snowflakes to Vitreous Glass

In the 5th century BCE, the Greek philosopher Democritus formulated the atomic theory of matter. Even before that, the existence of atoms had been universally accepted. The hypothesis that matter was the assemblage of hard small balls attracting each other was the source for understanding the structure of the elements.

At the beginning of the 17th century, Johannes Kepler found himself drawn towards snowflakes. As the stars in Fig. 1 show, the majority of the snowflakes display hexagonal symmetry: the angle between two branches is 60°. This angle of 60° was already encountered in the compact packing of discs (Fig. 2 of Chapter 8), where the centres of neighbouring circles form equilateral triangles; in the same way as is found in the compact packing of hard spheres (Fig. 5 of Chapter 8). It is probably this observation that prompted Kepler to suppose that snowflakes are formed by microscopic balls in a compact packing arrangement. As we shall see, this hypothesis was not true. However, it opened the way to other theories that began with the study of the crystals and contributed to providing a solid basis for the atomic theory of matter.

The Crystalline Order

A French scientist, the abbot René Just Haüy (1743–1822), is considered the founder of the study of crystals, the science known as crystallography. In his *Essay sur la structure des crystaux* published in 1784, he suggested that all crystals are made by the repetition of small identical bricks (Fig. 2). This microscopic vision had been inspired by the macroscopic appearance of the large number of crystals he studied, cut and observed (Fig. 3).

Figure 1. Some snowflakes. The forms taken by the snowflakes are extremely varied: among other things, they depend on humidity and temperature.

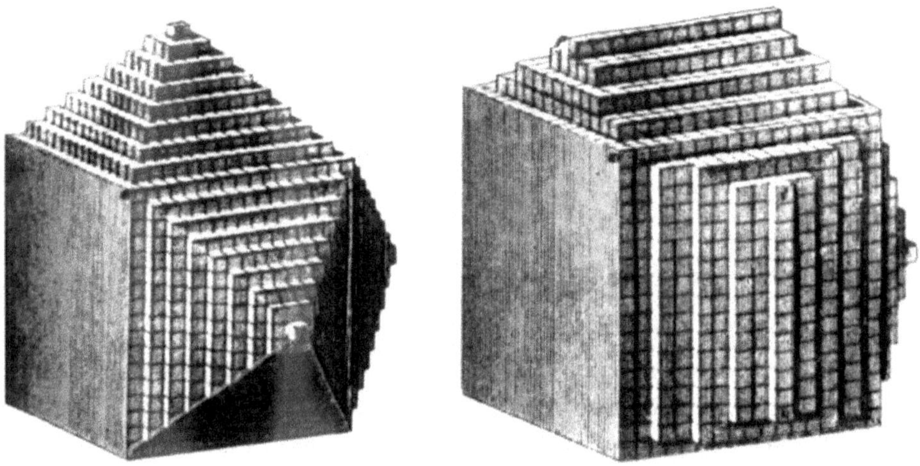

Figure 2. Drawings by René Just Haüy giving his interpretation of two forms of pyrite crystal (FeS_2). According to Haüy, the crystal is an assembly of small identical parallelepipeds placed with faces set in contact. These parallelepipeds being very small, their edges are not visible to the eye, and the faces appear smooth.

The idea that a crystal is purely a three-dimensional packing was taken up a little later (1824) by German physicist Ludwig August Seeber. In addition, he supposed that the basic unit of packing was small hard balls interacting with each other. In many materials, to a rather good approximation, we can indeed suppose that the atoms are hard spheres attracting each other. Even though atoms are not hard spheres, they have some analogous properties. In fact, it is rather difficult to push two atoms together beyond a certain distance, which is the sum of their so-called "atomic radius". This expression, commonly used in laboratories and textbooks, indicates that researchers have not entirely lost the habit of imagining

Figure 3. A pyrite crystal (FeS_2) also known as "fool's gold". This crystal naturally adopts very different geometries, from the simplest (a cube) to the more elaborate, which makes collectors happy.

atoms as small hard balls! These atomic radii are of the order of 10^{-10} m (a little smaller for the atoms of hydrogen, oxygen and carbon, and a little larger for the common metals).

Let us return to the crystals. Let us suppose the atomic hard balls are formed by attracting each other, in order to minimise their distance and so their attractive potential energy, to adopt a compact structure of the type studied in Chapter 8. In fact, there are several elements whose stable form at low temperature is the face-centred cubic arrangement of Fig. 7(b). For example, we can list silver, gold, copper and nickel, and around about 35 elements in total, with this form of packing. However, not all crystals are really compact, as we are going to address.

Crystallography and Its Instruments

Nowadays, crystallography can rely on very efficient experimental devices to investigate crystal structures. We have already shown that crystals are spatially periodic arrangements of atoms, molecules or ions (Chapter 8). One of these elementary units can describe the entire crystal: these units, called cells, are in fact periodically repeated along all three spatial directions. Furthermore, we can say that a long-range order exists. Electron microscopy (particularly the electron tunnelling microscopy) provides images where atoms can be observed individually (see Chapter 28), providing further evidence of their regular arrangement. The atoms appear as small balls, but we should not forget that they are far more complex objects, being formed of a nucleus with electron clouds around them.

Electron microscopy, on the other hand, allows us to study only the surface of the crystals. In order to study their insides, the most commonly used methods rely on X-ray diffraction (see Panel on page 129). On irradiating the crystal, the rays are diffracted and are creating interference phenomena that vary in intensity in specific directions. When a photographic plate is placed on their trajectory, we obtain an image, the diffraction pattern, that reveals the structure of the crystal.

The first diffraction patterns were obtained by German physicist Max von Laue in 1912 and by British physicists William and Lawrence Bragg in 1913: they justify René Just Haüy's theory. Thanks to the technique of X-ray diffraction, we can also determine the structure of very complex molecules, such as, for instance, the proteins in our bodies. The most celebrated example is the one for DNA, whose structure (see Chapter 28) was derived in 1953.

Inside our bodies, the proteins are obviously not crystals, but they can be extracted and then crystallised. The analysis of large molecules requires intense, well-collimated and monochromatic X-ray beams. These are generated by synchrotrons, which have been used for several decades. European nations have

Figure 4. The ESRF synchrotron ring at the confluence of the Isère (above) and Drac rivers. Electrons under the influence of a magnetic field move in a ring at a speed close to the speed of light. When the direction of speed changes, they emit intense X-rays. On the right in the cylindrical domed building is the nuclear reactor of the Laue-Langevin Institute.

X-ray diffraction from crystals

What are the principles of X-ray diffraction by a crystal? When a radiation of wavelength λ *of the same order of magnitude as the distance between atoms (a fraction of a nanometre), a phenomenon occurs similar to that when light is sent through Young's two slits (see Chapter 3): the atoms of the crystal diffract the X-rays similar to how Young's slits do for visible light. The crystal behaves as a regular lattice made of a large number of Young's slits (see figure). The rays diffracted by the atoms can then interfere.*

In any nonperiodic medium, in general, the interference is destructive: the diffracted waves can be approximately wiped out, apart from in the direction of propagation of the incident ray. In a crystal, because of the periodicity, there are other specific directions, different from the propagating ray, along which the waves emitted by the atoms are in phase. These directions reflect the existence of crystal planes to which many lattice points belong: the reticular planes (from the Latin word, ***rete****).*

Constructive interference occurs when the diffracted rays obey a condition known as the Bragg relation

$$2d \sin \theta = n\lambda,$$

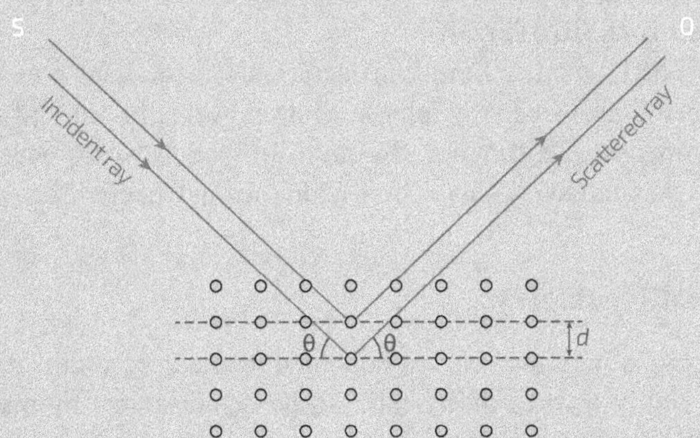

Principle of X-ray diffraction from a crystal. A source S emits X-rays directed towards the crystal at an angle θ with respect to the lattice planes. An observer at the point O measures the intensity of the beam that is received. Two lattice planes are separated by the distance d involved in the Bragg relation.

(*Continued*)

(Continued)

where d is the distance between the two lattice planes, θ the angle between the incident ray and two lattice planes, and n any integer. Experimentally, we can explore the Bragg relation by keeping the crystal fixed with respect to the X-ray beam and by varying the wavelength. By placing the photographic plate perpendicular to the trajectory of the diffracted X-rays, we obtain a pattern similar to the one reported in the following: the brilliant spots indicate the directions for which the Bragg relation is verified. This method was devised by Max von Laue (1879–1960) and was awarded the Nobel Prize in Physics in 1914.

For any direction of the lattice planes, there is always a wavelength that satisfies the Bragg condition. By studying the diffraction pattern, we can derive the distance between lattice planes, thus deriving the structure of the crystal, when rather simple. In more complex cases, besides finding the directions along which we obtain constructive interference and when the intensity is non-zero, we must also measure the intensity of the various diffraction lines.

jointly built such a generator in Grenoble, called the European Synchrotron Radiation Facility (ESRF) (Fig. 4).

Near the ESRF, another European facility, the Laue-Langevin Institute, has been installed and provides very intense neutron beams in order to study crystal structures by means of diffraction. Neutron diffraction works according to the same principles as for X-rays, as will be addressed in Chapter 22.

Crystals and Geometry

Crystallography is not just an experimental technique. Mathematics is also involved: several properties of crystals can be demonstrated by means of theorems. One of these is the crystal lattice, namely the three-dimensional arrangement in which the atoms or the molecules of the crystal are placed. Well before the research by von Laue, a large number of mineralogists devoted their attention to the classification of the various types of crystal lattices. A particular case is when the cell behaves as a single atom: such a lattice is defined as a Bravais lattice. Each Bravais lattice is characterised by the geometrical transformations (rotations and symmetries) that leave its configuration unchanged. There are only 14 types of Bravais lattices (Fig. 5). This was proved mathematically by the French scientist Auguste Bravais in 1848.

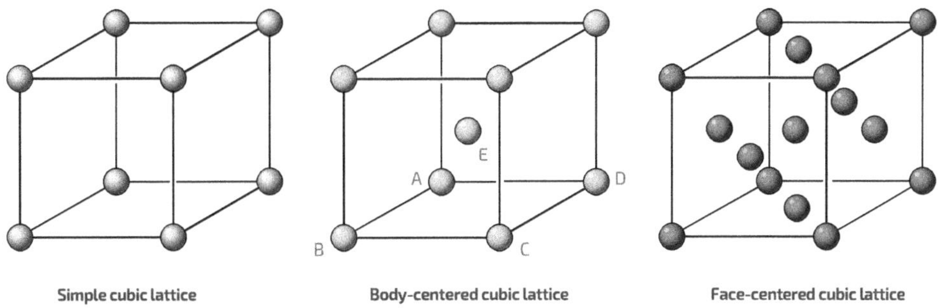

Figure 5. The three elementary cells forming three different types of Bravais lattices by multiple repetition in space. The two forms on the right are not the smallest units whose spatial repetition generates the complete lattice. In the cubic-centred lattice, for instance, the elementary cell is a non-rectangular parallelepiped having as its base the square ABCD and one of the vertices E, the centre of the cubic lattice.

More generally, for all crystal lattices, the geometrical transformations that leave them invariant are limited. For example, a rotation by an angle of order 5 (namely by an angle 72° = 360°/5) is forbidden for an infinite crystal that must obey translational invariance: we shall address this topic in the following section.

On the contrary, by definition, crystals are invariant for an infinite number of translations: in fact, it is sufficient to choose distances and directions that respect the periodicity of the crystal itself. This property is nowadays considered the most characteristic property of a crystal, rather than the macroscopic symmetry that can be appreciated even by external observation. The beauty of crystals, with their shining faces and their various forms, continues to move us regardless, for example, a small diamond on a ring or the gorgeous minerals that can be admired in natural history museums.

What is the link between the microscopic structure of a crystal and its macroscopic geometry? The presence of plane and shining faces, both in the natural case as well as resulting from cutting, are connected to the regularity of the microscopic structure. In fact, it is rather easy to obtain well-defined and smooth planes from a crystal by the operation of cleaving: if an atomic bond is fragile, all the bonds in analogous positions in the lattice are also fragile. This property is used by lapidaries in order to cut raw diamonds (Fig. 6). In an analogous way, we can understand how the formation of plane faces can be obtained during crystal growth.

Symmetry of Order 5

Crystals can display symmetry of orders 2, 4, 6... but not of order 5. The exclusion of the number 5 from this list is a direct consequence of their mathematical

Figure 6. In order to obtain one cut stone, the lapidary initiates with a small cut in the raw diamond, then they strike a knife in it: the diamond splits into two pieces along a plane of cleavage. By smoothing, the gemcutter creates many small faces (58 for a brilliant diamond).

Figure 7. The corolla of the campanulas displays a symmetry of order 5.

character: it is impossible to fill the space with a basic unit displaying a fivefold symmetry. In two dimensions, we noted that it is possible to tile a plane by means of rectangles, squares, hexagons or equilateral triangles but not by using pentagons.

This anomaly worries many researchers, and even nature seems to accept it only reluctantly, adopting it rather frequently elsewhere. In flowers, the symmetry of order 5 is very frequent (Fig. 7). The same happens in small aggregates of atoms obtained in laboratories. Researchers have been able to create aggregates with the chemical formula Al_{13} or $Al_{12}C$ arranged with a central atom and 12 peripheral

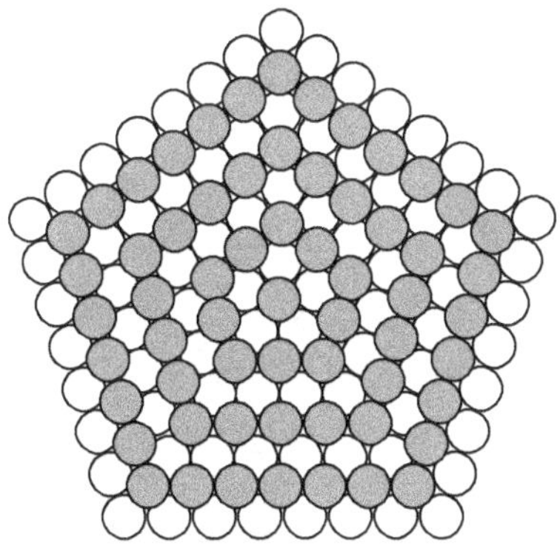

Figure 8. A very compact arrangement of the hard balls with the 5th-order symmetry (the first two layers are represented in white and grey). On each pentagon, adjacent balls touch each other, being slightly shifted with respect to two neighbouring pentagons. The sides of pentagon alternately consist, depending on the layers, of an even or odd number of balls.

atoms: they form regular icosahedrons with axes of symmetry of order 5 and not the cubic octahedron that is the basic unit for the compact arrangement.

Even for an infinite medium, symmetry of order 5 is not totally impossible. If we accept the limitations of building up a periodic assembly, we could be happy with the arrangement in Fig. 8, where the balls in each layer are placed onto the sides of regular pentagons. The packing factor that results is 72% which is not too bad compared with the maximum possible of 74% (see Chapter 8).

Quasi-crystals

Even before the work of crystallographers in the 19th century, the theoretical impossibility of finding a crystal with fivefold symmetry had already been demonstrated. Thus it came as a surprise in 1984 when a team led by Daniel Shechtman was analysing the structure of an aluminium and manganese alloy synthesised in their laboratory. Their diffraction pattern displayed localised brilliant spots, as for any crystal (Fig. 9).

But to their surprise, the alloy had symmetry of order 5, in spite of the laws of crystallography. Indeed, the alloy was a "quasicrystal", with a "quasiperiodic" structure (see Panel on page 135).

The International Union of Crystallography (IUCr) gave the formal definition in 1991: a "solid without three-dimensional periodicity" but displaying a diffraction pattern "essentially discrete" as the one of crystals. The brilliant spots of the diffraction pattern are indeed much more numerous than in crystals: in reality, there are infinitely many for any portion of the diffraction pattern, the majority

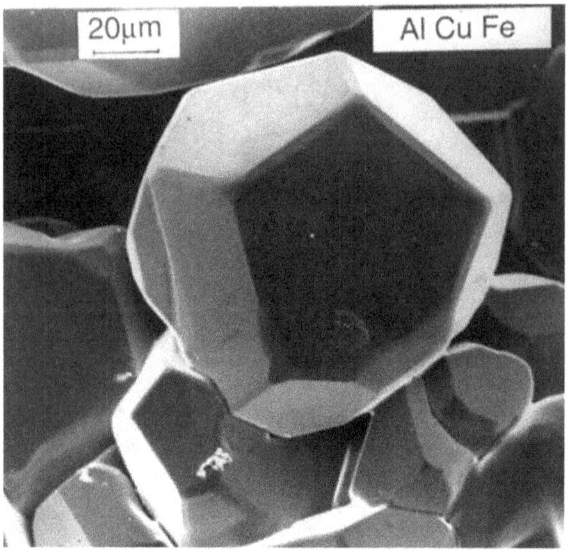

(a)

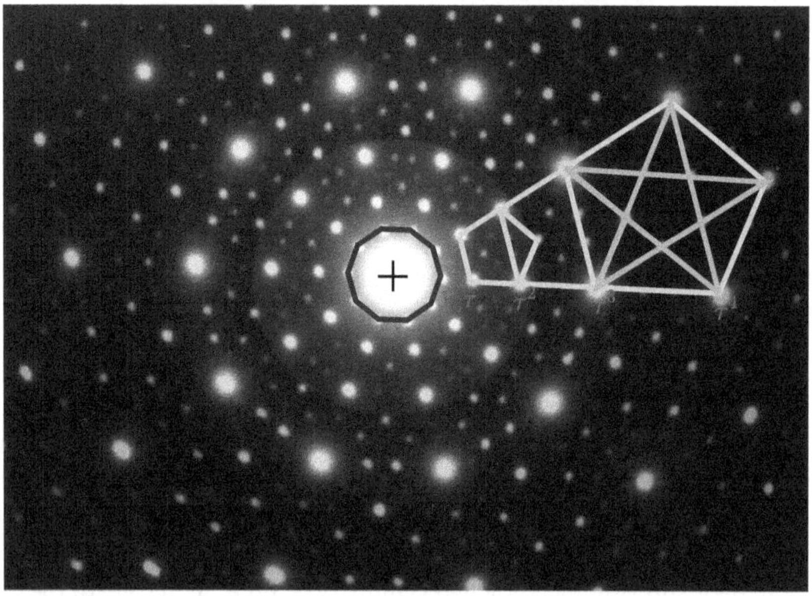

(b)

Figure 9. (a) AlCuFe quasicrystal of icosahedral morphology (courtesy of Dr. Annick Quivy, CECM-CNRS, Vitry, France). (b) Its X-ray diffraction pattern (courtesy of Dr. Jacqueline Devaud, CECM-CNRS, Vitry, France). The symmetry of order 5 that appears in this pattern is not possible for crystals.

Original tiling for a bath hall

Quasiperiodic structures are an unexpected boon for those people who dislike monotony and would prefer to tile the floor of their bathroom in a more original way than the usual paving by hexagons, squares or triangles. They could find inspiration in the scientific work of the English mathematician Roger Penrose. In 1974, he suggested tiling the floor by means of two types of rhombuses, thus "locally" obtaining symmetry of order 5. The reader can play in the attempt to form nonperiodic larger tiling. It is not so easy as it would seem!

being so weak as to be almost invisible. That astonishing discovery was rewarded with the Nobel Prize in Chemistry to Shechtman in 2011.

Amorphous Matter

We have seen that in general a crystal is the state for a system of atoms or molecules that minimises the energy at low temperatures, even if the size of the crystal is small. Is it true that all the atomic assemblies display the typical crystalline order at long range? The answer is no: some materials are present in an amorphous form, namely they do not exhibit any symmetry, not even locally. The diffraction pattern does not show any spot or any structure. Their structure is similar to that of a "frozen" liquid.

A familiar example is the glass of our windows that are basically formed by silicon dioxide (SiO_2) (Fig. 10(a)). This silica also exists in crystalline form, for example, as quartz that we can find in magnificent forms in nature (Fig. 10(b)), as well as cristobalite or tridymite.

All these crystals are "built" of SiO_4 tetrahedra connected to each other by their vertices in different ways, depending on the crystal form. In view of the different ways of placing one tetrahedron near another, it is not surprising that the various positions are taken at random, and this gives rise to amorphous silica or vitreous glass. Locally, the glass retains the structure of a crystal: thus the amorphous materials are characterised by "short-range" order. It is noted that both in a quartz crystal, as well as in silica, the configuration is far from compact. Each atom, instead of having 12 nearest neighbours, has only two (for the oxygen) or four (for the silicon). Thus there is still much room for placing a number of supplementary atoms.

Amorphous materials have many technological applications. Thanks to a long-lasting capability, strong resistance to corrosion and an optimal combination of electrical and magnetic properties, amorphous metallic alloys, for example, are employed in orthopaedic prostheses or in tools for surgeons.

Conclusion

In this chapter, we have been involved with the structure of matter, using the assumption that the constituent atoms behave as hard spheres attracting each other

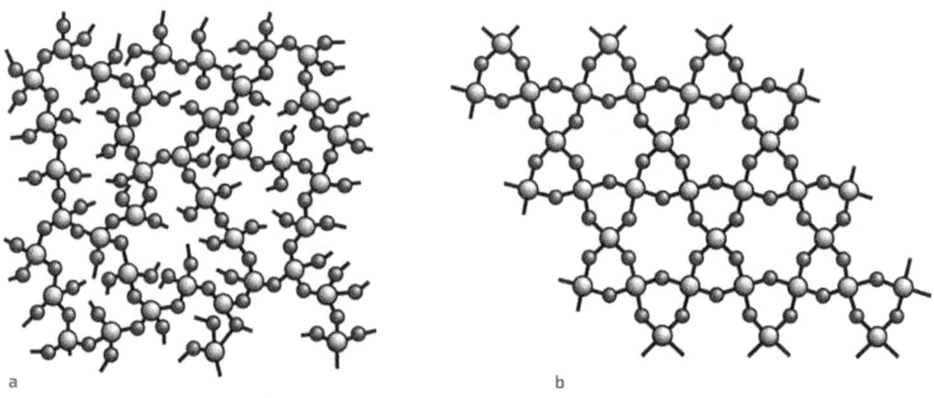

Figure 10. Structure of glass (a) and quartz (b). These two forms of silica are composed of interconnected SiO_4 tetrahedra. For simplicity, the image is projected onto a plane: each oxygen atom is bonded to two silicon atoms, and each silicon atom is bonded to four atoms of oxygen.

and forming a compact assembly. This hypothesis can be wrong, and amorphous or strong crystalline materials are far from being compact. When Kepler, in his attempt to explain the form of snowflakes, assumed the existence of compact packing, he was rather far from reality! In fact, the structure of ice at the atomic level locally resembles silica (Fig. 10), by replacing the silicon with the oxygen and the oxygen with the hydrogen. Each water molecule tends to have four other molecules nearby, analogous to the silicon atoms in silica that tend to have four oxygen atoms nearby. Therefore, as for silica, the structure of ice is not very compact, at variance with the assumption by Kepler. However, the hypothesis by Kepler and other scientists, such as Robert Hooke (1635–1703) or Mikhail Lomonosov (1711–1765), who followed the same idea, contributed to the advancement of Physics. Often progress is achieved through errors that other scientists attempt to correct.

Part 2

Everyday Physics

The reader is already aware of the importance of Physics in the modern world: our technological devices are ever more complex and numerous and require an increasing amount of energy. However, we may be less conscious of the presence of Physics in a variety of phenomena related to the daily life, for example, while eating.

Why does the French aperitif "pastis" become opaque when mixed with water? What is the origin of the bubbles escaping a bottle of sparkling wine? Why is a sound produced when fingers slide around the rim of a glass? All these questions, and many others, will have responses in this part of the book.

We will also take a train ride, during which we shall discover that the change in pressure the passengers call "ears popping" when entering a tunnel can be explained by resorting to a property discovered in the 18th century by a Swiss scientist. Do you know whether this variation of the pressure is an increase or a decrease? You will discover it in the following pages.

Chapter 10

A Conversation on a Train Ride

What do a train entering a tunnel and a liquid flowing through a narrowing pipe have in common? The response involves pressure. The area of physics called fluid dynamics, created in the 18th century, mostly due to the work of Daniel Bernoulli, finds many applications in our everyday lives.

Once upon a time, when high-speed rail services did not exist, three physicists were travelling by train. A few minutes after departure, the train suddenly entered a tunnel. The entrance was accompanied by an unpleasant sensation in the ears, similar to what happens when rapidly descending from a mountain (Fig. 1).

One of them complained that such an overpressure should not happen in a modern train. A second physicist, disagreeing, pointed out that it was not an overpressure but rather a sudden decrease of the pressure. The third one claimed that it had to be the result of the compression of the air when the train suddenly enters the tunnel. It appeared that they had forgotten the principles worked out long ago (18th century) by the Swiss scientist Bernoulli. In 1738, he published a book, *Hydrodynamics*, yielding fundamental results about the behaviour of fluids. It is noted that his name was Daniel (1700–1782) since several members of that family obtained significant successes in various scientific fields (his father Jean and his uncle Jacques are equally famous for their work in Mathematics and Physics). It is amusing to observe that Daniel was also the name of the physicist in the group travelling on the train who was an expert in hydrodynamics, and who decided to provide a lecture during our trip.

Figure 1. When the train enters a tunnel at high speed, an unpleasant sensation occurs in the passenger's ears due to a variation of the pressure.

About the Train Travelling in the Tunnel

Daniel addressed the problem in the following way. Let us assume that the train is moving at constant velocity while the air in the tunnel is effectively stationary. To be precise, the air is not at rest everywhere (Fig. 2(a)). In front of the train, at Section A, the air isn't moving. At the back, in Section B, it is also at rest. As the train proceeds, the air in front of it must vacate the place taken by the train, while, at the back, the air moves to fill up the place the train has left behind. Thus an air current moves in the direction opposite to the direction of train travel, a current that forces the air to pass from the front to the back of the train. To cause such a movement, a force must be acting. Is this force due directly to the train? No! The train does create a force against the air, but it is rather the friction force that acts to keep the air in contact with the train. In order to have a flow of air in the opposite direction to the train's motion, there must be a pressure difference! The pressure is weaker in section C than in section A in Fig. 2, thus explaining why the air moves towards C. In the meantime, it must be stronger in B than in C in order to push the air arriving from C. On the other hand, the pressure in A is the same as the pressure

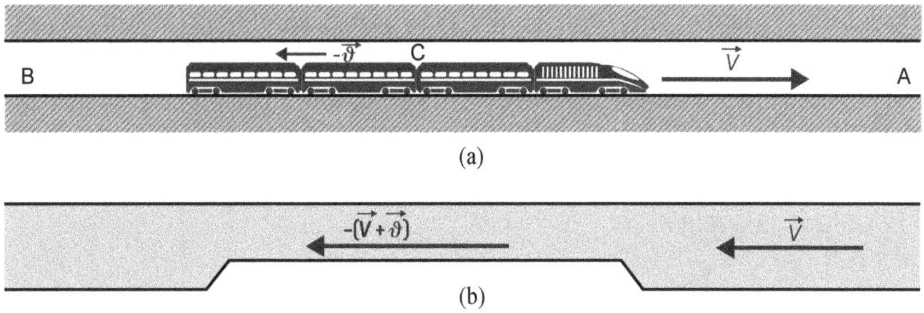

Figure 2. (a) The train entering the tunnel at velocity V causes an air current at velocity $-\vartheta$ in the opposite direction. A reduction in the pressure is happening. (b) An analogous decrease in pressure is induced when a fluid exits a narrowing pipe at velocity $-V$.

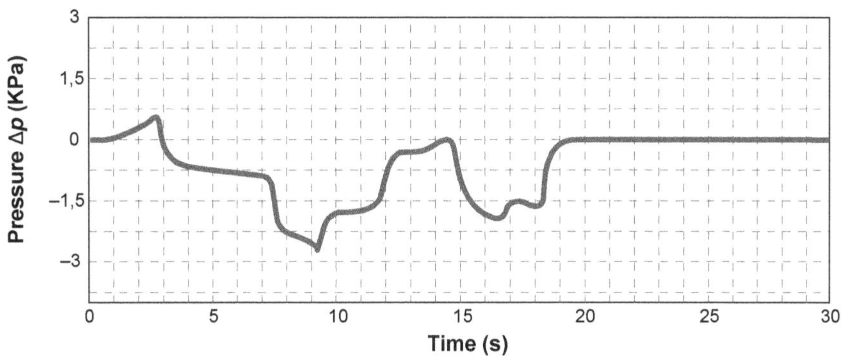

Figure 3. Evolution of pressure registered by a measuring device placed 72 m from the front, while the train is moving into the tunnel. On the whole the pressure is smaller than atmospheric pressure, during the passage of the train. Thus a decrease in pressure is occurring. The fluctuations present in the curve will be explained at the end of this chapter.

in B, both simply being at atmospheric pressure. Therefore, the train passing into the tunnel creates a decrease in the pressure around it (Fig. 3).

The Flow of Fluids in Pipes

During the 18th century, Daniel Bernoulli evidently could not know about trains. The problem he was analysing was the flow of fluids in pipes (Fig. 2(b)). The analogy with the train in a tunnel is evident. The pipe is the tunnel or, better, the tunnel plus the train: the presence of the train inside the tunnel is the equivalent of the narrowing of the pipe. For the reason that we shall address in what follows, the narrowing in the pipe causes the decrease of the pressure.

Bernoulli evaluated the drop in the pressure P by taking into consideration the flow of an incompressible liquid with no viscosity. His formula was very simple. Since the flow remained the same in any section of the pipe, the velocity V of the flow is necessarily higher in the narrow part than in the wider parts. If ρ is the density of the fluid, then for a stationary and non-turbulent flow, at any point in the fluid is valid the equality:

$$P + \rho V^2/2 = \text{constant.}$$

Experimental proof of low pressure at the pipe narrowing

The variation of the pressure for liquids flowing in a pipe where the diameter changes can easily be experimentally verified. Let us have the fluid circulating in a pipe with one wide and one narrow section, with open vertical tubes connected to the pipe as shown in the figure. Assuming that the presence of the tubes does not perturb the flow much, these tubes can indicate the value of the pressure at the different parts of the pipe. Once the flow is established (stationary state), the height of the water in the tubes is different: in the tubes connected to the wide sections, it is higher than in the part of the pipe which has narrowed. By measuring the difference in height Δh *in the tubes in A and C or in B and C, the difference in pressure can be deduced:* $\Delta P = \Delta h g \rho$*. By knowing the rate of flow and the diameter of the pipe, one can also deduce the velocity and thus verify Bernoulli's theorem.*

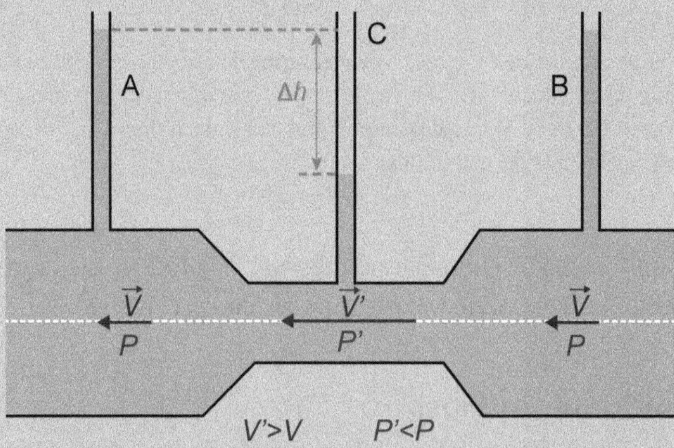

In the presence of flow from right to left in a horizontal pipe, the levels of the liquid in A, B and C evidence the different pressure present corresponding to the narrow section, due to the different velocity of the fluid, which is larger in B with with respect to A and C ($V' > V$). The same value of the pressure in A and C ensures that no change in the rate of flow has occurred, namely that the energy dissipation of the fluid due to the friction is negligible.

This equation is exact when gravity plays no role, namely when all the points being considered are at the same height. We shall see later on how to take gravity into account.

Compressible and Incompressible Fluids

As has been pointed out during the conversation with our colleague familiar with hydrodynamics, the Bernoulli theorem holds for an incompressible liquid, such as water. The air in the tunnel is not incompressible. In fact, the quantity $P + V^2/2$ cannot remain rigorously constant along the tunnel. However, the general argument remains valid: the rate of flow is the same, so where there is a narrowing, the air necessarily moves at a higher velocity. To induce this acceleration, a force is involved, and this force is related to the pressure difference and in the region of narrowing a decrease in pressure is induced. This phenomenon is known as "Venturi effect" after the Italian physicist who studied some devices based on the flow of fluids along pipes of different diameters around the time of Bernoulli.

One of the other two physicists on the train, Paul, was still arguing: if the velocity of the flow is different in the front and in the back of the train, why is the pressure in the inside and outside the train the same?

Daniel admitted that the velocity of the air is different. On the other hand, if one of the windows of the train is badly fitted, the pressure has to be about the same. This phenomenon is used in the *Pitot* tube, a device that allows us to measure the velocity of a flow (see Panel on page 146).

It can detect the value of the pressure by means of two *ad hoc* holes, and through the use of Bernoulli's theorem the velocity is deduced. These holes are the equivalent of the windows in the train if not hermetically sealed.

Our train has the windows well closed, the glass being fixed in place and not openable. The train attendant, who has just arrived, can complete the explanation, confirming that the windows are hermetically closed. Thus the effect of an increase or decrease in pressure can only be felt with a certain delay, of the order of 10 or 15 s.

One Simple Experiment

One of the other passengers remained somewhat dubious. The Bernoulli theorem is contrary to intuition. To let an object enter an orifice, for instance, a cork in a bottle, one must push. Yet, in the case of the train, a decrease in pressure occurs? It is true that some real results appear contrary to intuition, says Daniel. However, the cases of the cork in the bottle and the train in the tunnel are very different in

Measuring the velocity of an aircraft

How do we measure the velocity of an aircraft with respect to the air around it? Bernoulli's theorem (with some modifications to take into account that the air is not incompressible) reduces this measurement to an estimate of the pressure difference. This estimate is carried out using a Pitot tube installed outside the fuselage. The principle of such a measurement was proposed by French physicist Henry Pitot (1695–1771).

Pitot tube installed under an aircraft wing. This device became infamous in June 2009: the pilots of the flight from Rio de Janeiro to Paris reacted inappropriately after ice froze over the Pitot tubes, thus causing the fall of the aircraft.

The Pitot tube consists of two Γ-shaped tubes with parallel (C) and perpendicular holes drilled (B and B′) (see figure).

The quantity to be measured is the speed of the airplane with respect to the air, which we denote by −V (it is convenient to refer to the aircraft so that the air speed immediately in front of the airplane, at point A, is equal to V). When it enters the tube, the counter flow is split. A small part of it, C, is filled with air, the velocity of which is zero here since the tube is filled and closed at the other end. The air velocity is also zero at D: thus the pressure at points C and D coincides. Another part of the air flows around the pipe and reaches the holes B and B′ practically without changing its speed V, which makes it possible to determine the pressure that is established at point F.

(*Continued*)

(*Continued*)

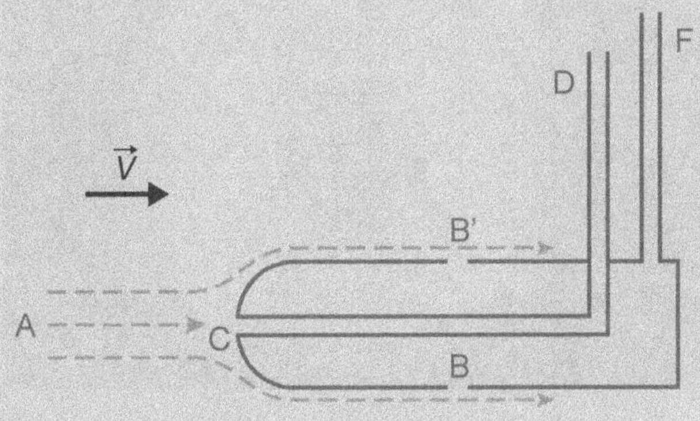

Next, the pressure gauge measures the pressure difference $P_D - P_F$ between points D and F, which is equivalent to the pressure difference $P_C - P_B$ between points C and B and coincides with the pressure difference $P_C - P_A$. Now we just have to calculate the speed of the plane using Bernoulli's theorem. In the case of an incompressible flow, the latter leads to the equality $P_C = P_A + \rho V^2/2$. Since the measured pressure difference $P_D - P_F$ is equal to $P_C - P_A$, the aircraft speed turns out to be

$$V = \sqrt{\frac{2(P_D - P_F)}{\rho}}.$$

reality. The next time you are travelling by train, just bring a barometer or even a simple device used to measure the altitude when climbing in the mountains (Fig. 4).

It will be possible to verify that a pressure decrease occurs during the passage of the train through the tunnel.

This experiment is really worth setting up since Bernoulli's theorem assumes that the fluid is incompressible, and its flow is laminar (i.e., without vortices), while the train bursting into the tunnel creates a significant disturbance of the air located there.

In reality, a certain additional pressure is registered right at the entrance, and then the decrease predicted by the Bernoulli theorem is followed by an increase in pressure (see Fig. 3). Entering the tunnel produces a wave of extra pressure in front of the train: this wave propagates through the tunnel at close to the speed of

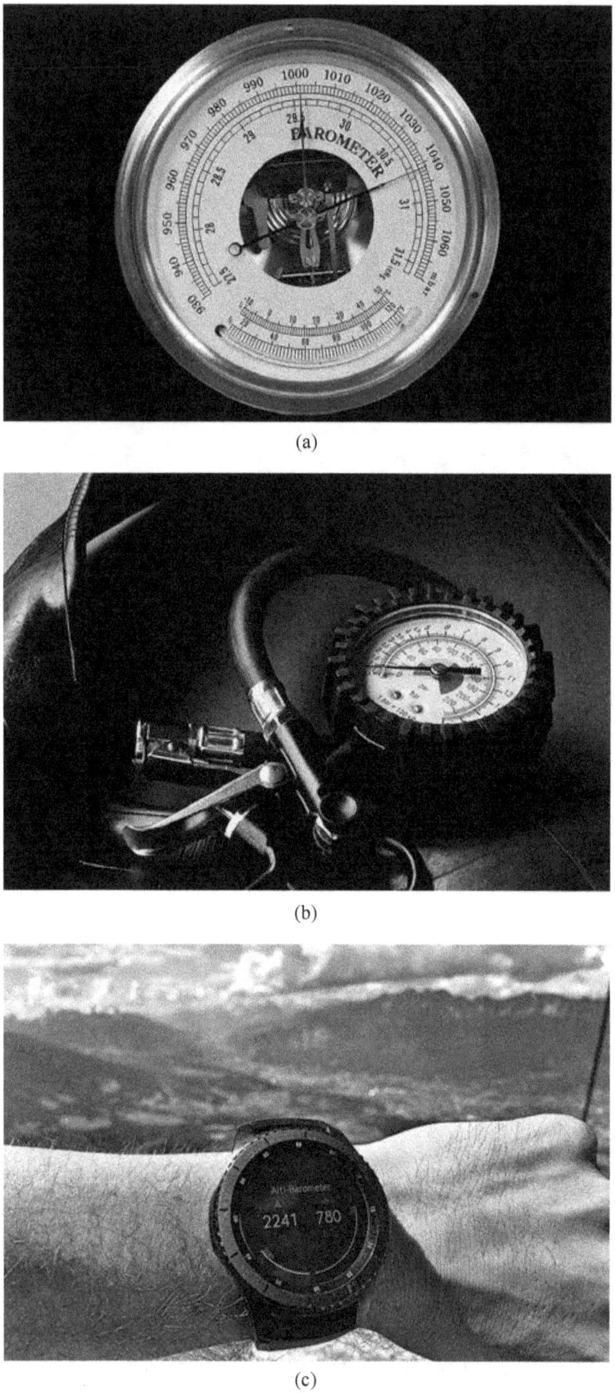

Figure 4. Some devices for measuring pressure. A barometer (a); a manometer used for the pressure in car tires (b); and an altimeter (c). This latter device measures the decrease in pressure that occurs on increasing altitude in the atmosphere.

sound and arrives at the end of the tunnel well before the train. Furthermore, through the capriciousness of fluid mechanics, that wave is reflected instead of exiting the tunnel.

Thus, it meets the train again, is reflected again, and so on. Similarly, after entering the tunnel, a low-pressure wave forms behind the tail of the train.

The Effect of Gravity

One further observation by a passenger: another consequence of the Bernoulli theorem goes against our intuition. To elaborate, we assume that a fluid flows due to the pressure difference when it would seem that it is the flow of the fluid itself that induces such a pressure difference. Often the flow of a fluid is due to its weight.

True, admits Daniel, and indeed the experimental setup used by Bernoulli to derive his equation implied a role for gravity, with the flow occurring in a vertical tube connected to a container (see Fig. 5).

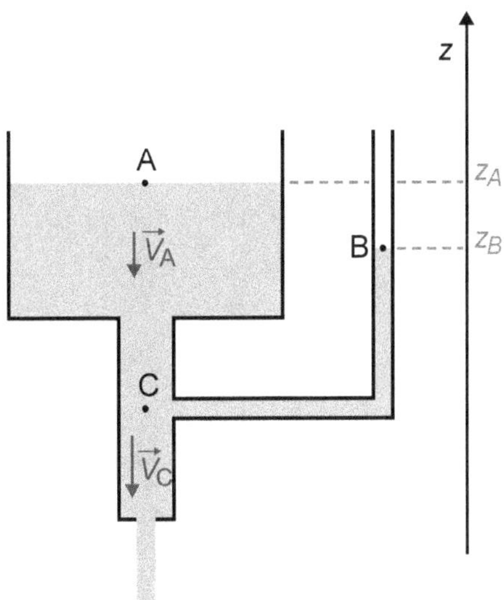

Figure 5. One of the Bernoulli experiments, described in his book "Hydrodynamics". The water inside the large container flows through a narrow tube. Without any flow, the levels in A and B are the same. Once the water flows, a difference in the levels in A and B is detected. In the container, $P_A + \rho g z_A + \frac{\rho V_A^2}{2} = P_C + \rho g z_C + \frac{\rho V_C^2}{2}$, while in the narrow tube of uniform cross section, $P_B + \rho g z_B = P_C + \rho g z_C$. Since $P_A = P_B$ is atmospheric pressure, we deduce $g(z_A - z_B) = V_C^2/2 \quad V_B^2/2$. From the rate of flow, the velocities V_C and V_A can be derived (the widths of the container and of the tube are known).

The flow is not horizontal as in the tunnel, and the Bernoulli theorem must be formulated in the form

$$P + \rho g z + \frac{\rho V^2}{2} = \text{constant}$$

with g being the acceleration due to gravity, while P and V are the pressure and the velocity at the height z. Historically, Bernoulli paid special attention to a specific problem: the duration of the outflow of liquid from a vessel through a small opening (see Panel below on this page).

The astuteness of a plumber in emptying a container

By considering the case of a container emptying, we shall realise that the Bernoulli theorem needs to be applied with some care. The reader can easily carry out the following experiment (see figure).

Simply fill a plastic bottle that has a hole at the bottom with water and measure the time taken for it to empty, repeating with different starting levels of the water. What is the relation between the height H of the water and the time τ required for emptying?

With various provisos, we could write that the quantity $P + \rho g z + \frac{\rho V^2}{2}$ is the same at the surface of the liquid and at the exit from the hole, or that the pressure is the same and given by atmospheric pressure P_0. For simplicity, let us assume that the diameter of the bottle is much larger than that of the hole. Thus the speed of the water exiting the bottle turns out to be

$$V = \sqrt{2gh},$$

h being the height of the water (having initial value H). This would be the same speed acquired by an object falling (with zero initial velocity) from the height h under the action of gravity.

In practise, this result is only an order of magnitude estimate since it is based on some assumptions that are only approximately valid. The pressure at the exit of the hole is indeed given by P_0, but this is not the real pressure inside the jet that is coming out. In fact, at the bottom of the container, we should add the additional pressure ρgh to P_0, this being the weight of the water column. The total pressure cannot suddenly change to equal P_0 at the exit of the jet, instead it will reach this

(Continued)

(*Continued*)

latter value progressively. Thus at the exit, the speed of the jet is smaller than V. This can be taken into account by reducing V by a coefficient C, smaller or equal to 1 and depending on the geometry of the hole. The value of C is known to be 0.6 for a perfectly circular hole made at the bottom of the container. The time required to empty thus turns out to be

$$\tau = \frac{S}{Cs}\sqrt{\frac{2H}{g}}.$$

The coefficient C can be modified by adding a kind of device to the hole such as the tube sketched by the two black parallel segments in the figure. These devices are commonly used by plumbers.

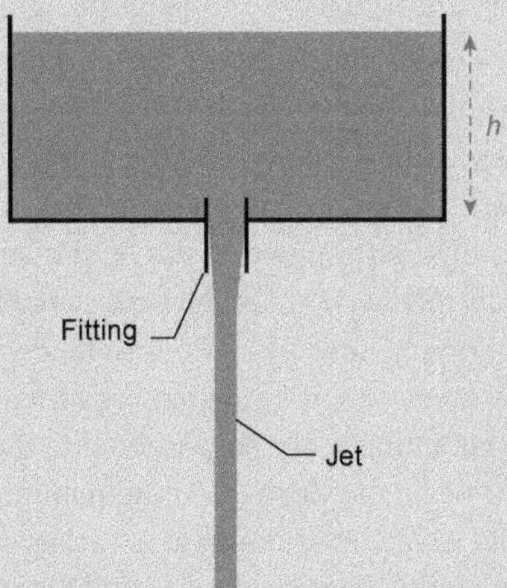

Emptying a container filled with liquid up to height *h*. The Bernoulli theorem implies that the speed of the flow exiting the container is proportional to $h^{1/2}$, the square root of the height. Thus the speed decreases as the container empties.

Viscosity and the Boundary Layer

The problem of emptying a container through a hole had already been studied, a century before, by an Italian scientist, Evangelista Torricelli. He figured out that the speed of emptying was independent of both the type of fluid and the shape of the container, and was instead proportional to the square root of the height of the fluid in the container. This result is only approximate, for several reasons that we are going to address. We displayed some regret — had Daniel pulled our legs?

He said: "I have only simplified some aspects, as physicists should do." When applying the Bernoulli theorem to the emptying process, we have assumed that the velocity of the liquid flowing in the tube is the same regardless of the distance from the walls of the tube being used. That it is the same, for instance, in the middle of the flow as it is very close to the walls. This is not correct, the velocity has to vary depending on the distance from the wall of the tube, with additional conditions imposed by thinking about continuity. In particular, very close to the wall, the velocity of the liquid must be the same as the wall, namely zero. Therefore, when the flow occurs in a tube of radius R that is very small with respect to the length L, the speed of the flow is weak all along a section of the tube. The rate of flow Q is given by an equation derived in 1844 by physicist Jean Léonard Marie Poiseuille:

$$Q = \frac{\pi}{8\eta} \frac{\Delta P}{L} R^4,$$

where ΔP is the pressure difference between the two ends of the tube and η the coefficient of viscosity of the fluid (measured in kg m^{-1} s^{-1}). The rate of flow will be very different for water and oil since the latter is more viscous. Furthermore, it is noted that, for a given difference of pressure, the rate of flow varies very strongly (proportional to the fourth power) with the radius of the tube.

By taking these effects into account, in the case of flow in narrow tubes (which are not too short), the emptying speed turns out to be dependent on the nature of the liquid and is determined not by the square root of its level height, but by this value itself. In addition, being inversely proportional to the viscosity of the liquid, the velocity of the liquid begins to depend on the nature of the latter. That is the reason why we originally considered the case of a large tube, in which we can assume to a good approximation a uniform velocity of the flow, independent of the distance from the wall. The region where the velocity changed significantly, decreasing to zero upon the direct contact with the tube wall, is only a thin boundary layer.

What is the thickness of this layer, asks one curious passenger. This depends on the viscosity of the fluid, which does not allow the velocity of the fluid to change too dramatically when moving away from the wall. Therefore, the thickness of the boundary layer is much larger in oil in comparison to that in water. In general, the quantity that plays the most significant role in fixing the boundary layer is the ratio η/ρ. It is of the order of 10^{-6} m^2 s^{-1} for water and 1.4×10^{-5} m^2 s^{-1} for air at room temperature.

Daniel initiated some calculations. For a tube of length L in which V is the velocity of emptying, the maximum thickness of the boundary layer is of the order of $\sqrt{\eta L/(\rho V)}$. Therefore, for a length of the tunnel around 250 m and a velocity of 180 km h^{-1}, the thickness is just a few centimetres. This value is almost negligible in comparison to the distance of the train from the wall of the tunnel, and the Bernoulli theorem can safely be applied. The same holds for the flow of water in a tube of length less than a metre where the rate of emptying is 1 m s^{-1}, and thus the thickness of the boundary layer is of the order of a millimetre.

A Problem in the Water Supply… The Water Hammer

Paul added, "I want to be able to guess the reason why the emptying process in a tube could lead to an increase in pressure, regardless of our previous conversation. I remember the year I left my house during the winter after having turned off the water supply. In the spring, when I returned to my house, I suddenly turned on the supply without taking the precaution of opening the taps in advance. Serious damage occurred in the pipes, and I had to call a plumber. He told me I was the victim of a 'water hammer' (Fig. 6)."

It is remarked that the circuit of pipes distributing water around the house is at a pressure >1 atm. However, when the general valve at the entrance to a house is opened, the water acquires such a high speed that the pressure can increase significantly over the original value, by compressing the air inside the pipes.

Daniel again began some calculations in his booklet to derive the equations controlling the phenomenon: too late, the train was entering the railway station.

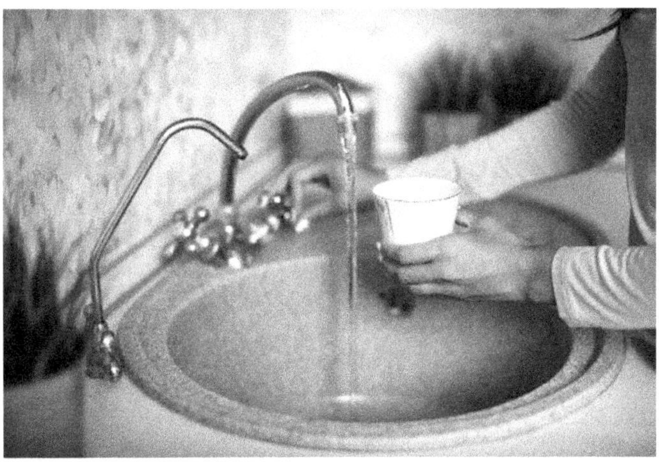

Figure 6. The sudden opening of a tap could lead to extra pressure in the pipes with related damage: the water hammer. This definition also applies to the series of characteristic noises when a tap is suddenly turned off and the circulating water stops.

Chapter 11

The Stradivari Legacy

Musical devices, at least for some aspects, belong to the realm of typical physics instruments. Even if they have been improved throughout the centuries through the works of lute makers or musicians who possibly did not have such a relevant scientific knowledge, indeed they represent wonderful applications of the laws of physics. In this chapter, we shall deal with the king of the string instruments, the violin.

Following previous musical instruments such as the lute or the viola, the violin was born in the 16th century in Italy, precisely in the town of Cremona, where the Stradivari, the Guarneri and the Amati families heightened the making of string instruments to an artform. Even nowadays, Cremona maintains its tradition. Tourists visiting this town can explore a number of shops displaying violins. (Fig. 1).

There is also a town of violins in France; this is Mirecourt, in the Vosges. The high school is named after and dedicated to a great lute maker of the 19th century, Jean Baptiste Vuillaume, born in Mirecourt, who founded a unique lute-making shop in France.

How the Bow Excites Vibrations

The violin is an instrument with rubbed strings: it has four of them, wound tight and strung onto a support (Fig. 2).

This support transfers to the resonance case the vibrations of the strings which result from the friction of the bow. When playing, the violinist makes a slow "go and come" movement with the bow. The direction of this motion is changed after about 1 s, even less. The string of the violin vibrates several hundred times per

Figure 1. The cathedral and the baptistery in Cremona, Italy. The renowned lute maker Antonio Stradivari (1644–1737), known as Stradivarius, was born and died in Cremona.

Figure 2. The violinist rubs the bow over a string that enters in vibration. This vibration is transmitted to the harmonic table (the upper part of the resonance case) through the support. The musician selects the height of the sound by blocking with the fingers the string against the handle. At the end of the handle, there are four pegs that allow the musician to adjust the tension of the strings.

second; in other words, in terms of the frequency unit, at several hundred hertz. This frequency fixes the emitted note (do, re, mi…), and it depends on the tension of the string (tuned by the musician by means of the pegs) and on the position of the fingers of the artist.

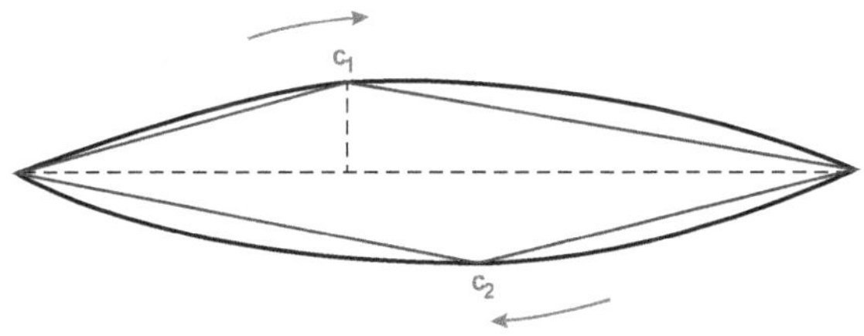

Figure 3. Motion of the violin string under the friction of the bow, following the study by Helmholtz. The angular point c_1 propagates from the support to the finger of the violinist, from the left to the right, designing an arc of parabola and vice versa from right to left (c_2).

The vibration of violin strings has been described by the German scientist Helmholtz in the year 1862. Thanks to instruments that he devised and made, he was able to demonstrate that the vibrating string, to a good approximation, takes the form of two segments of a straight line, with an angular point that propagates at constant velocity from one side to the other of the string and then back in the opposite direction (Fig. 3).

Leaving aside musical technique, we are going to address the following question: how does sliding a bow on the string cause its vibration? If the reader should think that it occurs spontaneously, we would invite them to fix a thin cord to two points and to slide a ruler on it (Fig. 4).

The cord is shifted from its equilibrium position, but it will remain in that new position until the contact with the ruler is active, and does not vibrate when the contact is interrupted. Thus a cord would not vibrate after contact, regardless of the object one uses for that. At variance, why does this happen if one uses the bow of a violin? We shall see that the solution to the mystery is just in the pitch, a resin that is smeared on the vegetable fibres of the bow.

Static Friction and Dynamical Friction

The interaction of the bow and the string is the physical process known as "friction". Friction effects are often disturbing, as is the case when they are the cause of the loss of mechanical energy, which is being transformed into heat. For example, a sizeable part of the energy spent to move our car is lost because of the various frictional effects involving the different parts. Friction effects are indeed useful in other cases, for instance, when initiating a movement: a proof is when we walk on an ice layer onto the ground.

There are two types of friction: the *static* friction involves two objects in contact, where one is immobile with respect to the other, while the *dynamical* friction

Figure 4. Experiment of rubbing a ruler on taut cord. The cord is shifted from its equilibrium position but does not vibrate unless one keeps the contact with the rule. On the other hand, one can get a sound "pizzicato" by briefly plucking the cord.

The laws of dynamical friction

The dynamical friction of solid bodies sliding onto another solid in general obeys the laws established in the year 1699 by Guillaume Amontons and subsequently (1781) specified by Charles Coulomb (better known for his works in the field of electrostatics). These laws state that the force related to the dynamical friction does not depend on the velocity. Instead, the dynamical friction depends on the force applied perpendicularly to the surface, for example, on the weight of an object resting on the ground, or in our case on the pressure applied to the string by the bow. The ratio of the two forces is the "friction coefficient", and this is independent of the force applied by the violinist. In a way somewhat unexpected, the friction coefficient is independent on the size of the contact area of the surfaces: it depends only on the nature of the two materials.

French physicists Amontons and Coulomb were the first to publish the friction laws. However, it appears that Leonardo da Vinci (1452–1519) was aware of those

(*Continued*)

(*Continued*)

laws about two centuries before, as it is suggested by his drawings (see figure). The brilliant painter from Tuscany was also a valuable physicist.

Drawing by Leonardo showing that the friction coefficient does not depend on the contact surface (Atlanticus codex, f532r). In the case of samples of different forms but of the same nature when the plane is progressively inclined, they initiate to slide at the same time. From D. Dowson, *History of Tribology*, Longman, New York (1979).

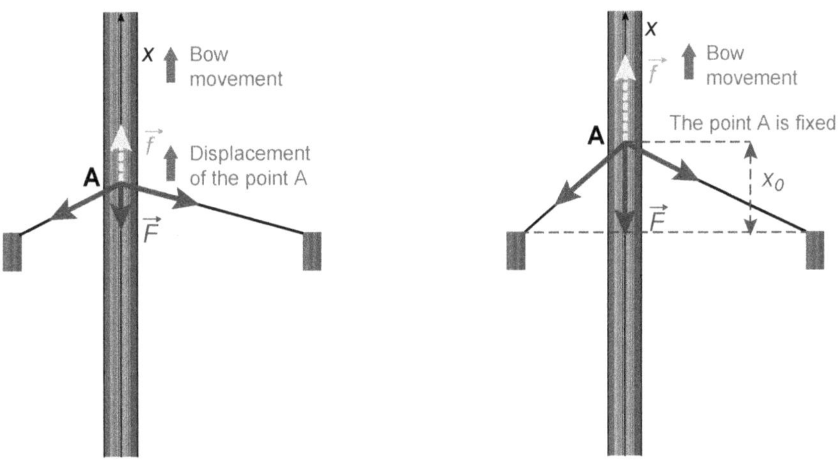

Figure 5. (a) Static friction of the bow against a violin cord. The bow moves up. The force due to the static friction **f** equilibrates the resultant **F** of the forces due to the tension; the point A of the contact of this with the bow stays immobile with respect to the bow. (b) When the tension of the cord is too strong, the friction changes to the dynamical one: the contact point A is in motion with respect to the bow, but its position x_0 with respect to the violin is fixed. The force related to the dynamical friction equilibrates the tension due to the cord.

is active when the two objects have different velocities. The frictions contrast the relative motions of the surfaces which are in contact (see Panel on page 158).

The friction force appearing in the process of sliding of the bow on the string is alternatively static and dynamical! When the violinist initiates to play, he/she puts the bow, initially immobile, on the string exerting a pressure that in the following we shall assume is constant. After a brief time, the displacement of the bow at velocity v_0 pulls the string at the same velocity: this phase of adherence is due to the static friction that the bow applies to the string in the same sense as the bow itself. On the other hand, the tension of the string exerts a reaction force of opposite sense. The two forces have to compensate each other so that the string follows the bow (Fig. 5(a)).

This first phase does not last too long: the force due to the tension of the string increases rapidly when the string is moving away from its equilibrium position, and it soon reaches the maximum extent compatible with the static friction.[1] The static friction can no longer compensate the force due to the tension; the adherence is lost and the string begins to slide with respect to the bow. Now the friction is

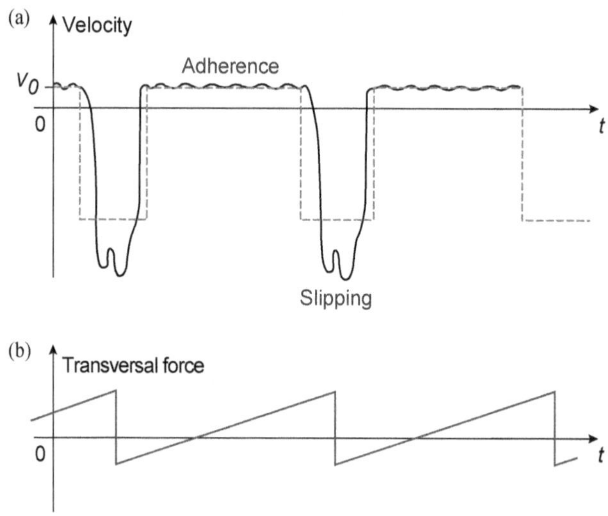

Figure 6. (a) Velocity of the violin cord at the point of contact with the bow, along the axis of its motion (dotted line: approximation; solid line: real behavior). The velocity changes very rapidly in between the adherence and the sliding phases. This sudden variation cannot be understood in the framework of our simplified description. (b) The alternation of adherence and slipping phases is described by a saw-edged curve for the transversal force due to the cord onto the support (that transmits the excitation to the resonance cage, main sound source).

[1] The existence of a maximum value of the static friction (for a given value of the pressure that the violinist is applying to the string by the bow) is a fundamental law for this type of friction.

reduced to the dynamical one, which is much weaker than the static friction. Then the string begins an oscillatory motion, as would happen to a spring pulled out from its equilibrium position: it immediately springs back because of the tension force until its sense (also known as its direction) takes the inverse sign; the string contra-sweeps in the wrong way, and then it moves in the same sense of the bow. Its velocity increases, and, at the end, it reaches the velocity of the bow; the "anchorage" again occurs, and a new phase of adherence starts over. This situation lasts until once again the tension becomes too strong and the string strays again. If this cord emits the note *la* of frequency 435 Hz, this means that it switches between two strays in 1/435 s. During the instant in which the cord and the bow have the same velocity, the bow could provide to the cord the energy required to trigger the vibration.

Thus, the contact bow–violin alternates adherence phases and slipping phases (Fig. 6).

Role of Resin

We have described how the cord begins to follow the bow and, then abandoning it, how it starts moving in the opposite sense. Then, again, it moves in the same direction as the bow, its velocity finally matching the one of the bow. Is there or is there not the chance that the cord cannot adhere to the bow, and then the friction

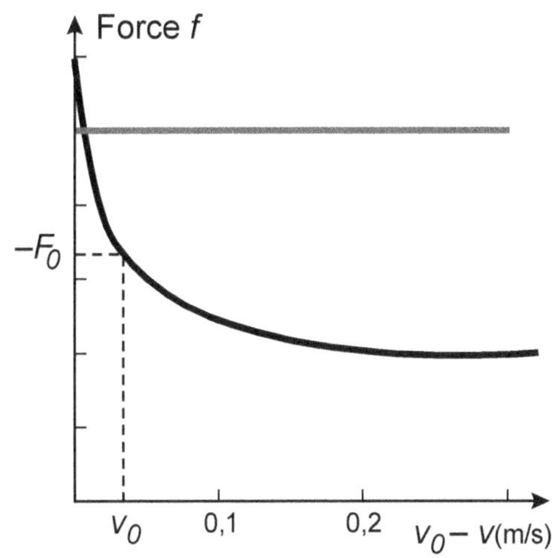

Figure 7. Behaviour of the dynamical friction force between the bow and the string of the violin as a function of their relative velocity (this velocity $u = v_0 - v$ is the difference between the velocity v of the cord and the velocity v_0 of the bow with respect to the violin). The horizontal line corresponds to the Amontons-Coulomb law, which is well verified by most solids. The violation of this law is what allows the violin to operate.

remains dynamical? This would be the condition that we have described above in dealing with the experiment of the ruler rubbing on a cord. Thus, on the bow, the string could take a fixed position instead of initiating the vibration. In this case, due to the tension of the cord, the resultant force F exactly equilibrates the force related to the dynamical friction (Fig. 5(b)). The cord would be immobile with respect to the violin in the position x_0 (and no longer with respect to the bow). Does there exist a position of the cord obeying this condition? The sum of the forces acting on the cord would be zero: it is an equilibrium position.

The existence of a possible equilibrium position is worrying! If the string should set in such a position, could the violin stop sounding forever? We are going to see that there is not that risk, and this is because of the wonderful property of the resin. When it is rubbed on the bow, again assuming that the musician exerts constant pressure on the string, the force of dynamical friction f decreases as a function of the velocity of the string with respect to the bow (Fig. 7).

In the framework of a simplified description (in particular, the torsion of the cord is neglected), we are going to show how the reduction of the friction does not affect the vibration of the cord, while, to a certain extent, it could amplify it. The equilibrium position x_0 of the cord then is unstable, and the violin never breaks down! Let us recall that an equilibrium can be stable or unstable. Most familiar is the stable equilibrium: it corresponds to a minimum for the potential energy (see Chapter 6). For instance, a ball at the bottom of a well, if slightly displaced from its equilibrium position, will return to its initial position. This is not what would happen to an object with a tip placed on a floor in a way that it remains vertical just resting on its tip. If one moves it even a bit from its equilibrium position, in contrast to the ball described above, it will fall down towards another final position.

The case of the violin during the slipping phase is similar. Let us suppose that the string from an equilibrium position x_0 (see Fig. 5(b)) makes a little displacement backwards, possibly because of some irregularity in the bow. The tension of the cord, proportional to $(x - x_0)$, would decrease, as would the friction force since it experiments an evolution contrary to the relative velocity $(v_0 - v)$. If one has a small decrease related to the tension, then the little displacement ahead is amplified.

In conclusion, the friction tends to increase the displacement from the equilibrium position: thus the equilibrium is unstable! This amplification effect is not ordinary. On the other hand, the vibrations are amplified by the friction only to a certain extent. When the velocity of the string becomes equal to the one of the bow, the relative velocity becomes zero, and the regime of dynamic friction is substituted by the static friction.

Until now, it has been assumed that the forces acting on a string of the violin are only tension and friction. Obviously, there are other forces, say air resistance, since the violin creates sound waves. The related force is rather negligible in comparison to the ones we have considered. This force evidently tends to dampen the vibrations of the cord. One could take that force into consideration with a little more complication, but things would not change that much.

Other Effects of Friction

Friction can induce other surprising effects, as is shown, for instance, when a billiard ball comes back after a strike against another ball (see Panel on page 164). The mastery in dealing with the friction laws is crucial for several industrial processes. Thus, when operating with a lathe on a piece of metal (Fig. 8), the blade that is cutting the piece could be affected by undesired vibrations. These vibrations are induced by the friction of the blade on the surface of the metal. These can be avoided by giving the blade appropriate forms. In a number of systems, for instance, combustion engines, it is very desirable to reduce the friction in order to improve the duration of the pieces. The direct contact of the pistons and the cylinders is avoided by means of lubrication, namely by interposing a layer of oil. In this case, the friction force becomes proportional to the relative velocity between the fluid and the piece (unless it is very small, see Chapter 15).

Cords and the Resonance Cage

As addressed above, the length and the tension of the cord determines the note emitted by the violin. However, by itself, the cord could not produce a sound of sufficient intensity: this is the role played by the resonance cage of the violin. The vibrations of the wood walls are the main source of the sound.

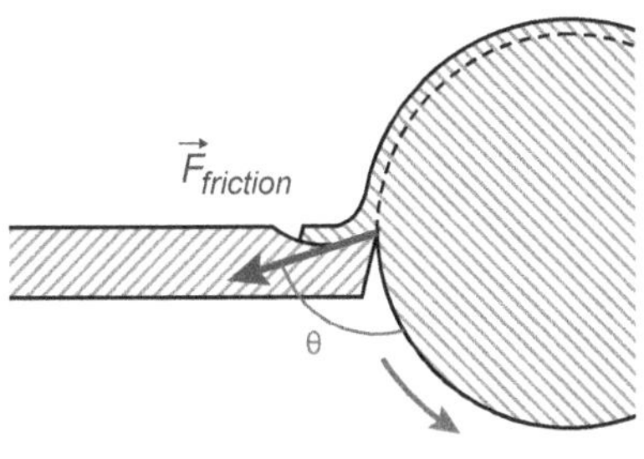

Figure 8. One of the effects of friction. During the lathing of a metallic piece (right), the blade vibrations (left) due to the friction against the piece can be avoided by an appropriate choice of the cutting angle θ through a proper form of the blade.

Tricks and astuteness of billiard players

If the violinists exploit friction without necessarily fully knowing their secrets, the billiard players can also make miracles. By hitting the ball under particular angles of incidence, they can obtain some effects that are really surprising for the unlearned. How do they work? Let us consider a ball hit by the billiard cue along an axis directed at the backwards of the ball centre and of the point of contact with the carpet (see figure). The hit pushes the ball ahead at the same time causing its rotation. This happens in a counterclockwise sense: if the ball rotates without sliding, then the rotation is in the opposite sense. During the motion of the ball, one has a decrease in the translation velocity and the rotation. If the cue imparts an initial rotational velocity sufficiently strong in comparison to the translational velocity, this latter goes to zero before that rotation stops. Thus, since the ball continues its rotation, the friction with the carpet forces it to come backwards!

Another technique consists of hitting the ball in a way that it comes back after the strike against another ball. It is enough to hit the ball below its centre (see figure). As above, the ball rotates in the counterclockwise sense with a reduced rotational velocity. The strike against another ball stops the translation, while some rotational energy still remains, and thus the ball comes backwards. Without any strike against another ball, then the translational motion continues without sliding, and the ball does not come back.

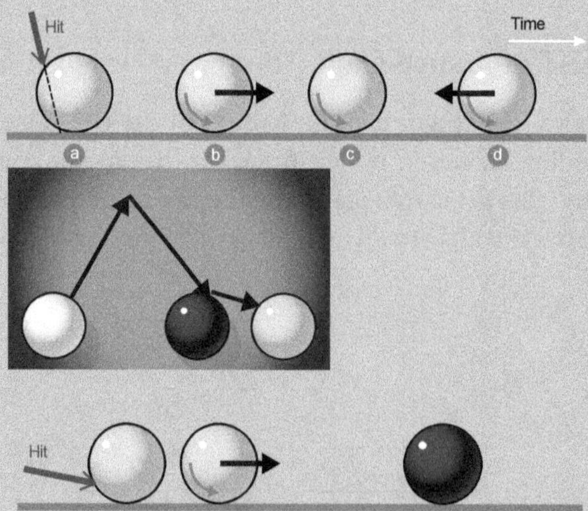

A billiard ball hit in a proper way, according to the direction indicated by the dash line in (a), starts its longitudinal motion at low speed (b), and, after some time, it stops, while a marked

(*Continued*)

(*Continued*)

rotation is still active (c). Thus because of the rotational motion, the ball can return backwards (d). It is also possible that the white ball strikes two other balls.

A billiard ball hit as shown by the gray arrow slips ahead while rotating, as shown by the curved arrow on the ball. After the strike against another ball, it can move backwards. By hitting the ball with a given angle, the billiard player can obtain a coming-backwards effect, surprising the layperson.

At variance with the cord of the violin and the bow, the friction of the ball against the carpet obeys the law of Amontons–Coulomb. According to it, the friction force (directed horizontally) is given by the product of the constant friction coefficient times the vertical component of the reaction force of the table, that, in absolute value, is equal to the weight of the ball. The reader being familiar with mechanics can write the equation of motion and solve it and obtain the conditions controlling the dynamics we have described.

What does "resonance" mean? In physics, for a given phenomenon, it means an acute maximum in the intensity around a particular frequency. A very simple example of resonance is provided by the electronic circuit with a resistor, a capacitor and an induction coil, with an alternate generator (Fig. 9).

Depending on the frequency of the generator, the intensity of the current can be almost zero or on the contrary very high. The function of the circuit is just to

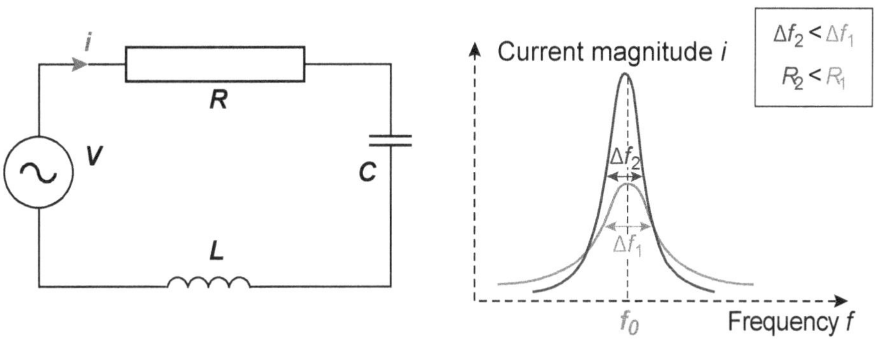

Figure 9. Example of a resonating electric circuit. The inductive coil L stores the energy when the current is flowing and then returns it when a variation of current occurs. Also, the capacitor C stores the energy that it could subsequently provide. The energy can be transferred from the capacitor to the coil and vice versa, in this way, creating oscillations in the circuit. The current obtained by a given potential difference V as a function of the frequency f of the alternate current displays an acute maximum for $f_0 = \frac{1}{2\pi\sqrt{LC}}$ (in assumption of smallness of the resistance R). The maximum is more acute when the resistance R is small.

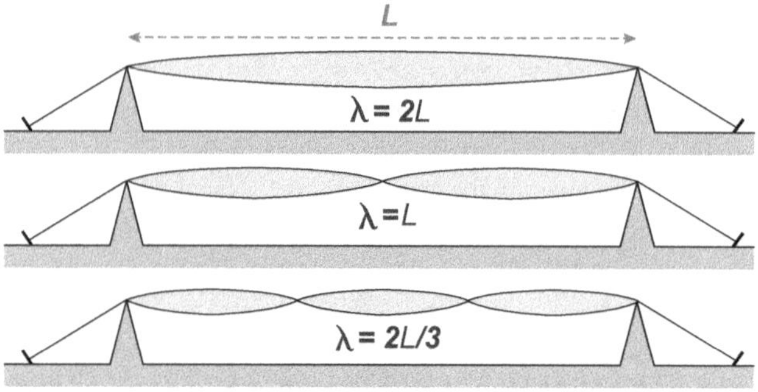

Figure 10. Vibrational modes of a string at fixed extremities.

select a particular frequency, and one defines the quality factor of the circuit Q to be the ratio $f/(\Delta f)$ of the resonance frequency (the one corresponding to the maximum) over the width of the resonance peak. For a radio emitter or receiver, it is important to have a narrow resonance, in order to emit or receive at a well-defined frequency.

The violin string is a resonant system. When the two extremities are fixed, the resonance is obtained for a wavelength λ of the vibration two times[2] the length L (Fig. 10).

However, there are other resonance modes in correspondence to $\lambda = L$ or $\lambda = 2L/3$, etc. In fact, the violin simultaneously emits all these modes in proportion, characterising the tone of the instrument. These cord vibrations are called *stationary waves*. By referring to the violin resonance cage, the word resonance has a rather different meaning. One does not wish that the instrument has an acute resonance to a certain frequency; on the contrary, for any vibration of the cord transmitted to the cage by means of the support, the violin must resonate by creating stationary waves characterised by vibrational minima, called *nodes*.

[2] For a sinusoidal vibration the displacement u of the cord from the equilibrium position at the abscissa x at the time t is given by

$$u(x,t) = u_0 \sin(2\pi x/\lambda) \sin \omega t.$$

Since the extremities at $x = 0$ and at $x = L$ are fixed, one must have $\sin(2\pi x/l) = 0$ or $L/\lambda = 1$ or 1/2 or 3/2, etc. The most general vibration corresponds to a sum of these sinusoidal vibrations.

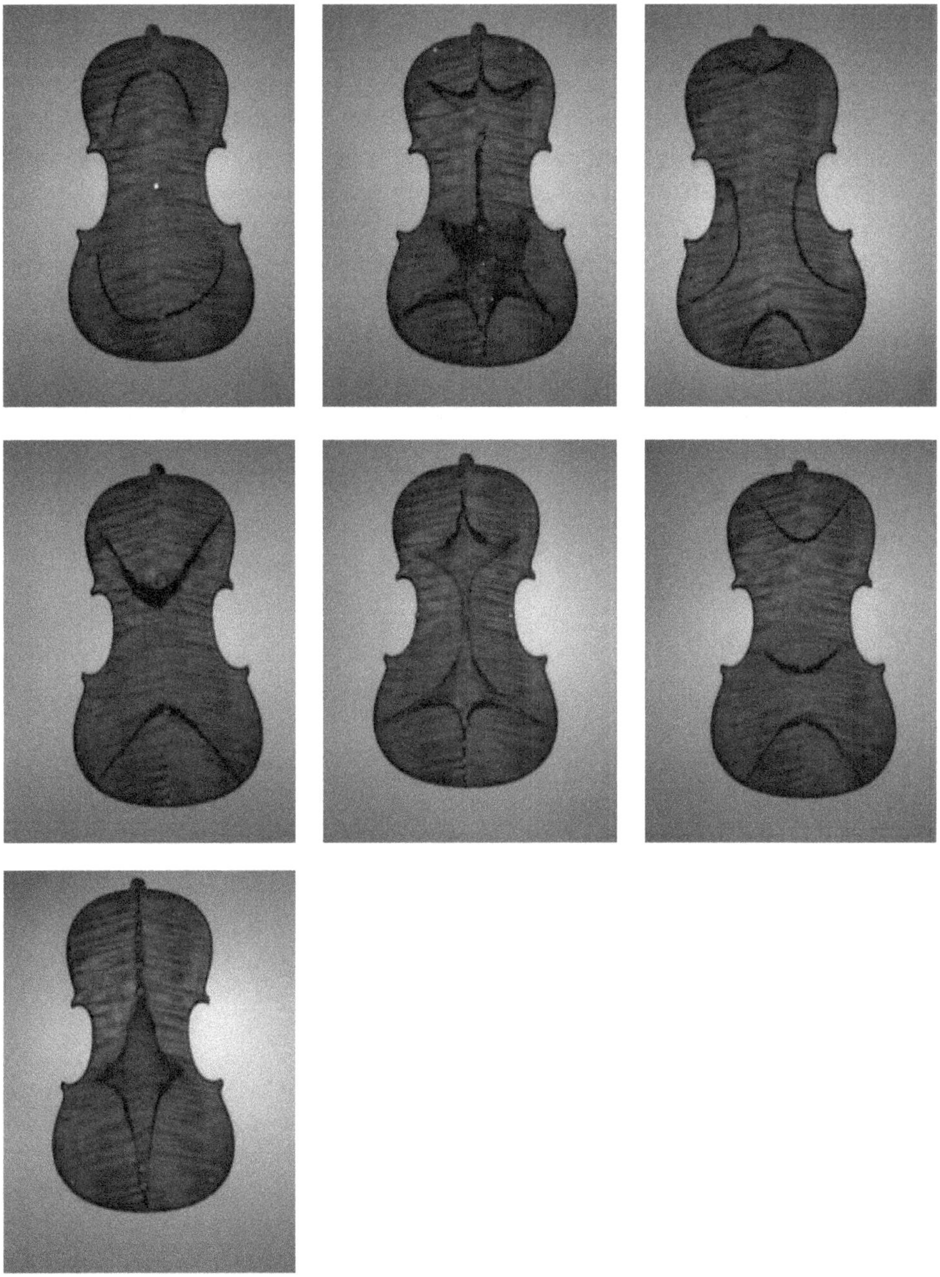

Figure 11. The Chladni patterns corresponding to the vibrational modes of the bottom of the resonant cage of violin, in correspondence to seven different frequencies. The black sand gathers in the places of minimum vibration. Courtesy of Emmanuel Bossy and Renaud Carpentier.

Chladni Patterns

The vibrational nodes of the cage can be evidenced by turning the violin and covering the cage with sand. If one excites the instrument at a given frequency, provided for instance by loudspeaker, the sand will be forced to settle at positions where the vibration is minimal. One will observe the so-called Chladni patterns (Fig. 11), named after the German physicist Ernst Chladni (1756–1827).

At variance with the electronic circuit described above, one violin has several resonance frequencies, but the vibrational nodes are clearly noticeable corresponding to the various resonances. As we mentioned above, this is not what the lute maker wishes to get. The latter, in fact, is not dealing with a flat table but rather with a complex instrument: the resonances are more complex when the violin is assembled.

Chapter 12

Singing and Silent Glasses

During a dinner, while we are waiting for the courses, we can create a particular orchestra: the participants could create special sounds from the glasses. It is easy to obtain harmonious songs from a glass of wine, while it is more difficult when the glass is full of champagne. We are going to reveal the mysteries of singing glasses, and that will bring us to musical balls…

It is known that glasses of wine can sing. Use a finger slightly wet with the liquid and let it slide on the edge of the glass. At the beginning, the sound could be unpleasant, but as soon as the edge is uniformly wet, the sound will become harmonious. By changing the pressure of the finger, one can vary the height of the sound. This latter also depends on the size of the glass, its shape and thickness. The mechanisms producing the sound are comparable to the one generating the sound in vibrating strings: it involves adherence and slipping sequences that excite the wall of the glass. During the vibration, the glass periodically compresses the air around it, creating the sound waves (see Chapter 2). By filling the glass, you will see some ripples at the surface of the liquid while the walls are excited.

Not all glasses are singing, and the search for a suitable glass might be rather time-consuming. The best singers are the glasses having thin walls and the shape of a paraboloid of revolution, possibly with long support (Fig. 1). In other words, the ones that more frequently we break! One of the authors of this book could hardly find one of those glasses, since in his home all had been broken and substituted by very thick glasses that were not suited for making vibrations.

Singing glasses have been the object of serious studies. It was found that the vibration of the wall essentially involves two elliptical configurations of the glass edges (Fig. 2).

Figure 1. Singing shaped as a paraboloid of revolution (namely the shape obtained by rotating a parabola around its axis). By rubbing one wet finger on the edge, sound is produced. The water facilitates the sliding.

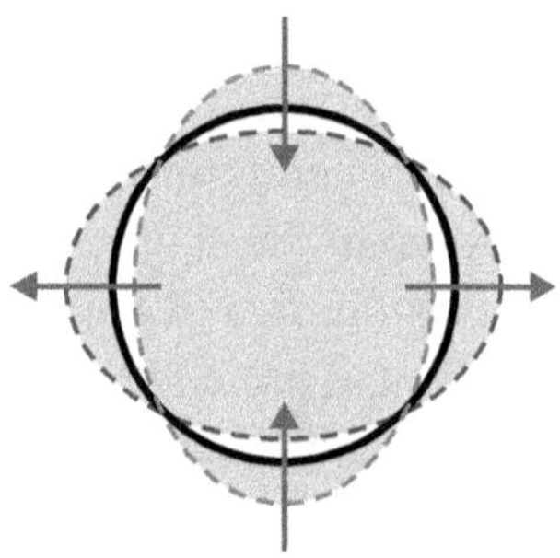

Figure 2. When sound is generated, the walls of the glass vibrate (the edge of the glass is shown here, top view): they alternately take an elongated shape in one direction, then in another, perpendicular to it.

These vibrations occur at a well-defined frequency that depends on the shape and the properties of the material; therefore, the sound produced is very pure. The level of the liquid in the glass is a relevant factor for the sound production: the higher the level, the lower the sound. By choosing a suitable type of glasses, selecting different sizes and different levels of liquid, all the notes can be obtained: with a little dexterity, it could be possible to produce one musical piece. A little-known rod instrument, the glass piano, is entirely based on that principle (see Panel on page 171).

The glasses-based piano of Benjamin Franklin

A great American scientist, as well as a great politician (qualities that are not commonly found in a single person), was strongly interested in the phenomenon of the "singing glasses". He was Benjamin Franklin (1706–1790), most known for his experiments on the atmospheric electricity and as the inventor of the lightning rod. He exploited the phenomenon in a musical instrument he had designed: the glasses-based harmonium. This instrument was based on a series of glass cups, each with a hole in its middle and placed at a regular distance along a staff. The engine had a pedal driving the rotation of the system, similar to the one in the sewing machine popular in the 20th century. By letting the wet fingers slide on the edge of the rotating cups, the musician could achieve crystalline sounds, more acute in the smallest cups and more grave in the cups with a large diameter.

Benjamin Franklin plays his glasses-based harmonium.

In the year 1793, Franklin gave the instrument to an English lady, madam Davies, as a present. She used the instrument on tours while giving demonstrations in several countries. Then the instrument apparently disappeared: one can guess that the same happened to other glasses-based pianos, possibly broken as a

(*Continued*)

(*Continued*)

consequence of clumsiness... Other instruments could remain in operation somewhat longer. Their musical qualities have been variously judged, possibly dependent on the mastery of the operator. Its marvellous sounds were claimed to have caused premature deliveries of babies or madness of the musicians! On the other hand, the instrument has been very well appreciated by some audiences, in particular by the great violinist Nicolò Paganini. The popularity decreased year by year: the instrument was completely removed from musical halls around the year 1830. After some decades, the instrument was again built with an electric engine in charge of the cups' rotation, while wetting of the fingers of the musician was performed by a humidifier. Using the internet, readers can evaluate the quality of sounds: it is possible to find a performance by the French artist Thomas Bloch of the K617 Mozart rondo using a glasses-based piano.

To Your Health!

A less elaborate way to get sound from glasses is just during a toast. When the glasses are colliding against each other, part of the wall is displaced from the equilibrium position. The glasses quickly return to the initial configuration after some oscillations that are progressively damped. Therefore, for a short time, sound is emitted. This sound in general is pleasant, but it is less pleasant when the glass is full of champagne: in this case, the sound is inexpressive and rather hard. Why?

The human ear can perceive the sound of frequency within about 20 Hz and 20,000 Hz. The crystalline character of sound emitted by a glass is due to the high-frequency components, say between 10,000 and 20,000 Hz. When the colliding glasses contain gas-free liquids or are empty, the oscillations can persist for a rather long time after the wall excitation. In contrast, the sound is quickly damped out when the glass is full of champagne. The explication is related to the carbon dioxide (CO_2) bubbles present in the liquid (see Chapter 14). After knocking the glasses, the sound waves propagate in the champagne: compression and depression waves are alternating. The solubility of gas in liquids depends on the pressure, being high when pressure is strong. When the pressure is small, degassing of the champagne occurs: bubbles are created in the glass, and they can dissipate the energy of the oscillations. Then the sound is damped much more quickly than in the absence of the bubbles (Fig. 3).

These observations made during a drink are only qualitative. Experiments performed at Paris Diderot University have allowed more precise studies of the sound propagation in gaseous media. The scientists did not use champagne for their experiments, nor did they use Vichy water, but simply water with air bubbles. They have found that even a small concentration of bubbles (say a bubble of 1 mm

Figure 3. Bubbles are spontaneously formed in a glass of champagne. Their average diameter is around 1 mm. Because of their presence the sound produced at the knocking is rather hard.

diameter for a cubic centimetre of liquid) could sizeably affect the velocity of propagation, by a factor of about 10, resulting in inducing a large effect on the sound in the audible range. The effect on the sound velocity is easy to understand: it is given by $1/\sqrt{\chi\rho}$, where χ is the compressibility (adiabatic) and ρ the specific density of the medium. This latter is equal to that of the liquid (namely of the water with alcohol for champagne). The compressibility is small in the absence of bubbles, while it is considerably increased in their presence. As regards the damping in the audible domain, to a major extent, it is due to the single bubble resonance in the liquid. This is what we are going to address in the following section.

Musical Bubbles

If the bubbles can dampen sound in the liquid, they are equally capable of emitting sounds! The murmur of the creeks, the large part of the noises during the flows of liquids and as well the singing of boiling water (see Chapter 15) are indeed due to bubbles. To understand this property, let us refer to one bubble of 1 cm diameter, in a large volume of water. When leaving the injection device, the bubble vibrates till they are entirely damped, generating sound waves. The fundamental frequency of vibration is known as Minnaert frequency, after the Belgian scientist Minnaert (1893–1970) who in the year 1933 pointed out the musical capability of the bubbles. For bubbles of air having a radius of some millimetres, in water and at atmospheric pressure, that frequency is in the range of the order of several kHz, in the audible range (see Panel on page 174).

Bubble vibration and spring oscillations

At the equilibrium, a small bubble in water is spherical. The radius is fixed by the condition that the internal pressure is equal to the pressure created by the water. Due to some perturbation (for instance, a sound wave), the bubble can be deformed and then it starts to oscillate around the equilibrium position.

These oscillations can be decomposed into an infinite number of deformational modes. Some of them correspond to relevant displacements of the bubble surface: then one deals with resonance. One of them, the Minnaert resonance, is characterised by a particularly low frequency, corresponding to very large wavelengths (of the sound in water and in air) with respect to the bubble size. These oscillations are similar to the ones of a mass hanging from a spring when displaced from its equilibrium position (see figure); in the absence of damping, the ball oscillates at a well-defined frequency f *depending on the mass and the elastic constant* k *of the spring. In order to understand the equivalence to the bubble, one should remark that the mass* M *in the case of the bubble corresponds to the mass of the liquid involved in the motion, while the role of the restoring elastic force of the spring in this case is played by the excess pressure force which tends to return the bubble surface to its equilibrium position. For the spring, this force is proportional to the lengthening* x *of the spring:* $F = -kx$*, where the restoring constant* k *is a characteristic of the spring itself. The oscillation frequency of the mass is given by*

$$f = \frac{1}{2\pi}\sqrt{\frac{k}{M}}.$$

Therefore, this frequency is increased when the rigidity of the system is increased or when the mass is small.

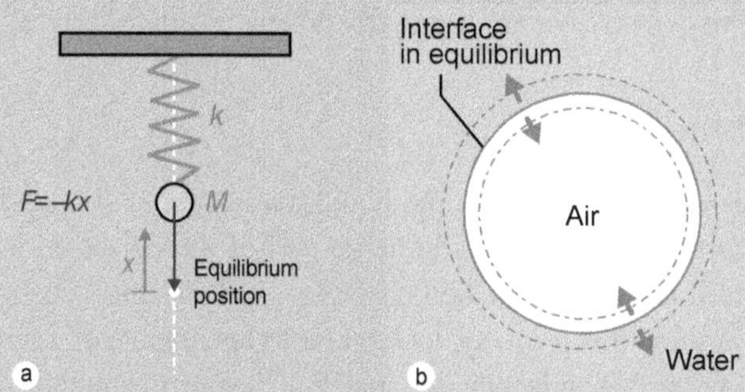

The oscillations of a mass hung by a spring (a) are analogous to the ones of the bubble (b) oscillating around its equilibrium position (continuous line) within two extreme positions (dotted lines).

(*Continued*)

(*Continued*)

What is the corresponding equation in the case of the bubble? We are going to figure out the quantities that must replace M and k by means of dimensionality arguments. In this way, we shall avoid a complex calculation, which on the other hand would be required for a quantitative derivation. The characteristic quantities involved in the problem are the density ρ *of the liquid, the equilibrium radius of the bubble, and the pressure P in the liquid. In fact, the restoring force acting on the surface mainly results from the pressure inside the bubble which at equilibrium must be equal to the pressure in the liquid. The mass M that one should take into account has to be of the order of* $(4\pi/3)R^3\rho$, *namely the volume of the bubble times the density of the liquid. As regards the constant k for the restoring force, it must be of the dimension of a force divided by a length. The only length in the problem is the radius R while in order to obtain a force one must take the pressure and multiply it by the square of a length, namely* R^2. *Thus, in this way, k is of the order of PR. By using these values in the previous equation, one obtains a result close to the formula developed by Marcel Minnaert:*

$$f = \frac{1}{R}\sqrt{\frac{2\gamma P}{\rho}},$$

where $\gamma = 7/5$ *for an air bubble.*

The surface tension σ *is not involved in this equation: its role is in fact negligible unless the bubble is very small.*

Chapter 13

Energy: Obedient Maid or Oppressive Mistress?

The energy consumption by mankind is continuously increasing, for transportations, thermal heating as well as industrial requirements. Electrical devices have invaded our everyday life; the electricity, such a docile maid, is always at our service, just from one switch or from one plug. However, will we always have sufficient production capability for a request that does not give rise to decrease? What will be the price paid by the environment? The requirement to provide energy for more and more devices could become an unbearable tyranny. Let us see some ways to remain safe.

Countries Facing an Energy Challenge?

In the middle of the 20th century, our grandfathers could learn at school that France had coal mines and hydroelectric dams. Sixty years later, the mines were closed, and the dams, which have grown in number, could provide only a small fraction of the energy required for economic growth. A similar situation has occurred in other countries. Cars require an amount of fuel that weighs heavily on a trade deficit balance. These fuels are generated from fossil sources that are diminishing, and moreover their combustion contributes to the increase of the greenhouse effect (see Chapter 7). In particular, in France, electricity to a large extent is produced by nuclear power plants. Also, these use an imported "fuel": uranium. The worldwide reserves of uranium are also limited, and they could possibly be sufficient for a century…

Thus many countries are in a situation of complete energy dependence and with the progressive decrease of the sources, this can be somewhat worrying.

Figure 1. The breeding reactor nuclear plant Superphenix (1,200 MW) in Creys-Malville, in between Lyon and Chambery, which was operational for some years at the end of the 20th century.

Where can we look for other energy sources? The first source to take into account is solar energy (see Chapter 28). However, it is not available during the night. Wind energy, which Denmark is looking toward, is also rather intermittent. A solution that has been attempted in some countries is to produce "nuclear fuel" *in situ* in the so-called *breeding reactor* plants (Fig. 1). The interest towards this quasi-miraculous way is obvious, but it has some difficulties or disadvantages, as we are going to address.

How to Exploit Nuclear Energy

Nowadays, the exploitation of nuclear energy in general is attained in the following way. In the reactor, the uranium is transformed into two more light elements: more precisely, each uranium nucleus is divided into two light nuclei, this process is called *fission*. This reaction generates a great deal of heat and that is the point! If one compares the energy obtained by fission with the one from burning oil, one uranium gram corresponds to more than a ton of oil.

In the nuclear reactor (Fig. 2), the heat produced by the fission is transferred to a first fluid, called heat carrier, that in turn transfers it to water; this latter is vaporised.

This water is pulled into the turbine by pushing on the shovels: the principle is the same used at the beginning of the 20th century for the steam engine where

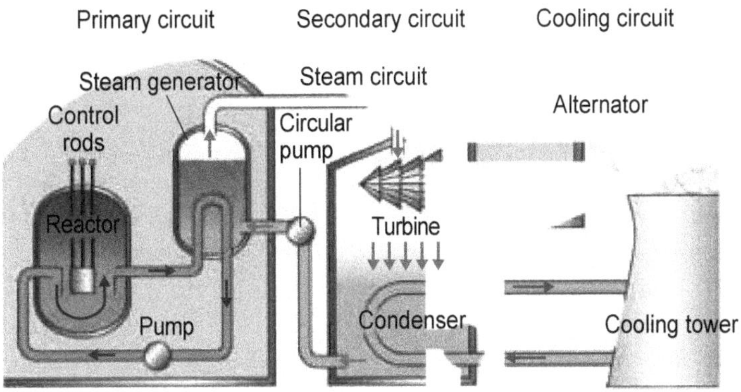

Figure 2. Principle of the procedure in a pressure reactor. On the left, in the heart of the reactor, nuclear reactions produce heat. This heat is transferred by means of a fluid which transfers the heat to the water. From its vaporisation, one can have an operating turbine connected to an alternator that produces electricity. By means of the condenser, the water can return from the vapour to the liquid state. The heat that is not converted can be discarded through a proper cooling system.

the engine was moved by the vapour pushing on a piston! Finally, the mechanical energy of the turbine is transformed into electricity by the alternator. Then this energy is delivered to the users, who may be hundreds of miles away.

Only part of the uranium is suited for fission; that part is represented by the isotope ^{235}U (see Panel on page 180). In other words, the natural uranium as it is found in the mines has only 0.71% of the isotope, the large majority being instead represented by ^{238}U that is not suited for fission. Thus, before being set inside the reactor, the natural uranium must be enriched: this step is obtained in refineries for the isotopic separation.

Controlling the reaction in a nuclear reactor

We have seen that in a reactor the energy is provided by fission, i.e., decomposition of uranium 235 nuclei into lighter nuclei. Fission in general is not spontaneous: it is triggered by the collision of the nuclei with a neutron (Fig. 3).

The fission is accompanied by the emission of other neutrons (in general two or three). These neutrons can cause new fission processes that in turn emit other neutrons and so on: this is a chain reaction. Each neutron created by fission gives rise, on average, to a number k of fission neutrons which can be larger than 1. This occurs in an atom bomb. The explosion of an atom bomb is obtained by assembling a sufficient mass of fissile matter (^{235}U, for example) which must be larger than a certain critical mass (see Panel). Indeed, if the mass is smaller, most of the

Elements of nuclear physics

The nuclei of the atoms are composed of particles called nucleons. One has two types of nucleons of almost the same mass: the protons, having positive electric charge, and the neutrons, which do not have any charge. The number Z of protons (called atomic number) characterises the chemical properties of the element. The total number of nucleons, usually labelled A, is the mass number. Two nuclei of the same element that differ in the A number are called ***isotopes****. Thus, for example, the carbon atom* (Z = 6) *has several isotopes, all having the same number of protons but different numbers of neutrons. Two of these isotopes are stable. They have respectively mass numbers 12 and 13. One writes them as* ^{12}C *and* ^{13}C*. Another isotope, carbon-14, i.e.* ^{14}C*, is unstable and spontaneously decomposes into nitrogen* ^{14}N $(Z = 7)$ *by emitting an electron: it is indicated as "radioactive".*

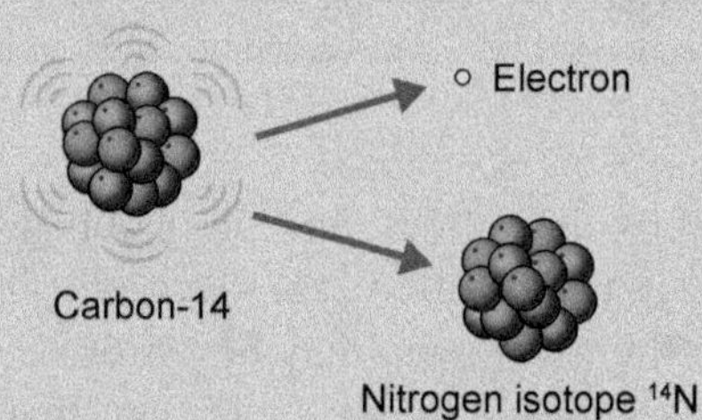

Sketch of the radioactive disintegration of carbon-14. The emission of neutrino occurs, which is not indicated.

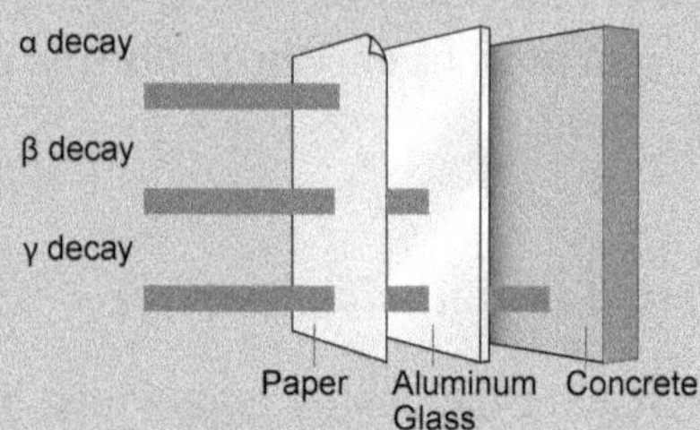

Penetration power of the different particles emitted in nuclear transformation.

There are many types of nuclear disintegration, depending on the nature of the particles that are emitted as a consequence of the nuclear transformation. The emission of an electron or a positron (a particle equal to the electron but having positive electric charge) is called beta radioactivity. Another type of radioactivity involves

(*Continued*)

(*Continued*)

the nuclei known as "heavy" and corresponds to the emission of helium nuclei: the α *radioactivity. Finally, the* γ *radioactivity corresponds to the emission of high-energy photons.*

The particles emitted during the radioactive transformation are very different in regards to their penetration power through the matter. Depending on their nature and intensity, they can be dangerous for one's health.

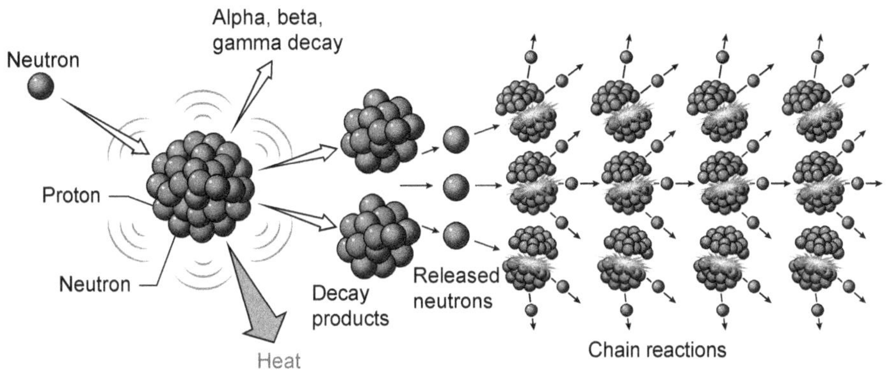

Figure 3. Principle of a chain reaction. By bombarding a fissile nucleus with a neutron, one causes the splitting into two lighter nuclei. This process is accompanied by the production of heat and at the same time by the emission of one or more neutrons. These latter can induce new fission processes.

neutrons escape without creating fission. The critical mass for pure ^{235}U is about 52 kg.

In a nuclear reactor, on the other hand, the power must be kept constant, which implies that the number of neutrons created per unit time must be constant, and the multiplication factor k must be equal to 1. To maintain this value, the reactor should be carefully designed. In particular, control rods made of a material that absorbs the neutrons (for instance, cadmium) are inserted inside the reactor at a tuneable depth, thus maintaining $k = 1$ or, if necessary, $k < 1$ if one wishes to stop the reactor.

Besides these absorbing control rods, in the majority of the reactors presently in use (and in the totality of the reactors in France), around the fissile material there is a moderator (in general, water) with the role of slowing down the neutrons in order to increase the probability of the fission processes. In fact, this slowing down of the neutrons increases their capability to produce new fissions so that one can work with uranium only weakly enriched in the ^{235}U (by an amount of around 4%). This weakly enriched uranium cannot be used to make an atomic bomb.

In an atomic bomb, what matters is to get a strong explosion, as devastating as possible. In case of control loss in nuclear reactors operating with slow neutrons, the fissile material is dispersed by the first explosion of moderate intensity, and the dispersion can cause the end of the chain reaction. This is exactly what happened at Chernobyl in 1986: after a series of human errors, an uncontrolled reaction that started in the reactor caused its destruction, fortunately without releasing all the energy stored in the fuel. The case of the disaster in Fukushima, related to a tsunami in 2011, was different (Fig. 4): the chain reaction was safely stopped by the safety devices, but the lack of cooling due to the breakdown of extra pumps due to the tsunami-related water waves could not prevent the explosion.

Plutonium: Fuel for 1,000 years?

As we have mentioned, before being set in the reactor, uranium must be enriched in the isotope 235, at the expense of uranium 238. Is this latter material devoted to being a useless residue? Not necessarily! In fact, when the enriched uranium is inside the reactor, the 238 isotope does not stay inactive. By absorbing neutrons,

Figure 4. Fire at a power plant in Fukushima Daiichi, Japan, in March 2011. The various engine failures due to the tsunami caused the breakdown of the cooling system of the reactor. The intense heat produced by the fission reaction induces chemical reactions leading to the decomposition of the cooling water with the emission of hydrogen. This hydrogen then exploded, and radioactive products were emitted in the atmosphere.

The birth of the atomic bomb

During the Second World War, physicists were concerned about the estimate of the critical mass of Uranium-235 required for achieving the chain reaction. It would seem that the Germans, in particular Werner Heisenberg, had overevaluated it so that they did not recommend the making of an atomic bomb since it had to be too heavy to be carried by aircraft. In the meantime, on the other coast of the Channel, Rudolf Peierls (of German origin as well but who settled in Great Britain after Hitler took power in the year 1933) had pointed out the correct way to get the estimate. He had published the calculation without being aware of the military implications! It was Otto Frisch from Austria, who also immigrated to Great Britain, who realised the consequences of the calculation. In the year 1940, Frisch and Peierls wrote down a memorandum, this time strictly confidential, to be sent to the British authorities: it described the procedure to create the atomic bomb, emphasising the possible devastating effects. In the USA, it was taken very seriously, and a huge scientific program of nuclear researchers was launched in 1942, the Manhattan project, with the collaboration of very distinguished scientists such as Enrico Fermi and Robert Oppenheimer. The project finally produced two atomic bombs, used on the Japanese cities Hiroshima and Nagasaki in August of the year 1945, with well-known consequences.

it partially transforms to plutonium ^{239}Pu. This plutonium is as effective for a fission reaction and therefore to provide energy. This property is used in the so-called breeder reactors where the fuel can be a mixture of ^{239}Pu (at least 10%) and ^{238}U. These reactors not only use plutonium as fissile material but also generate more nuclear fuel than the amount they use for the chain fission. For instance, if France were to just count on fissile material obtained in this way to obtain energy, then it would have the possibility to generate effective fuel for several thousands of years. In fact, the amount of uranium 238 presently stocked and potentially used for the transformation of plutonium would grant that long-lasting storage.

A breeder reactor is a reactor comprising *fast neutrons*: then the removal of moderating material is required. The use of fast neutrons implies a rather delicate design of the reactor. In the majority of the reactors at work in several countries, and definitely in France, the heat produced by the nuclear reaction is collected by means of water. Therefore, in the breeder reactors, one cannot use water: it would slow down the neutrons. Which heat transfer fluid could be used? Among the ones considered, the least difficult to handle seems to be sodium. Unfortunately, this element has an unpleasant property: it reacts violently with water and it bursts into flames spontaneously in the air. Even in the absence of water and air, the chemical reactivity of sodium limits the choice of the materials, in particular, when considering that they must be employed at high temperatures.

A few breeder reactors have been active in France. The last, the Superphenix (see Fig. 1), was closed in 1997 after a career that can be considered honourable for a prototype but otherwise affected by rather serious accidents. The development of dependable breeder reactors would require a long-range exploration of novel types of reactors.

The problem of nuclear waste

Other problems, common to all the reactors in use, involve the managing of the nuclear wastes. After use, the fissile material is a mixture of radioactive products, in addition to being very hot at the exit from the reactor. So it is stored in a "pool" fed by freshwater. Once the activity of the radioactive isotopes has been sufficiently reduced, two choices are possible: to treat them once again in order to obtain, for instance, plutonium, or immediately put them underground. When the wastes have been treated, the decrease in radioactivity as a function of time is

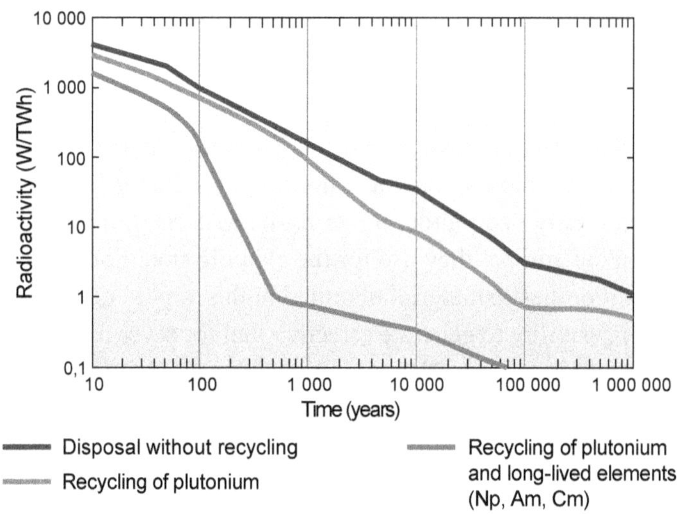

Figure 5. Comparison of radioactivity produced by the wastes, by the same amount of electricity, with or without the treatment by a breeder reactor. The quantity reported on the y-axis is the thermal release in watt as a function of terawatt–hours of electricity provided. The methods of treatment considered are putting underground without any recycling (upper line), the treatment by recycling to plutonium by a breeder reactor (middle line) and the treatment by recycling to plutonium and other radioactive elements at long time-life (Am, Np, Cm) induced by the fission (lower line).

considerably enhanced (see Fig. 5). Other elements (such as neptunium, americium and curium) can also be used as fissile material. According to this futuristic hypothesis, the radioactivity can be considerably reduced. Thus the breeder reactors are a partial response to the problem of handling the wastes.

In a more far distant future, the production of energy by means of *nuclear fusion*, namely the fusion of two light elements to obtain a more heavy nucleus, is a road presently under study (see Chapter 25).

Economic Estimates of Thermal Heating

Electricity produced in a plant is used several hundred kilometres away, for instance, in refrigerators or by operating laundry machines or in electric radiators. To convert again, electricity in heat implies unavoidable wastage since a large part of the heat initially produced by the fission is poured into rivers or into the sea (Fig. 2). On the other hand, electricity can be transferred at a large distance more easily than heat. How can we use less energy to avoid that wastage? Buildings consume the greatest portion of energy in France: three quarters of energy is spent only on their heating. Some progress can be achieved through a better insulation, but still some evolutions in the heating methods are indeed necessary. In the past, the traditional method to get heat was to burn wood, since, in this way, the chemical energy contained in wood through the combustion is extracted in the form of heat. During the 19th and the 20th centuries, wood has been substituted by combustibles such as coal, natural gas or fuel oil, obtained from petroleum. Then the electric radiators appeared: here the heat is obtained because of Joule effect (see Chapter 16) from electric current circulating in resistors.

Thermal pumps, again based on electricity, are a completely different heating method. Heat is extracted from a place where it is not needed and is pushed where it is useful. In order to obtain such heat transfer according to the second principle of thermodynamics (see Chapter 7), one has to spend energy. The surprise of the operation is that the energy spent is less than in traditional heating! How much exactly?

Having in mind to keep an apartment at the temperature T_2, while the temperature at the exterior of the house is T_1, it can be proved (see Panel on page 186) that in order to deliver the heat amount Q_2 the heat pump must use an electric energy W given by

$$W = Q_2(1 - T_1/T_2).$$

W and Q_2 are measured in joules and the temperatures in kelvin.

Heat transfer from a cool to a warm source

In most favourable cases, what is the minimum energy required in order to transfer energy from a cool source at temperature T_1 *to a warm place at temperature* T_2*? To find it, let us refer to an ideal cycle. Let us consider a cylinder with a piston with a fixed amount of fluid (see figure). This device will allow us to transfer heat from a cool room to a warm room. The cylinder is set in the cool room, and the temperature of its contents becomes* T_1*. Then an expansion is induced by pulling the piston, while the temperature stays* T_1 *by keeping the thermal contact with the environment. For the fluid to keep this temperature, a certain amount of heat* Q_1 *has to be taken from the cool room. Outside the cool room, an adiabatic compression is made (no exchange of heat with the environment occurs), and the temperature of the fluid becomes* T_2*. Then we place the device in the warm room, always keeping the temperature* T_2 *by compression. In order to not increase its temperature, the system must deliver the heat amount* Q_2 *to the warm room. Outside the warm room, a further adiabatic expansion is made and the temperature of the fluid is brought to* T_1*, and the cycle is completed.*

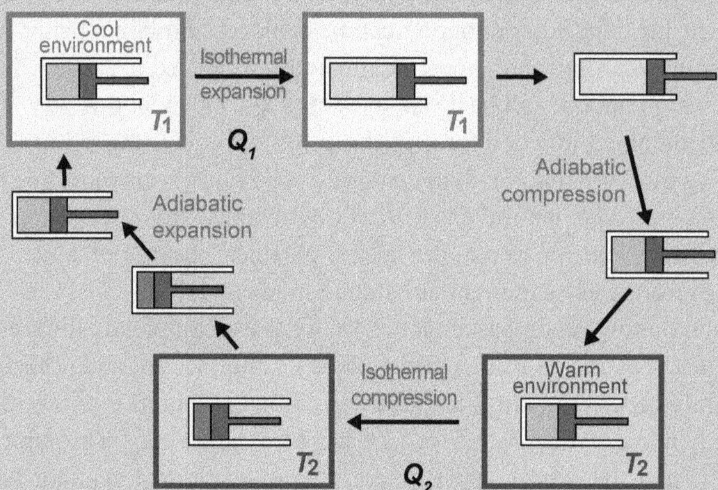

Refrigeration device. The fluid is placed in a cool environment and the amount of heat Q_1 is taken and then brought in the warm environment where it releases the heat amount Q_2. Isotherm means "at constant temperature" while adiabatic means "without heat exchange with the exterior". The device is a refrigerator if it used to cool a cold environment or is a heating pump if used to heat a warm environment.

(*Continued*)

(*Continued*)

The second principle of thermodynamics states that $(Q_2/T_2) - (Q_1/T_1)$ *is always positive. It would be zero in a certain theoretical limit. That corresponds to the case in which the transformations are made with such precautions to be reversible, namely that one could, in any situation invert the sense of the arrows in the scheme. This condition in practise cannot occur.*

In order to drive the piston, a certain mechanical energy W must be spent. The variation of that energy being zero in a closed cycle, the energetic balance yields

$$W = Q_2 - Q_1.$$

In the ideal case that all the transformations are reversible, one would have $Q_2/T_2 = Q_1/T_1$ *and then*

$$W = Q_2(1 - T_1/T_2) = Q_1(T_2/T_1 - 1).$$

The cycle described above is the most effective in order to transfer heat from a cool source to a warm one by providing mechanical energy. The inverse cycle (obtained by changing the sense of the arrows on the scheme) is called, after the physicist Sadi Carnot, the Carnot cycle (see Chapter 7).

Let us compare the performance of the heat pump with the one of a classical electric radiator for which the heat amount provided is equal to the electric energy W required for the heat pump. For a heat pump operating at $T_1 = 0°C = 273$ K and $T_2 = 20°C = 293$ K, the electric energy required is $W = 0.07Q_2$. Thus one has to pay only 7% of the thermal energy obtained. In practise, due to the losses, the expenditure is more than 7% but still much less than the 100% corresponding to the thermal energy provided by the electric heater.

Let us comment on, in a few words, the heat pump and its analogue refrigerator (Fig. 6). In both cases, the heat transfer is obtained by means of a fluid that moves along a pipe sometimes from the room to be kept cold or warm and sometimes from the other room (this fluid plays the same role as the one inside the mobile cylinder in the example addressed in the Panel above). This fluid experiences a cycle of transformations from the gaseous to the liquid state and vice versa. This increases the performance of the process since the transformation liquid–gas involves sizeable energy.

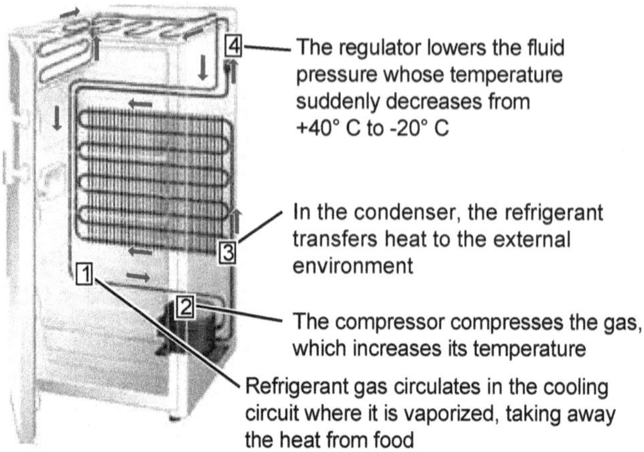

Figure 6. The operational principle of the refrigerator.

From Thermal to Electric Engines

Instead of wasting mechanical energy in order to get heat transfer, as in the heat pump, the thermal engine in our cars requires heat amount Q in order to provide mechanical energy W. The performance of an engine is estimated by resorting to a number defined by the ratio of the usable energy over the energy that one has to spend: this number is high when the engine has great efficiency. The second principle of thermodynamics limits the efficiency W/Q of all thermal engines. When the heat is transferred from the hot source at temperature T_2 (for example, the combustion room in the cylinder for the fuel-operating engine) to the cool thermostat at temperature T_1 (the exterior), the ideal efficiency is $(1 - T_1/T_2)$, namely the one of the Carnot cycle (see Panel on page 186). Nowadays, the best efficiency is around 35% for engines burning petrol, since a large part of the energy associated with the combustion of the fuel is dissipated in the form of heat. Thus the fuel-operating engine makes a poor performance in comparison to the electric engine that has an efficiency of around 95%. Nowadays, the limited autonomy of electric cars restricts their diffusion: their future is related to the development of efficient batteries.

Storage of Electricity in Chemical Form

The storage of electricity is a major aim: the solar or the wind sources provide energy in an intermittent way. Therefore, one has to be able to store the electric energy in case of surplus production and to provide it when required.

(a) (b)

Figure 7. (a) The alkaline batteries typically present in electronic toys convert chemical energy into electrical energy. (b) The batteries of our mobile telephone are made of several accumulators working according to the same principle. At variance with batteries, the accumulator is rechargeable: with a proper supply of energy, the chemical reaction is inverted, and the reacting elements are regenerated.

Providing electric energy on request? This is what the batteries do, having stored the energy in chemical form (Fig. 7).

When the battery is delivering current, chemical reactions of the constituents are occurring. These reactions, called oxidation–reduction reactions, imply exchanges of electrons among the chemical entities. Once the elements producing the reactions are exhausted, the battery is wasted. A fuel pile is not much different from those batteries since it is based once again on oxidation–reduction reactions. In theory, the fuel pile could last indefinitely since the reagents that feed it are continuously supplied. In practise, the lifetime, although much longer, still is limited.

The hydrogen pile, for example, has two compartments and a porous membrane in between (Fig. 8). In one compartment, the molecule of hydrogen H_2 in the gas state is transformed by a catalyser into two ions H^+ thus releasing two electrons.

These two ions filter through the membrane that must have the remarkable property to allow the ions to pass while the electrons are stopped. The electrons are forced to enter the electronic circuit open for them thus causing a current that can feed, for example, an engine. On the other side of the membrane, the electrons find the ions H^+ and, in addition, an air current, namely oxygen, thus being able to combine and form water.

The equation of the chemical reaction is $2H_2 + O_2 \rightarrow 2H_2O$ (two molecules of molecular hydrogen with one molecule of molecular oxygen provide two molecules of water). In this fuel pile, the chemical energy is transformed into electrical energy with good performance; far from being 1, anyway. Part of the energy obtained from the chemical reaction is indeed lost as thermal energy.

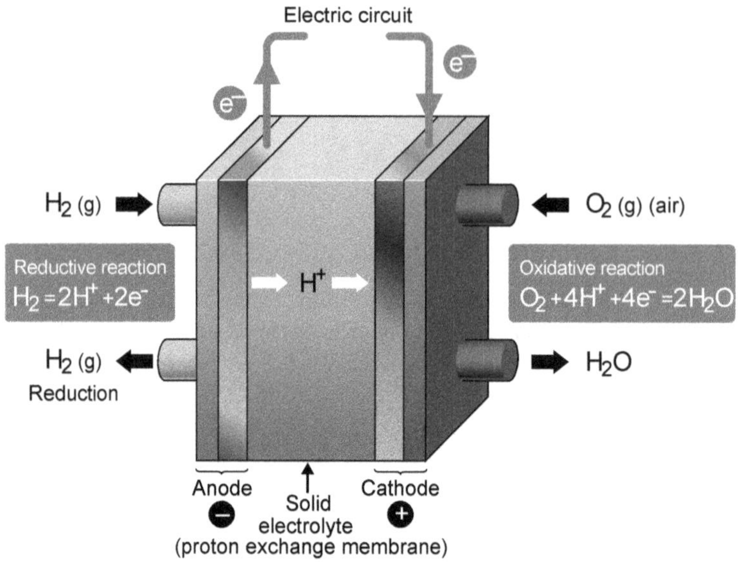

Figure 8. Principle of a fuel pile with Hydrogen and Oxygen as reagents. The only exhaust of the pile is water.

Figure 9. The electrical engine of this car is fed by combustible pile that uses hydrogen stored in its reservoir.

The hydrogen pile has the great merit to discharge only water! Thus, cars equipped with an electric engine using this pile do not yield any pollution (Fig. 9): it can go around simply by resorting to its hydrogen reservoir that has to be fed regularly. On the other hand, from the point of view of environmental preservation, one should take into account the global balance of the production of hydrogen. In reality, hydrogen is combustible and not so easy to obtain and store (see Panel below on this page). Furthermore, the catalyser required for the pile functioning, most times platinum, is expensive. Research is underway aim to overcome these two difficulties.

Water: The fuel of the future?

Hydrogen is a suitable energy transporter. How to get it at a reasonable cost and in a "green" way? An idea could be to take advantage of aeolic sources or solar panels since they provide energy in an irregular way, and thus temporary storage is required. During the production peaks, one could obtain hydrogen by means of electrolysis processes. It is indeed possible to obtain the process inverse of the one in the hydrogen pile by providing electrical energy: water is decomposed into molecular oxygen and molecular hydrogen and then stored.

However, hydrogen storage implies some difficulties. In a pure form it is explosive! Furthermore, a container of hydrogen in the gaseous state is very big, while it can be stored in the liquid state only at low temperature and high pressure. An innovative process has been suggested by a small French company, McPhy; to store the hydrogen in metals. Its General Manager claims to have been inspired by a prophecy of the scientist Cyrus Smith, the hero of "The Mysterious Island" by Jules Verne: "What will burn instead of coal?", he asked. "The water," answered Smith, "but the water decomposed by electricity in its constituents, namely oxygen and hydrogen. Thus, water will be the coal of the future", he concluded. Coal that when burning does not produce gases which contribute to the greenhouse effect...

To conclude, it should be emphasised that what we addressed in this chapter only provides a fragment of information. The reader could conveniently integrate it by resorting to specialised books.

Chapter 14

Nunc est Bibendum

"Nunc est bibendum", now is time to drink, wrote the Latin poet Horace in the first century BCE. He was wiser than his predecessor Alceo of Metilene who, in his lyric seven centuries before, claimed that the joy of wine up to inebriation: nmν χρη μεθμσθην, is time to get drunk. Before presenting other physical properties of drinks rich in alcohol, we shall begin with a brief history of the viticulture and the procedure to make wine.

Chemists without Knowing It

According to a tale, wine was discovered at the court of a Persian king Jamshid, many centuries before our time.[1] Due to serious depression, a court lady was planning to kill herself. Following the indication of a priest in the court, she decided to drink that strange liquid at bottom of the jars where the grapes were stored, a liquid that was supposed to be poison. The unexpected result was that the depression disappeared and the bad mood gave place to joy and happiness! The positive role of that liquid convinced the king to promote regular use of it. That fortuitous discovery had great impact; the cultivation of the vineyard and the wine consumption gained many adherents around the world. The Greeks celebrated wine with the addition of a new god, Dionysus, who the Romans subsequently adopted with the name Bacchus (Fig. 1). Among wine's attributes were its curative properties, as a good antiseptic and for caring the wounds. Christians

[1]Another story is the account given in Genesis that Noah was, probably, the first man to cultivate vineyards for making wine.

Figure 1. Fragment of the mosaic "Icarius and Dionysus" from the atrium of the house of Dionysus (Paphos, Cyprus). The god of winemaking, Dionysus (left), and the nymph Acme with a cup of wine.

have given wine a major role, being a crucial component during religious celebrations.

How did the juice of grapes become wine at the court of the Persian king? Without knowing it, those amateur chemists had discovered alcoholic fermentation, namely the chemical reaction that transforms sugar into alcohol, the molecule where carbon atom carries OH groups. In the grape juice, this reaction occurs thanks to micro-organisms: yeasts called *Saccharomyces cerevisiae* present in the skin of grapes that can transform the sugars contained in the fruits. In the simplest transformation of the juice, the glucose (and its isomer fructose), produces ethanol (C_2H_5OH) and carbon dioxide (CO_2):

$$C_6H_{12}O_6 \rightarrow 2C_2H_5OH + 2CO_2.$$

For the chemists, wine, like all the alcoholic drinks, is basically a mixture of water and ethanol. In small proportions, many other compounds (up to 2000 types of molecules) are also present in wine. The latter are the ones yielding characteristic fragrances and colours.

The Art of Winemaking

Winemaking is a difficult art. One of the major problems is that in the presence of oxygen the wine tends to transform into acetic acid (CH_3COOH). In other words, the wine becomes vinegar! Agronomists and Latin naturalists, like Columella and Pliny the Elder, provided recipes to avoid wine turning into vinegar. For instance, they recommended covering the amphora with pitch, probably in order to stop air entering the container. Our modern bottles have a watertight seal and do not need such treatment. However, it could be useful to avoid the penetration of air through the cork with a wax layer. On the other hand, if the winegrowers have given up the pitch, they often add sulphur dioxides with antioxidant properties to the wine.

Great biologist Louis Pasteur (1822–1895) devoted much interest to the wine; he had noticed that the oxidation of the alcohol in vinegar was due to a bacterium. The works by Pasteur gave scientific basis to the technique of winemaking, which previous generations of winegrowers had to learn by trial and error. For example, in order to avoid the fermentation going on so long that the transformation to acetic acid occurs, it is appropriate to divide pretty soon the must (namely the juice obtained from the grapes under pressure) by the yeasts. This is the main purpose of the filtration of the must. On the other hand, fermentation is an exothermal reaction (meaning that it generates heat) so the temperature of the must can increase up to about 42°C. Such a temperature could cause the evaporation of volatile compounds, including the precious fruity or floral aromas characterising high-quality wines. To preserve those natural treasures, nowadays, winemakers resort to fermentation in an ambient atmosphere (at about 18°C), and this means that the chemical reactions take considerably longer; typically three weeks instead of 7 or 8 days for the natural fermentation. It is useful to resort to filtration at low temperatures, for example, 4°C, in order to get rid of unwanted products: at this temperature the latter solidify while the wine filters and one achieves the required separation. In fact, the temperature of the wine solidification decreases by about a third of a degree for 1 percent (in volume) increase of ethanol[2] (Fig. 2).

The considerations recalled above explain why the cellars where wine is produced look like scientific laboratories. We could write at length on this subject, but we prefer to devote attention to addressing some physical properties of wine.

Wine Tears

When a glass of wine is slowly rotated, with slight inclination in order to get the walls wet, one can observe a curious phenomenon (see Panel on page 197). Inside the glass a slight layer of wine appears, and, at the top of it, small viscous rivulets, called tears of wine, can be seen (Fig. 3).

These rivulets gently slide down along the walls of the glass, while further tears appear. It is the volatility of the alcohol, greater than that of water, that is

[2] When the wine is defined to have alcoholic degree 10%, it means that the ethanol concentration is 10% in volume. From 1 litre of this wine, one could extract 1 dl of pure alcohol (at 20°C). The mass density of ethanol being 0.787 g l^{-1}, a fourth of such wine contains a little less than 20 g of alcohol.

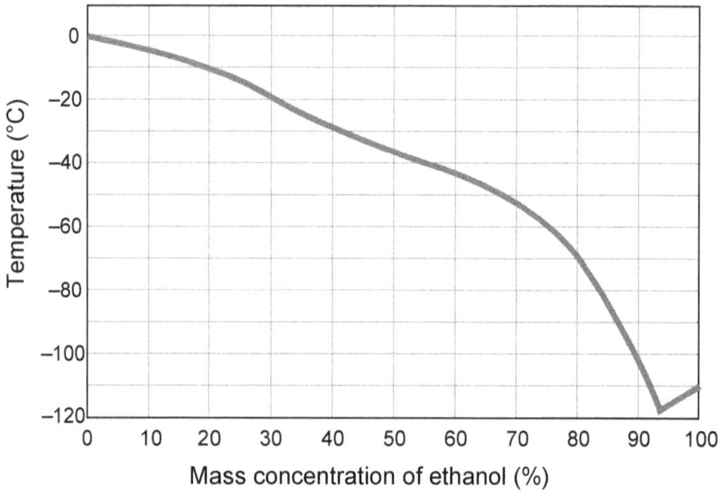

Figure 2. Temperature at which the solution water–ethanol initiates the solidification as a function of the mass concentrations of ethanol, given in percent. It is noted how, for concentration of alcohol smaller than 93%, the solidification temperature of the solution is considerably below the one for pure water (0°C) and decreases on increasing the alcohol concentration. Alcohol is a kind of antigel!

Figure 3. The tears of wine, sometimes known as "legs".

responsible for this phenomenon: a flux of liquid climbs along the layer in order to replace the alcohol that has quickly evaporated in that thin and wide area. It is not only related to equilibrating the concentration, but also in order to decrease the surface energy, namely the surface tension, as established by Italian physicist Carlo Marangoni in 1865. The surface tension of the water is in fact larger than that of alcohol. So what happens to this upstream? At a certain height, the force associated with the gradient of the surface tension coefficient, pulling the film upward, and the gravity force of the already raised volume of the film, are balanced. The film stops rising, and the incoming liquid accumulates in the rim formed along the perimeter of the glass. Just as the flow of liquid from the tap breaks up into drops, so, due to the Rayleigh–Plateau instability (see Chapter 6), this rim is also unstable. Thicker and thinner areas are formed along it, resulting in an imbalance: the liquid from the former spills downwards and the wine "tears", flow down the surface of the glass. The upward flow of liquid constantly feeds the whisk, so that the process continues as long as the concentration of alcohol in the wine remains sufficient.

Condition for the occurrence of wine tears

What are the conditions for the occurrence of wine tears? We have discussed the relevant role of the film of liquid on the walls of the glass after rotation. The film has to be sufficiently stable, and therefore the surface tension has to be relatively weak. When the glass contains pure water, its surface tension σ_{lg} *is rather strong, so the film can hardly be formed. For wine, or simply for a mixture of water and alcohol, the surface tension is smaller, and this favours the formation of the film.*

For a solution with a high concentration of alcohol (say more than 20%), one does not need to rotate of the glass in order to have formation of the film: the liquid climbs the wall of the glass spontaneously (one speaks of total wetting) before the tears are formed. This has been pointed out by two French scientists, Jean-Baptiste Fournier and Anne-Marie Cazabat.

For their experiments, the two researchers took the precaution of placing an alcoholic solution, rather than in a glass of wine, inside an expanding cup, so to avoid the saturation of the atmosphere by alcohol vapour. In atmosphere too rich in alcohol, the evaporation might stop since the molecules of alcohol evaporated are

(Continued)

(Continued)

equilibrated by the ones that return to the liquid. The reader can check: if a coaster is set over the glass of wine where tears are occurring, then the formation of the tears will stop within a few minutes. If the coaster is removed, the air will be renovated, and the process will start over. It should be remarked that the experiments by Fournier and Cazabat have been carried out strictly in a solution of water and ethanol.

If one should repeat the experiments by using real wine, then other factors might play a significant role and affect the formation of the tears: say sugars, tannins and others. On the other hand, at variance with popular belief, glycerine did not much modify the experimental observation.

Champagne Bubbles

Champagne is the most northern wine-growing area in France. Already in the Middle Ages there was wine production: red wine, not sparkling, and most likely not so good. Due to the cold climate in the winter, alcoholic fermentation could stop well before the sugar had entirely turned to alcohol by the yeast's action, and the activity could resume only in the spring. At that time, the wine might already be inside the bottles. The carbon dioxide produced by the fermentation increased the pressure and could cause the explosion of the bottles! With the help of one Benedictine monk, Dom Pérignon (1639–1715), winegrowers in Champagne learned to take advantage of that secondary fermentation in order to create a novel and now famous sparkling wine: champagne. Nowadays, after a preliminary fermentation in the cellar, the second fermentation is artificially induced by the addition in the wine of the "liqueur de tirage", namely a mixture of sugar, yeasts and other wine. The product is enclosed in thick bottles, strong and well sealed. After some months, one notes the appearance of the deposit of un-active yeasts inside the wine. It is the sign that the sugar has been consumed and that the second fermentation has been achieved. By keeping the bottles inclined upside down, the deposits are collected in the neck, and they remain there for a certain time (Fig. 4).

During an ageing period that can last up to three years, the rich flavours typical of champagne develop. When the deposit has to be removed the neck of the bottle is frozen by plunging it in liquid nitrogen, which opens the bottle: the pressure from the carbon dioxide drives out the deposit contained in a piece of ice. To complete the preparation of the bottle, one adds the "liqueur d'expédition", namely a mixture of sugar and champagne. This last process will fix the type of

Figure 4. Bottles of champagne in the aging process. The bottles are kept inclined upside down and are frequently moved to make sure that the deposit of the yeasts properly occurs inside the neck of the bottles.

champagne: brut, demi-sec or sec. At this point, one has only to fill up, close the bottle with a proper cork and a " muzzle" and sell it — at a price that has to take into account the attention spent and the work done!

Once the bottle is sold, it will be opened soon thereafter. This operation causes the development of a certain amount of foam. Why? The carbon dioxide produced during the second alcoholic fermentation accumulated inside the bottle that was hermetically sealed. The pressure on the cork can reach up to 6 or 7 atm at the end of that process. This high pressure explains why some bottles used to explode before the strength of walls had been adequately increased by the producers. When we open the bottle to enjoy the champagne, the pressure inside the liquid suddenly drops to 1 atm. The amount of gas stored inside the liquid is increasing with pressure. By opening the bottle, the solubility of carbon dioxide inside the wine strongly drops, and the gas comes out, forming the bubbles. A bottle of 0.75 l of champagne contains about 9 g of carbon dioxide, which corresponds to 5 l of gas at room temperature and pressure. Even if the fraction of the gas taking part in the effervescence is only about 20%, there is still plenty of gas to cause a large number of exploding bubbles!

Experimenting with sparkling wine is not recommended for children, but it offers a good way to let them play. It is enough to take a piece of chocolate for them to observe a nice movement inside the glass, from top to bottom and vice versa. The chocolate initially sinks, but approaching the bottom of the glass it collects bubbles which then lifted it up, due to Archimedes' buoyancy, and then at the surface, they evaporate so that the chocolate again sinks.

The bubbles not only play with a piece of chocolate as if it was a ball, but they even play music when they pop: a characteristic crackling that has been studied by means of very sensitive microphones. For convenience, it has been studied by using soap water, where the foam has longer lifetime. What has been discovered by the recording? When bubbles explode occurs at random inside the foam, one can hear a buzz similar to the noise of a radio badly tuned or near a waterfall: a so-called at random "white noise", having the same intensity at all frequencies. This is not what one really hears. Each explosion, lasting about 0.001 s, usually triggers a sequence of explosions that our ear experiences as a single sound signal, even in the case that some bubbles blow up separately without the capability of the ear to catch it. If the explosions occur at random independent events, then they would be produced with probability $1/\tau$ per unit time; consequently, the distribution of the time intervals t separating two explosions shall be centred around an average value τ defined in that way. This is not the situation occurring for the bubbles in the champagne: since their explosion is a collective phenomenon, there will be correlations among successive explosions, and a characteristic time between two explosions can hardly be defined. It can be proved that the probability that a time t separates two explosions is proportional to $1/t$. Several natural phenomena follow analogous laws, for instance, the avalanches and landslides. These phenomena do not have characteristic time and characteristic intensity. Thus an avalanche can involve a few grams of snow or, though less likely, can be large enough to wipe out a road.

On the Clouding of the Pastis

Pastis is the traditional aperitif in South-East of France, with strong flavour of anise. Its preparation is very simple, and it is accompanied by a surprising phenomenon. When the glass is filled with pastis, you can admire a beautiful, slightly green transparency. Then the same amount of water is added, evidently transparent. By mixing two liquids, both transparent and opaque, almost white mixture appears. How to explain this surprising phenomenon? The explanation can be found in the composition of the pastis: it is a mixture of alcohol (45%) and water, with a little amount (0.2%) of anethole, an extract from grains of anise, present also in fragrances and some medicines. The anethole is very soluble in alcohol. On the contrary, it is only little soluble in water. In pure pastis, the concentration of alcohol is such that the molecules of anethole can form a solution. When water is added, the anethole molecules are no longer surrounded by a sufficient number of alcohol molecules: they cannot go into a real solution, and instead they collapse into little groups that remain as a suspension in the liquid. Thus the pastis mixed with water forms an "emulsion" (as is milk, a suspension of fats, or mayonnaise). The opacity is due to the light scattering by those relatively large groups of molecules (see Chapter 3): in pastis as well as in milk, the incident light is strongly scattered for any wavelength. The size of the little drops is indeed of the order of a micrometre, namely of the same order of the wavelength of the light, and therefore the scattered component is almost white (Fig. 5).

The size of pastis droplets has been studied at the Institute Laue-Langevin by means of neutron scattering. Some readers might think that the research on the pastis was motivated by the desire of drinking the correspondent glasses. It has to be said that this is not true: the study on the neutron scattering required the substitution of H_2O with the "heavy water" D_2O, where the hydrogen atoms are replaced by the heavier isotope deuterium. The heavy water strictly speaking is not toxic,

Figure 5. The pastis problem. Pure pastis is transparent. After the addition of a little amount of water, the pastis becomes turbid: the anethole molecules form small drops of micrometric size that diffuse the light. If some drops of a product for dishes are added, the pastis attains its transparency: the surface-active molecules of the product favour the contact among the anethole and water molecules, and the size of the drops decreases.

but it still does not have a good reputation. Thus the pastis used for the experiment had to be discarded!

Pastis is a relatively stable emulsion. If a glass of pastis is left at room temperature for about 12 h, the anethole droplets would progressively melt and lift up at the surface, creating a floating phase. The pastis would become almost colourless and transparent.

The Titration of Vodka

Winemaking is not so easy in cold countries. Wine is often substituted by alcoholic distillates obtained by apples — this is the case of the Calvados in Normandy — or by starting from cereals, for instance, barley, maize or rye for whisky, as it happens in several countries. Vodka, also known as "wine from bread", is obtained from cereals: often in Russia, vodka is drunk with a meal, as it happens in Italy and in France with wine. The present titration of vodka is around 40% of in alcohol.

This composition has not been so clearly stated in the past. Sometimes the innkeepers took the liberty of adding water. To stop such misuse, it seems that the tsar Ivan the Terrible issued a decree that allowed the unsatisfied clients to beat the innkeeper to death if the vodka served was not flammable. One can suppose that the concentration of 40% was right, the one to cause easy flammability, thus saving the life of the innkeeper.

Distillation: Technique of the alchemists

Alchemists were the precursors of modern chemists. Distillation is one of their ingenious inventions: it allows one to obtain the "aqua vitae" from the wine or, in more general terms, to separate the components in a liquid mixture. Let us refer to Fig. 7: at a given temperature, for example, 85°C, the gas phase in equilibrium with the liquid one is much more rich in alcohol than the latter. In order to increase the alcoholic titration of solution (for instance, wine), it is sufficient to warm it and let the vapour get out and condense in a cold container. This is the work of the alembic (see figure). For example, on warming progressively the mixture of water and ethanol at a molar fraction of 0.2 in alcohol, the mixture initiates its evaporation at about 83.5°C. The molar fraction of alcohol in the vapour is 0.54, and this corresponds to the one in the first droplets in the distillate. Step by step, when the

(*Continued*)

(*Continued*)

evaporation is pursued and the temperature of the mixture increases, the two phases lose alcohol: the alcoholic concentration in the liquid decreases. Therefore, it is convenient to stop the process rather soon while the vapour is still rich in alcohol.

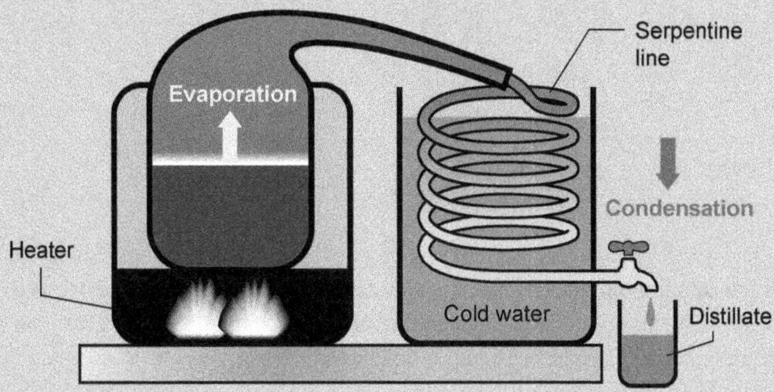

Distillation principle with the alembic. The mixture to be distilled (for example, wine) is heated in a vessel. The vapour rich in alcohol is collected in the upper part of the vessel. This vapour is progressively cooled down in the serpentine kept inside cold water. The liquid, namely the distillate, is collected at the end of the serpentine.

After the first distillation of wine, one obtains "aqua vitae" strongly alcoholised. If the distillation is repeated over and over, one can collect a solution more and more concentrated. On the other hand, one cannot go over the azeotropic concentration of 96°, i.e., 90% in molar concentration, for which the vapour has the same composition as the liquid: the distillate has the same composition as the hot solution. The applications of the distillation are not limited to making liqueurs: the distillation of lavender, for instance, is common in Provence. The fragrant oils present in the plant are brought by the water vapour in the alembic, and one can extract them.

A more simple method to test the concentration of ethanol in water does exist: to evaluate the density, which decreases when the concentration of ethanol increases. The measure is performed by means of the densitometer, essentially a glass tube, ballasted, that is put to float and sink more or less inside the liquid (Fig. 6).

The principle was probably already known at the time of Ivan the Terrible, however its real making is rather delicate and thus was not so diffuse in Russia at

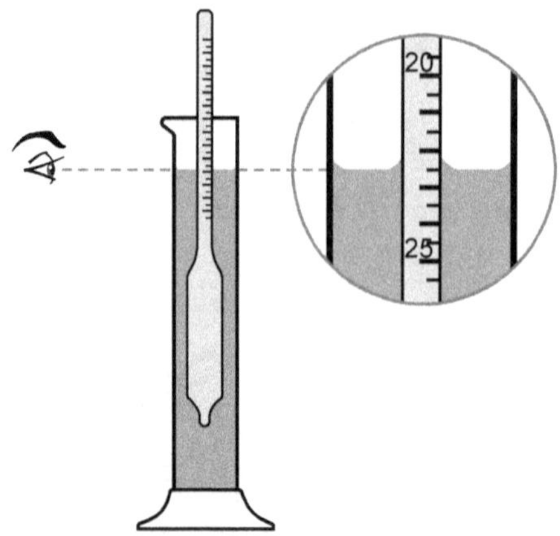

Figure 6. The device is a sealed glass hollow tube with a weighting compound at the bottom. The flask tapers towards the top where graduations are applied to it. When the device is lowered into a container of wine, it sinks more or less depending on the Archimedean force acting on it, and therefore depending on the density of the liquid. The graded scale allows one to estimate the density by reading the number at the surface of the liquid.

that historical moment. Another tzar, Alexander III, in 1894, decided to substitute the empirical criterion of Ivan the Terrible with a method based on the density. What to select? The tzar asked the advice of one of the most prestigious scientists of that period, Dimitri Mendeleev, who probably suggested a concentration of 38% and then rounded up to 40% by the tzar. At this concentration of alcohol, a glass of vodka can stand in the open air for many hours, and although the volume of liquid will decrease due to gradual evaporation, the concentration of alcohol in the drink will remain constant. Mendeleev was certainly aware that one has a concentration, much more high, where the solution evaporates progressively still preserving its alcoholic concentration (Fig. 7). At that concentration, the solution is said azeotropic, and it corresponds to an amount of alcohol of around 96%. That is why it is hard to buy in a pharmacy alcohol totally pure and produced with the traditional method of distillation.

Vodka has to be enjoyed cool. It is easy to have the vodka properly cool during the winter in Russia, and especially in Siberia where the average temperature is around –10°C: one has to simply leave the bottle outside. The same treatment applied to water would break the bottle, as it is well known. The solidification of water implies an increase of the volume of about 10% (see Panel on page 206). Why does a bottle of vodka not undergo the same destiny? From one side, we have seen that a solution of alcohol in water at about 40% keeps its liquid state well

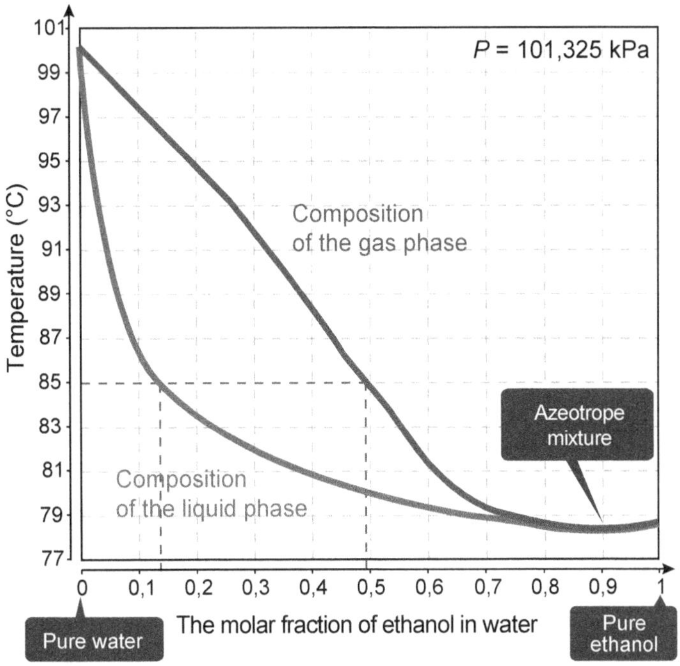

Figure 7. Composition of the phase liquid–gas in mixture of water and ethanol at the equilibrium at room pressure (about 101 kPa). The composition of each phase, along the *x*-axis, is given in molar fraction of alcohol, namely the number of alcohol molecules divided by the total number of molecules. Pure water boils at 100°C while ethanol at 78.4°C. In between the two temperatures, two phases coexist at the equilibrium: for example, at 85°C, a liquid phase has about 14% of alcohol in molar fraction and a gas phase about 49%.

below 0°C. A glass of vodka in the freezer will remain liquid unless your freezer should be so efficient to cool down at about –30°C. At that moment, when the vodka initiates to solidify, the solid phase being formed is almost entirely pure water. The remaining liquid is rich in alcohol and could stay liquid down to a lower temperature. On further cooling, the temperature of the liquid continues to decrease until the temperature is the same as your freezer: the liquid phase does not disappear. To have total solidification, one should reach the temperature of –120°C, which is not reached even in Siberia! What happens if you put also a glass of wine, rather rich in ethanol, in the freezer, at the same time that you are doing the experiment with vodka? This time you would see something that looks like homogeneous solid. However, touching it with a finger, you would realise that is not a unique solid block. In fact, one has a series of grains of ice hanging around a liquid rich in alcohol. This liquid adheres to the grains, and it is not easy to separate. Therefore, solidification is not a handy way in order to increase the amount of alcohol in a vessel: distillation is more convenient.

Water: An extraordinary liquid

The mass density of ice, at room pressure, is 917 kg m^{-3}, while the one of liquid water is about 1,000 kg m^{-3}. This property of water, to have a volume larger in a

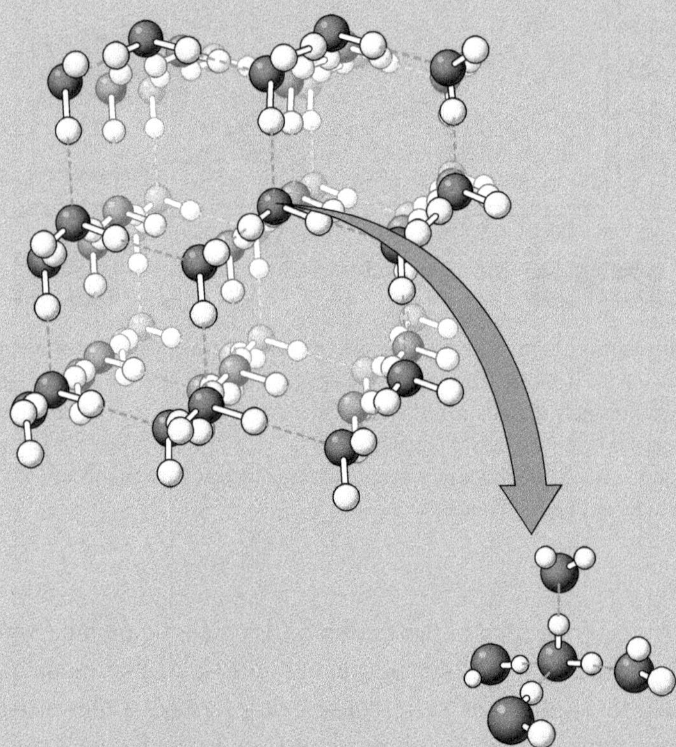

At normal pressure, ice takes the hexagonal lattice and is not very compact. The tops of the hexagons represented in the figure are occupied by oxygen atoms. On the places nearby, there are hydrogen atoms, as it is shown in the small figure. Each oxygen atom has four neighbouring hydrogen atoms, two of them nearest neighbours, thus forming a water molecule, and the other two more far away (that belong to another water molecule). The size of the hydrogen atom is very small, and the space is occupied most by the oxygen atoms. Each oxygen atom has only four nearest neighbours, instead of the 12 occurring in the more compact lattice.

(*Continued*)

(*Continued*)

solid state compared to one in a liquid state, is really exceptional: it is opposite to what happens with the majority of materials. Metals, for instance, increase their volume by about 3–4% when melting. The behaviour of water is related to the structure of ice at the atomic scale (see figure): it implies vacancies where it could be possible to set other water molecules. When the water crosses from solid to the liquid phase, the chemical bonds become weaker, and under the effect of pressure (for example, the atmospheric pressure), the molecules take advantage of the free space in order to set at smaller distance. It is necessary to warm water further for the vacancies have to be occupied by the molecules: in fact, it happens at about 4°C that the water has the minimum specific volume. On increasing the temperature, water increases the volume on heating as it happens for most materials: but in between 0 and 4°C, water reduces its specific volume under heating!

Wine, Alcohol Intake and Health

In this chapter, we have recalled the physical and chemical properties of wine, as well as the role that wine plays in the present civilization and the pleasure that it can bring to our existence. In this last part, we are going to mention the effects of the wine on our health. It is well known that an immoderate consumption of alcohol implies serious diseases, such as liver cirrhosis, cancers and mental illnesses, besides the risk of car accidents. Things are rather different for a moderate use of wine, especially red wine, as we are going to address.

Pasteur, in the year 1866, in a heavy work dealing with the wine-related diseases, wrote "wine can indeed be considered the most hygienic and healthy drink". But his assertion could not be based on real statistical data: many of them were obtained only at the end of the 20th century. Those data belong to a group of studies about the factors that influence the health and the diseases (not only the plagues).

In 1991, statistical data collected in a study called Monica[3] was published in the prestigious medical journal *The Lancet*. The data caused a great deal of attention from the press because their conclusions were somewhat unexpected: the wine had beneficent effects, favouring good health, at least in regard to

[3] Monica (Multinational MONItoring of trends and determinants in CArdiovascular disease) is an international study planned and coordinated by the International Organization for Health (OMS) having the aim to measure the tendencies and the determinants of deaths because of cardiovascular diseases and the correspondent risk factors for these illnesses.

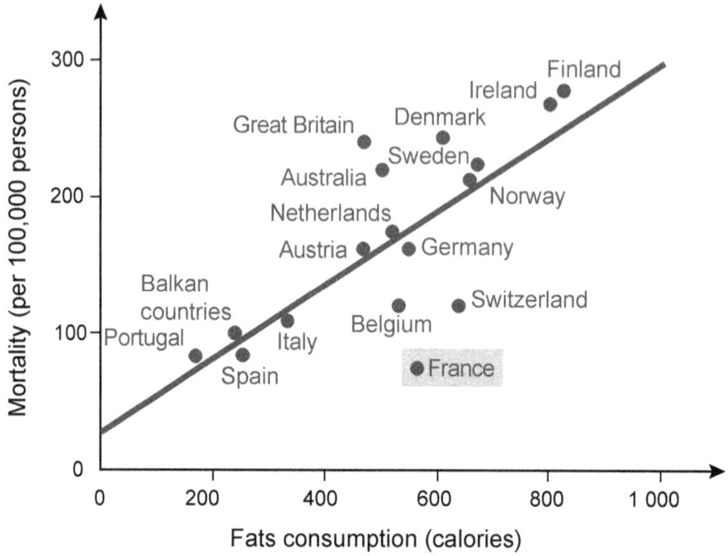

Figure 8. The more animal fat (and therefore cholesterol) we consume, the mortality due to cardiovascular diseases increases.

cardiovascular diseases. Figure 8 reports the rate of mortality for these illnesses as a function of the animal fats intake, in various countries. The correlation is noticeable: the more animal fat (and therefore cholesterol) we consume, the more the mortality due to cardiovascular diseases increases. The correlation is almost linear, and the data are reasonably well fitted by a straight line. A point is markedly out from the fitting curve: it is the one for France! In spite of a rather large intake of cheese and butter, the death rate for cardiovascular diseases is relatively low (although still remaining the second source of mortality after cancer). It is noted from the figure that French people eat more fat than the English yet still have around four times fewer heart attacks!

The US television station CBS was the first of the media to call attention to that result. In 1991, it was presented with the aggressive title "The French paradox". Could that be attributed to the daily intake of red wine, particularly in the Bordeaux area where most of the data had been collected? Further researches, performed in regions of red wine, confirmed that a moderate intake was beneficial in order to decrease the risk of cardiovascular accidents.

In Fig. 9, the result of a study more recently carried out in France is reported, in the form of a curve reporting the risk decrease as a function of the alcohol intake. One could remark that the reduction in that curve is smaller than the one reported in the Monica project, where for a daily intake, the reduction in the risk was significantly greater. This discrepancy could be related to the fact that in the

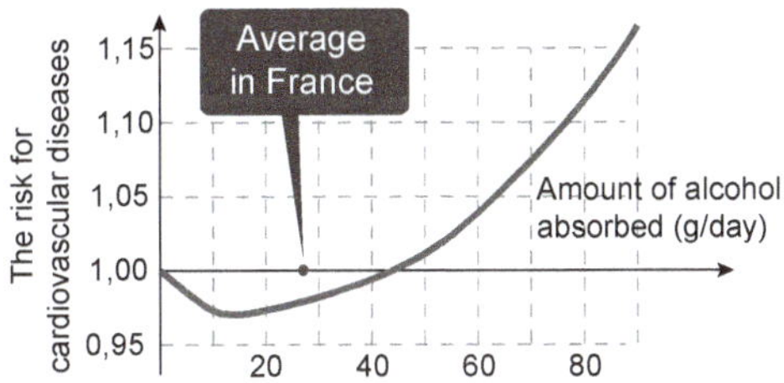

Figure 9. Risk of death by cardiovascular diseases as a function of the daily intake of alcohol (in grams per day), in comparison to the one for a teetotal person. The curve has been drawn by the authors of the study performed on the basis of the epidemiological data in France, in the year 2009. (S. Guerin, A. Laplanche, A. Dunant, C. Hill, Alcohol-attributable mortality in France, *Journal of Public Health*, **23(4)**, March 4, p. 1 (2013). For the data of the Monica project, see E. B. Rimm *et al.* Prospective study of alcohol consumption and risk of coronary disease in men, *The Lancet*, 338, pp. 464–468 (1991).).

Figure 10. The Chinon castle in Indre-et-Loire where King Charles VII received Joanne of Arc in 1429 and where nearby Franc Rabelais was born some decades later in a region of vineyards celebrated in the story Gargantua and Pantagruel.

Monica results the decrease was estimated in correspondence to the intake of red wine in particular areas, not of simple alcohol itself. In fact, the characteristics of the wine seem to play a relevant role. Wine contains more than 2,000 substances, and crucial to health benefits is the phytoalexins, the polyphenols and particularly the transresveratrol, cardioprotective due to antioxidants. These compounds should be present in a significant amount of the red wine of the Bordeaux region and obviously from other vineyards of similar characteristics.

Obviously, there are other, noncardiovascular, alcohol-related diseases. Besides liver cirrhosis, also certain cancers as the one involving the oesophagus could be favoured by the alcohol assumption. Thus, the advantage of regularly drinking even moderate amount of wine could be debated if one has public health in mind. Still, wine deserves credit: it helps in bearing the small or great troubles of human beings and yields a special atmosphere, a sensory pleasure, at social gatherings. With the stipulation: do not overindulge!

Glasses of wine are agreeable provided that the wine is of good quality (and furthermore if we want to get the benefit described above, it should be red and coming from certain vineyards). The quality is guaranteed by clever winemakers, and we can visit them and their cellars during touristic trips in beautiful wine-producing areas, which are also often rich in historical souvenirs (Fig. 10). Finally, Physics, specifically the Nuclear Magnetic Resonance, offers a nice and dependable method to determine the place of origin and the composition of the wine we buy, by analyzing the spectra of the deuterium (the isotope of hydrogen) in the ethanol molecule (see Chapter 27).

Part 3

The Sapiens Cooker

Homo erectus appeared about a million years ago and was probably the first to master fire. His cuisine was nothing other than roasting meat of elephant or buffalo that had been killed by means of knives obtained by polishing off stones or bones. The subsequent 10,000 centuries have seen dramatic changes in the way of living of the human being: our knowledge, expertise and creativity have evolved in parallel to our alimentary practises.

Nowadays, modern humans manipulate the aliments in sophisticated ways. Cooking has become an art, and dinners are pleasant, convivial events.

In this part of the book, we will address several subjects cooking-related; from the use of the microwave oven to the way to cook spaghetti and the preparation of tea or coffee. It will be shown that the kitchen is the scene of several phenomena of physical or chemical interest. It will be emphasised how mastering the temperature allows one to obtain special eggs Japanese style or to cook meat at the best. Furthermore, the so-called "molecular gastronomy" will be addressed, showing how novel methods can provide meals with special properties.

Chapter 15

Waiting for a Cup of Tea

In the famous cartoon "Asterix in Britain", the authors insinuate that a cup of tea is strictly similar to a cup of hot water. Without having in mind to continue the joke, this chapter about tea will be devoted to boiling water: where are the bubbles generated in the kettle? Why is some noise produced? What is their speed when the bubbles lift up?

There are several heavy manuscripts and chapters of specialised books devoted to the best way to serve and drink the tea. We shall limit this section to its preparation, that in practise is nothing other than bringing water to the boil, before adding the leaves to get the infusion.

Never Two Kettles without a Third One

Let us start with a very simple experiment. We take two kettles exactly the same and with the same amount of water, at room temperature. One of the kettles is left open, while the other is covered by the lid (Fig. 1). Which one will first arrive at boiling? It is not required to be an expert cooker to answer: the covered kettle will first arrive at boiling. The question is: why?

While waiting for the two kettles to get hot, let us take a third kettle and set it onto the heater. Same amount of water, same power, same temperature. We want this third kettle to get the water to boil more quickly than the others. What to do?

Someone could possibly suggest to add a certain amount of water already hot, thus increasing the temperature. This would be a poor idea. Boiling will be delayed! In fact, it would be necessary not only to provide the energy required to

Figure 1. Traditional kettles are set on a heater. Some have a beak with the whistle emitting a strident sound when the water reaches boiling temperature, and thus some vapour is given off. There are also electric kettles that incorporate their own heater.

boil the initial amount of water but also the energy to bring the water added to the boiling point. Evidently, if a certain amount of cold water is substituted with the same amount of hot water, then the boiling point will be reached more quickly.

Appearance of First Bubbles

Let us leave the third kettle alone and pay attention to the first, covered with a lid. It makes a vague noise. If one lifts up the lid, some bubbles being formed at the bottom of the kettle can be noticed, which then detach and move up towards the surface.

Why do the bubbles move up? Because of the Archimedes force, obviously, that acts on any body immersed in liquid: it is equal to the weight of the water that has been shifted. From the correspondent force, directed upwards, one should subtract the force acting on the bubble in the opposite direction due to the resistance of the water: resistance force having strength that increases with the increasing speed of the bubble. At the beginning of the ascent, the bubble velocity is small, and the water resistance is insignificant: the bubble rises faster and faster to the surface. Within a few centimeters, its speed becomes high enough for the resistance force to compensate for the force of Archimedes; the steady-state bubble velocity in this case depends only on its radius (see Panel on page 215).

Bubble motion and turbulence

Let us detail the equations describing the motion of the bubble during its ascent. We will assume that the bubble is a rigid sphere; this is far from being correct, however, it will lead us to a fairly accurate result. The bubble is thus subjected to the weight force (negligible), to the Archimedean force, and to a kind of friction force that resists its motion.

For a sphere of radius R, the Archimedean force expelled can be written as

$$F_a = (4/3)\pi\rho g R^3,$$

where ρ and g are respectively the water density of the liquid and the gravitational acceleration. The resistance to the motion, for small velocity v is given by the Stokes equation

$$F_s = -6\pi\eta R v,$$

where η is the viscosity coefficient of water and v the speed of the bubble. At large velocity, the viscosity no longer plays any role, and an approximate expression of the resistance reads

$$F_s' = -\pi\rho R^2 v^2/2.$$

The ratio of the two resistance quantities is given by 12 times the inverse of the "Reynolds number" Re, a dimensionless number very useful in the fluid mechanics

$$Re = \rho R v/\eta.$$

The Stokes equation is correct within error of 10% for Reynolds number Re < 1. For a sphere of radius about 1 mm in water, the Reynolds number is around 200. Then we shall refer to the second expression of F_s, yielding a satisfactory order of magnitude.

Thus the total force can be written $F = F_a + F_s$. According to the general principle of the dynamics (see Chapter 4), one can obtain the equation of motion of the object once its mass is known. The vector sum of the forces is equal to the acceleration times its mass, namely

$$F = m(dv/dt).$$

However, when we refer to the motion of a bubble in liquid, what is really moving are the water molecules outside the bubble. Should we apply the fundamental principle of the dynamics to the water molecules? This would be very difficult! Fortunately, the fluid dynamics shows that the principle can be applied directly to the body in motion in the liquid provided that a proper mass is attributed to it. In

(*Continued*)

(*Continued*)

the case we are dealing with, the mass we have to consider is the real mass of the object plus an "extra mass" (so called added mass) given by $\delta m = (2/3)\pi\rho R^3$ *namely the half of the mass of the volume of water being expelled! For a bubble, the total mass is practically reduced to the extra mass δm only. Just after detachment, till the speed is small, the resistance of the water is negligible. Then one has*

$$Fa = \delta\text{m}\ dv/dt,$$

which, after substituting the expressions for the Archimedes force and the added mass, leads to a surprising result: dv/dt = 2g. The motion thus is twice as fast as the one during the free fall, obviously in the up direction instead of down.

After some centimetres of ascent, when the forces F_a *and* F_s *become in equilibrium, the bubble continues its uniform motion with velocity that only depends on the radius R. The experimental estimate is around 20 cm* s^{-1}*, for a bubble of diameter around 1 mm, a value compatible with the equations given above.*

Why is the Stokes equation no longer valid when the speed of the bubble is rather high? It can be used only for the so-called "laminar regime" when the current lines flow as would be attached to the object (see figure). For sufficiently high speed, the disordered vortices along the trail of the bubble arise. This is the regime called turbulent. These vortices absorb energy, thus slowing down the motion of the bubble.

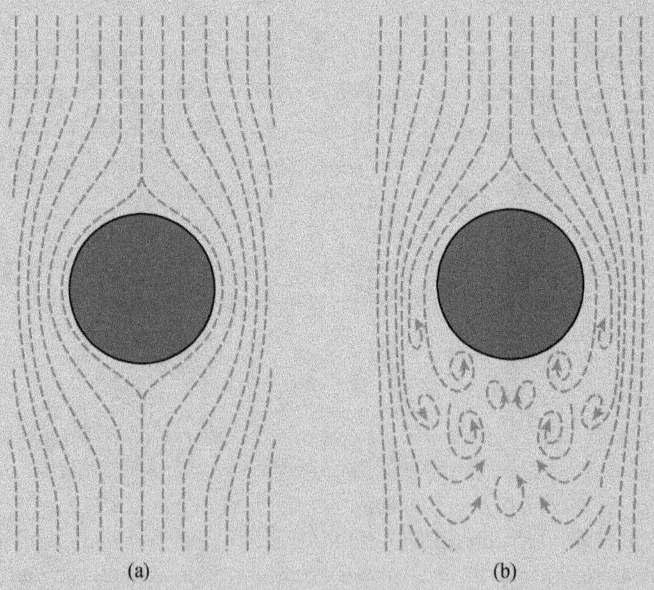

(a) Laminar flow around an object; (b) turbulent trail behind the obstacle. The dash lines indicate the current lines.

Evaluation of the bubble implosion time

Let us estimate the duration of bubble implosion. As above, the fundamental principle of the dynamics can be applied to the mass m of the water that enters the bubble during the implosion:

$$ma_s = F_b$$

where a_s is the acceleration of the border of the bubble while F_b is the force related to the difference of pressure. One has $F_b = S\Delta P$, where $S = 4\pi R^2$ is the surface of the bubble and ΔP the pressure difference between the interior and the exterior of the bubble itself. By substituting the mass by $m = (4/3)\pi R^3 \rho$ (namely the volume of the bubble times the water density of the water), one has $\rho R^3 a/3 = R^2 \Delta P$. We shall assume for ΔP (that depends only on the difference in temperature between the upper and the lower layers of water) a constant value. Let t_2 be the time required for the implosion and R_0 the initial radius of the bubble, the acceleration can be substituted by R_0/t_2^2, thus obtaining the order of magnitude

$$t_2 = R_0(\rho/\Delta P)^{1/2}$$

where the square root of 3 has been omitted in view of the approximate estimate we expected.

At standard pressure, the saturating pressure of the vapour decreases by about 3 kPa per degree (see Table). Thus we can assume that ΔP is of the order of 1 kPa. Then a bubble of radius 1 mm implodes in about 1 ms.

The bubble formation

Why do bubbles form when the water is heated? At first sight, it has to do with the gas inside the water that gets free (see Chapter 14), and when the temperature is sufficiently increased, it is due to bubbles of water vapour. For a bubble of radius R to be stable, it is required that the internal pressure is greater than the external pressure by the amount $\delta P = 2\sigma/R$, where σ is the surface tension of the water: it is the Laplace pressure (see Chapter 6). In our case, the external pressure is the one due to the liquid plus the atmospheric pressure. As regard the pressure inside the bubble, defined as *pressure of saturating vapour*, it depends on the temperature: it is the pressure of the vapour being in equilibrium with the liquid at a given temperature. It increases when the temperature grows (see Table 1).

Table 1. Pressure of water-saturated vapour at representative temperatures.

T (°C)	96	98	100	102	105	110	115	120	125
P (kPa)	87.7	94.3	101.3	109	120	143	169	196	232

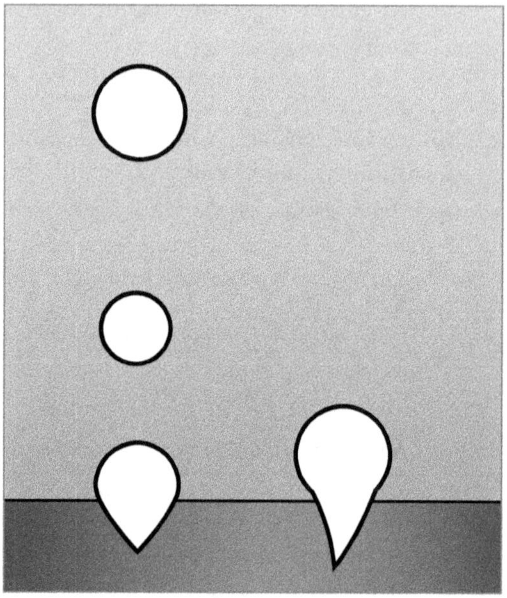

Figure 2. As soon as the temperature of the water becomes high enough, bubbles appear on the defects of the bottom of the kettle. Due to the action of surface tension forces, while their volume is small, the bubbles remain at the bottom and, being filled with steam, gradually grow.

Therefore, bubbles appear primarily on the bottom of the kettle, being hotter. But even there, the nucleation of a bubble is not easy: after all, at first, while the radius of the bubble is still small, the Laplace pressure which must resist the air and the vapour is very large! Experiments have shown that the appearance of bubbles most often occurs at defects in the bottom of the kettle, where conditions for nucleation are more favourable: the initial size of the bubble is determined by the characteristic size of the defect (Fig. 2). It is said that such nucleation is heterogeneous, in contrast to homogeneous nucleation when bubbles appear in the volume of the liquid.

The bubbles' ascent

Bubbles just born do not detach immediately: as they are small, they remain stuck because of the surface tension effect. In fact, a bubble of a given volume is more

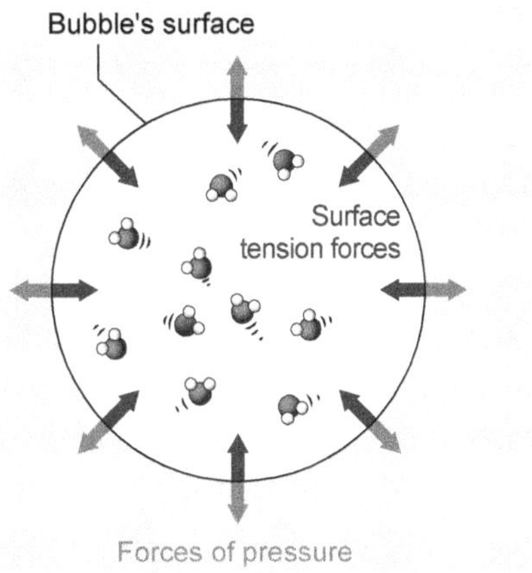

Figure 3. Bubble of water vapour before the real boiling starts. The water molecules are moving in a disordered way: a large number of collisions against the surface of the bubble occur. The resulting pressure has to equilibrate the surface tension. When the bubble lifts up, colder regions are attained: the water vapour condenses, and the bubble implodes.

"comfortable" when stuck onto the bottom of the container. When the nascent bubbles are resting on the bottom, there are two forces acting upon them: the Archimedean buoyancy force, which pushes it upward, and the surface tension that keeps it attached to the surface. As the bubble grows, the Archimedean force rises too, and at a certain moment, it exceeds the retaining force of surface tension. The bubble takes off, starting on its journey upward. During this phase, the bubbles have size of the order of a millimetre (while during the boiling process, they can reach a size of the order of a centimetre). One can evaluate the time required for the bubble to detach from the wall: it has to displace by a distance of the order of its radius R under acceleration of the order of g. Thus, the time required is of the order of $(2R/g)^{1/2}$, namely a time of the order of 0.01 s for a bubble of millimetre size (Fig. 3).

Let us follow the bubble during its ascent. Due to its lifting up, the bubble reaches a region where the temperature is decreased; the temperature distribution in the water being far from homogeneous during the boiling process. In fact, water is not a good heat conductor, and desiring to drink the tea, we have forced a strong heating: while the temperature at the bottom of the kettle is above 100°C, the water in the layers at various levels does not reach this temperature. During the ascent of the bubble, the vapour inside it quickly returns to the liquid state: the bubble collapses and disappears (see Panel on page 215), or, if it has a small amount of water inside, it strongly squeezes. It is only when the water boils in the entire volume that the radius of the bubbles increases with increasing height.

Song of water in the kettle

The appearance of the bubbles in water is accompanied by the emission of a regular noise. What is the source of this phenomenon? It seems possible to rule out the hypothesis that just the propagation of the bubble in the liquid can cause sound waves in the audible range. In fact, in the air, the high speed of a gun bullet can indeed cause audible sound waves, but this is not the case as, for instance, for a tennis ball. On the contrary, the two phenomena, namely the detachment of the bubble from the bottom and the implosion, could be good candidates. In fact, they induce oscillatory effects inside the liquid. What is their frequency? Through simple calculations, one can deduce that a bubble of radius 1 mm can detach in about 0.01 s, which corresponds to a frequency range around 100 Hz. The implosion, in about 1 ms, corresponds to a frequency around 1,000 Hz, a little higher frequency sound. A simple observation corroborates these estimates: just before the boiling process fully sets in, when the bubbles stop the implosions, a careful listener can identify that the sound becomes lower, being emitted now only due to the detachment from the bottom. During the real boiling, the tonality of the sound changes again, the noise being at this point just the one due to the burst of the bubbles at the surface.

The different frequency components of the sound emitted are also functions of the water level and the shape of the container. The song of the water being boiled was studied during the 18th century by Scottish physicist Joseph Black (1728–1799): his conclusion was that the sound results from a kind of duet between the bubbles going up to the surface and the vibrations of the walls of the container.

Kettle with a whistle

The whistle is a remarkable device. It can emit a sizeable noise in spite of its small volume. How does it work in the kettle? It is a rather difficult problem of fluid mechanics, as it is proved by the interest that arose in a group of researchers in Cambridge, UK. Here are their conclusions. The whistle fitting the beak of the kettle is formed by two metallic plates with a hole in their centres (see figure). The vapour comes from the cavity of the beak, of size of order of a centimetre, that fixes the wavelength of the sound that is going to be emitted. The longer the beak is, the lower the tonality of the sound. Then the steam jet exits the cavity through a narrow

(*Continued*)

(*Continued*)

opening and enters the whistle. In this case, the steam jet turns out to be unstable: it behaves like a trickle of water at the outlet from the tap. After walking a short distance, it breaks up into separate drops. Leaving the whistle at a sufficiently high speed, the jet forms numerous vortices, which are the sources of the sound waves we hear.

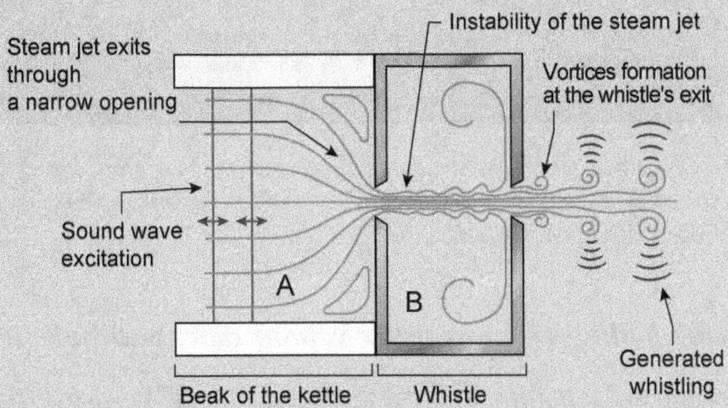

Operating principle of the whistle. The vapour crosses two cavities A and B turning into the two holes. Coming out from the whistle, the jet forms vortices that are responsible for the sound. From R. H. Henrywood and A. Agarwal, The aeroacoustics of a steam kettle, *Physics of Fluids* **25**, p. 107101 (2013).

Water is boiling

Now in one of the kettles, the temperature is so that the vapour bubbles reach the surface and there they break out (Fig. 4). As expected, it was in the covered kettle that the boiling point was first reached. The process is indicated by a jet of vapour exiting from its beak. By itself, the water vapour would not be visible, but at the exit from the bill, it condenses in small droplets of liquid water. These little droplets accompanying the jet get visible as a white cloud, thanks to the same diffusion process when the light makes the clouds white (see Chapter 3).

The vapour jet is sufficiently strong to cause harsh sound through a whistle (see Panel above on this page). Sufficiently strong as well to cause some burns if our hand intercepts the trajectory of the jet. Within half a second, we should receive a mass of vapour of around 0.3 g. This consideration leads us to pose another question.

Figure 4. Once the boiling process is stationary, the vapour bubbles grow during their ascent. Above a certain size, they lose the spherical shape. They arrive at the surface where they break out, and the surface appears rich of water-whirls.

Are burns from boiling water or water vapour more painful?

What burns the most: the boiling water or the water vapour? It is a question similar to the one: what is more heavy, the lead or the feathers, or the variant, 1 kg of lead or 1 kg of feathers? Thus one has to be more precise "what burns the most, a gram of boiling water or a gram of water vapour at 100°C?". We immediately state the water vapour! On a surface that is colder, such as our skin, the vapour starts to condense and after that cool down. Now the condensation downloads a lot of heat. A mass m of 1 g of vapour that condenses at 100°C downloads an energy given by mr, where r the latent heat of condensation that for water is 2,257 kJ kg^{-1} (see Chapter 18). This energy is much more than the one downloaded when the same amount of water is cooled down. In fact, this latter is given by $mC\Delta T$, where $C = 4{,}190\ J\ kg^{-1}\ (°C)^{-1}$ is the specific heat for water and ΔT is the difference with the final temperature (about 40°C, when the temperature becomes acceptable) and the initial one, 100°C. For $\Delta T = 60$°C, the ratio of the two energies being respectively received is around 10. Thus it is concluded that a mass of water vapour is 10 times more painful than the same mass of boiling water.

The Kettle without a Cover

While we were doing our calculations, the water in the kettle without any cover reached the boiling temperature. Why is it late? That is because part of the energy provided by the heater has been used to produce the evaporation of part of the water, which is released into the atmosphere. Even in the covered kettle, there was such evaporation, but, in this case, the vapour was trapped. Some condensation occurred onto the cover, thus returning the energy to the system. In fact, once the vapour concentration has reached in the covered kettle a certain value (dependent on the temperature), then the number of molecules that evaporate is exactly the same as the number of molecules that return to the liquid.

Some more explanation. It is not necessary to wait for the boiling temperature (100°C at the ordinary atmospheric pressure (see Panel on page 224)) for some water pass from liquid to gaseous state: in salt marshes or even for the water of the sea, one has slight evaporation. Water evaporation, at 100°C or even at lower temperatures, does require energy; the same energy is gained in the inverse transformation, namely during the condensation crossover to the liquid state of the vapour.

There is a small advantage of an uncovered kettle: when the water evaporates there is a certain reduction of the amount of water that must be warmed up to 100°C!

Meanwhile, as we have addressed above, the energy required to heat a certain amount of water of some tens of degrees is much less than the energy required to bring the same amount to the vapour state. The saving obtained in bringing only part of the water to boiling temperature is therefore small in comparison to the cost of the vaporisation. If we have, on average, heated the water in the uncovered kettle to a given temperature 100°C $- \Delta T$, the cost therefore is $(r - C\Delta T)\Delta m$, where Δm is the mass of the water evaporated and ΔT is of the order of 20–30°C: the latent heat r and the capacity C have been given above. As regards the amount of water evaporated Δm, it is proportional to the area of the water surface and how much the heater has been on, and therefore inversely proportional to the power of the heater; in practise, it involves some percent of the total amount of water. The reader can verify that the waste in energy having forgotten to set the cover is around 30 percent of the total energy provided.

Water Is Hot!

Physicists like to experiment using cages where the temperature is uniform, such as the calorimeter, in order to be able to control the various conditions. Cooks know that this state rarely occurs in their pans or their ovens. The water at the bottom of the container, in direct contact with the heat source, is evidently hotter than the one at the surface. This is particularly true before boiling but also remains this way during the boiling process (Fig. 5).

The pressure cooker and cooking at high height

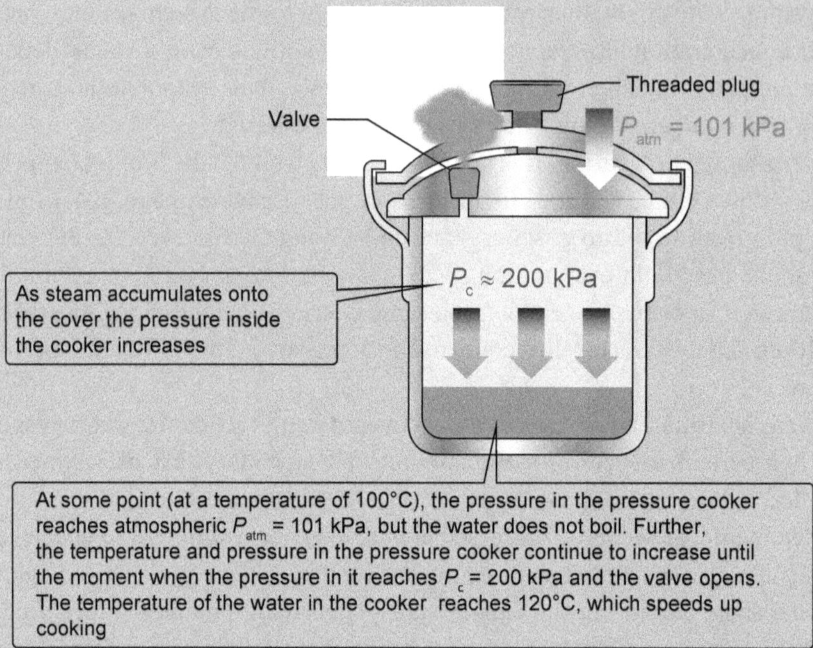

A pressure cooker. When some water is heated, the vapour is kept inside up to a given value of the pressure, controlled by the valve. As a consequence, one gets a temperature around 120°C, evidently much higher than in a usual pan.

The boiling temperature of a liquid is a function of pressure: it increases with increasing pressure. This property has been used in a kitchen device: the pressure cooker. This instrument elsewhere is known as an auto-cooker. Its cover is firmly closed so that the water vapour which results from evaporation cannot get out. As much as it accumulates onto the cover, the pressure inside the cooker increases above the atmospheric pressure, and thus the temperature of the water boiling inside increases as well. To avoid an explosion, the device is equipped with a valve that opens up when the pressure reaches a value of about 2×10^5 *pascal (Pa). The temperature inside the pressure cooker is typically around 120°C, and this allows a rapid cooking, much faster than in the usual pan. The device requires some precaution: after having completed the cooking, never open the cover without having first cooled the cooker with cold water and removing the cover with some care to avoid*

(*Continued*)

(*Continued*)

a jet of high-temperature vapour. In fact, the water at 120°C, suddenly brought at the atmospheric pressure. would start again to boil tumultuously! Furthermore, it is necessary to avoid cooking food that could clog up the valve.

Historical experiment for the measurement of the atmospheric pressure at the Puy de Dôme, from a carving of the 19th century. On September 19, 1648, the son-in-law of Pascal observed the height of the mercury column in the barometer and compared it with the one detected at low height, in Clermont-Ferrand.

The pressure cooker minimises the energy in all circumstances, but it is particularly useful at altitudes where it would be hard to cook the food in ordinary pans. For instance, at the top of Mt. Everest where the pressure is reduced to 3.5×10^4 Pa, the water would boil at about 70°C!

It is appropriate to mention that the pressure unit is called pascal in honour of Blaise Pascal (1623–1662). This French philosopher and scientist was the first to prove that the atmospheric pressure is reduced when climbing in mountains: he also demonstrated that the atmospheric pressure is related to the weight of the air column above the ground. The related experiment was not so easy to carry out: he had to carry to about 1,500 m of height a rather bulky instrument, the mercury-based barometer that the Italian scientist Torricelli had just invented (see figure).

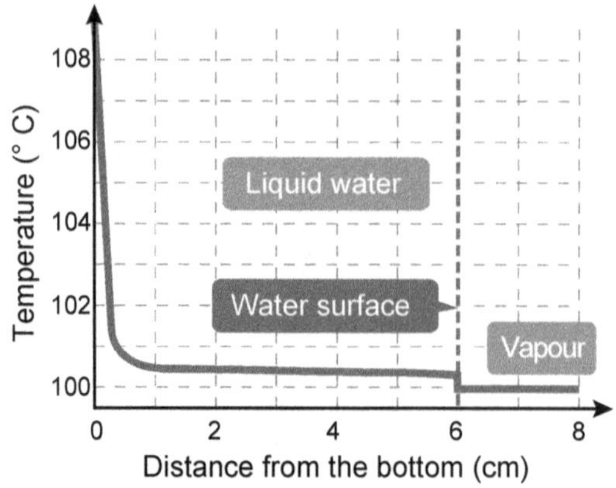

Figure 5. Temperature of the water as a function of the distance from the bottom of the container, during the boiling process, at the pressure of 1 atm. The temperature of the vapour is 100°C, namely the equilibrium value. That of the water is close to 100.4°C in the boiler, while towards its bottom reaches about 110°C. The temperature at the bottom of the boiler depends on the intensity of the heat source.

It should be remarked that all the water at this point has a temperature above 100°C, in a state defined as "metastable". Thus the water bubbles up at the surface increase their size, growing because of the water that is vaporised at the contact. After turning off the heater, the water continues to evaporate. Then let us pour water into the teapot through a strainer filled with tea and let the tea brew.

How to serve the tea

While the tea infusion is being formed in the kettle, let us choose a proper container to pour it. In the East, instead of the traditional china cups, small special containers are preferred. Possibly as the heritage of the Asiatic nomad tribes, these containers are easy to put away and less fragile. Their flared shape also implies the cooling down of the liquid at the surface (since the zone in contact with the air is greater), thus avoiding some burns on the lips. In Azerbaijan, the glasses *armudi* (pear) are as well characterised by a flared swan neck. Towards the bottom, the liquid stays warm, and the spherical form minimises the ratio of surface/volume and the cooling that occurs close to the surface of the liquid.

Nowadays, the use of cylindrical mugs is more popular; it may be because they are more easy to manufacture and to be personalised by various colour motifs.

Chapter 16

When Physics Invades the Kitchen

The chefs of the old days prepared dishes with very simple equipment. Today, culinary experts have at their disposal a whole arsenal of electrical appliances with multi-page manuals and the principles of their operation which, most likely, seem hazy to most users! Let's describe the physics that has invaded modern kitchens.

Traditional Methods for Heating

When the prehistoric human being learned how to handle fire, that day marked the beginning of a new era. Fire offered the possibility to work with metals and as a consequence… to create cars or to travel to the Moon. Some people think is that fire was significant because it allowed human being to no longer eat raw vegetables and meat. The development of civilization went on together with the evolution of cooking techniques. For good food critics, a steak properly cooked could be considered one of the symbols of civilization. Over the centuries, the connection between fire and food has changed, more precisely, it has been enriched with new methods; however, the fundamental process behind cooking has always remained the same. The fire on which the mammoth meat was fried changed at first to a family hearth in the fireplace and then a wood-burning stove (Fig. 1) or a coal stove, later being replaced by a gas stove, and finally, with the advent of electricity in the kitchens of the 20th century, electric stoves appeared, and in the 21st century, they are increasingly being replaced by induction.

Figure 1. Food preparation in a saucepan on a wood stove. Traditional culinary methods are usually based on heat in contact with hot metal (here — a plate), hot air (in the oven) or boiling water. Temperature throughout a pan is almost constant due to convection, which makes liquid or vapour circulate.

The Electric Plates in Cast Iron

The first electric devices for cooking have been the plates in cast iron including inside an electric circuit transporting electrons. In this way, the plates became hot due to the Joule effect. When an electric current circulates in a conductor, this gets hot! The electric radiators and the toasters function according to that principle. When the conductor is a thin wire, the heating is so strong that it emits light. This is what happens in incandescent bulbs, where the tungsten wire is heated above 2,000°C (see Chapter 7).

What is the power dissipated in the Joule effect? It depends on a characteristic of the conductor, the electrical resistance, and on a characteristic of the electric source, called electromotive force, approximately given by the tension (or potential difference) at the terminals of the conductor (Fig. 2).

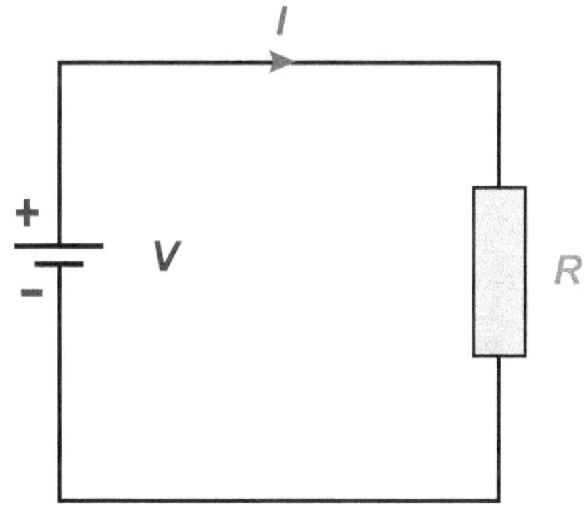

Figure 2. Sketch of continuous current generator at potential difference V (implying tension V between the two terminal + and –) connected to the resistance R. The electric current of intensity I that can be measured by the ammeter is flowing through the circuit.

The electrical plugs at home deliver a tension of 220 V in Europe and 120 V in the USA. If the conductor resistance is R (measured in SI units in ohms) and the intensity of the current is I (measured in SI units in amperes), then the electric power dissipated (measured in SI units in watts) because of the Joule effect is $W = IR^2$ or $W = V^2/R$. For a given electric tension V, the power dissipated is increasing when the electric resistance is decreased. The pan in contact with the plates gets hot by thermal conduction. The air in the room is also warmed to a moderate extent, but much less than in cooking devices with a gas fire. After having turned off the power, the plate remains warm for a relatively long time, implying the risk of being burned ourselves by touching it. Let us turn to another subject of this chapter: modern methods of cooking.

Induction Plates

When a conductive circuit is subjected to a variable magnetic field, an electric current is generated. This is the phenomenon called *electromagnetic induction*, and it was discovered by English physicist Michael Faraday (1791–1867). Electromagnetic induction finds countless applications in modern technology, including in induction stoves in modern kitchens. How do they work?

In such a stove, located under the plate (the so-called hob), is a device (Fig. 3) which generates an alternating electromagnetic field. Until nothing is set over the plate, the energy consumption is small: practically no heating occurs, and one can even touch it without injury. When a metallic pan is set over the plate, then the

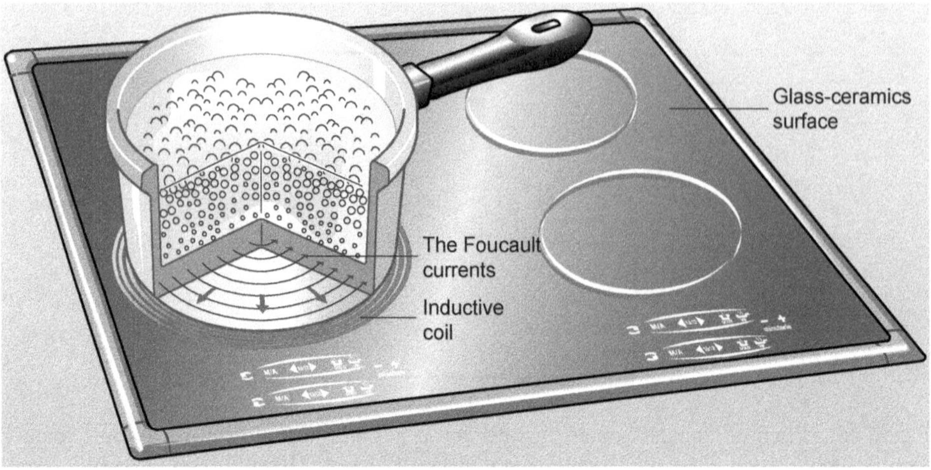

Figure 3. Principle of heating by induction. The pan is on the plate made in glass-ceramic material (that can resist high temperature variations) which is over the inductive coil. The coil is made of many turns of thin wires, and a variable electric current is circulating. It induces a variable magnetic field, the strength and the direction of the field oscillate at high frequency. Foucault currents are produced in the bottom of the pan, heating it through the Joule effect.

magnetic field generates an electric current inside the base of the pan itself. A small clarification is now appropriate since, at the beginning of this section, an electric circuit was mentioned. Where is it in our case? Simply, the pan itself is the circuit, and this explains why it has to be metallic. The electrons of the metal are being moved under the influence of an induced electromagnetic field, forming the contour which passes through the bottom of the pan. The currents circulating in a compact block of metal are called Foucault currents in honour of Léon Foucault, who used a pendulum to demonstrate the rotation of the Earth (see Chapter 4). Thus, the bottom of the pan heats up, thanks to the Joule effect. The heat generated is transferred to the food being cooked in the pan, such as soup, and as a result the contents boil quickly.

The induction electromotive force is proportional to the strength of the magnetic field and the rapidity of its variation, namely to the frequency if the field is due to the alternate current. The frequency of the power supply in homes is 60 Hz (50 Hz in Europe), and this value is not sufficient to cause an electromotive force sufficient to cook the food in the pan. Thus, inside the device, there is a frequency multiplier, often by a value by a factor of 500 or 1,000.

Using an electric current to produce another current, what for? Because, in this way, the current is generated exactly where it is required with no electric contact and no wires that might be dangerous, unlike other methods of cooking, such as the gas cooker or the metallic plate directly heated by the Joule effect. The air of the room does not increase its temperature. The glass–ceramic plate is indeed heated by the contact with the bottom of the pan but to an extent much less than for the plate in cast iron. Also, the energy consumption is decreased by the induction plate, furthermore it carries no risk of burning. The cooking is fast, the only negative remark being that not all the kitchen items can be used in that way of cooking. In fact, the pans not only must be metallic but ferromagnetic, for example, steel or cast iron. For various reasons, the bottom of the pan has to be thick, in particular so that it is not damaged by the high temperatures.

It turns out that while only metal cookware is suitable for an induction hob, another device popular in our kitchens does not accept it. Let's move on to discussing it.

The electromagnetic induction

Around the year 1830, physicist Michael Faraday carried out a series of experiments demonstrating the phenomenon of electromagnetic induction. A simple experiment can evidence it.

Let us consider a coil made by a conducting wire (for instance, of copper) forming a series of n loops each having area S (see figure). When a magnetic bar is introduced inside the coil, causing a flux of a time-dependent magnetic field H, then an electric current arises in the circuit. The intensity of the current, indicated by the ammeter, depends on the circuit (and particularly on its resistance), but the electromotive force V occurring at the extremities of it is given by a simple equation, $V = -n d\Phi/dt$, where Φ is the induction flux through each loop, which in turn is proportional to the area and the intensity of magnetic field H. Thus the electromotive force increases by increasing the number of the loops in the solenoid.

Nowadays, the majority of our electric devices work by exploiting the phenomenon of electromagnetic induction. One can mention the alternators at work in the nuclear, thermal or hydraulic plants providing the electricity we use in our houses (see Chapter 13), or the chargers of our mobile telephones. These devices include a transformer (see figure) that, starting from a source of alternate current of a given voltage, can provide current of another voltage. Furthermore, the device includes a coil that provides a magnetic field in a metallic cage. Due to this field, the iron

(Continued)

(*Continued*)

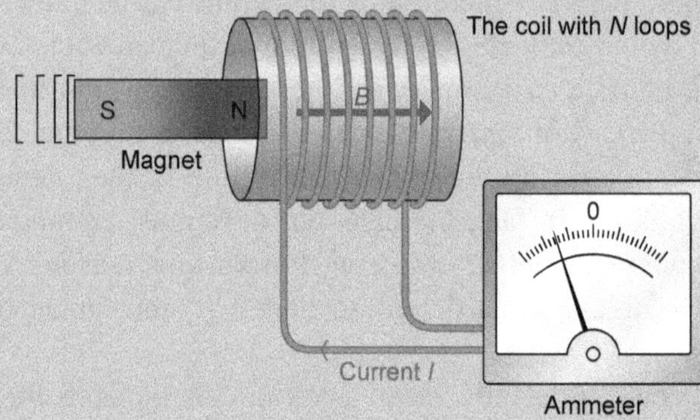

A variation of the magnetic field though the winding coil (for instance, due to the introduction or the removal of a magnetised bar, generating a (non-homogeneous) magnetic field causes a current that can be detected by the ammeter.

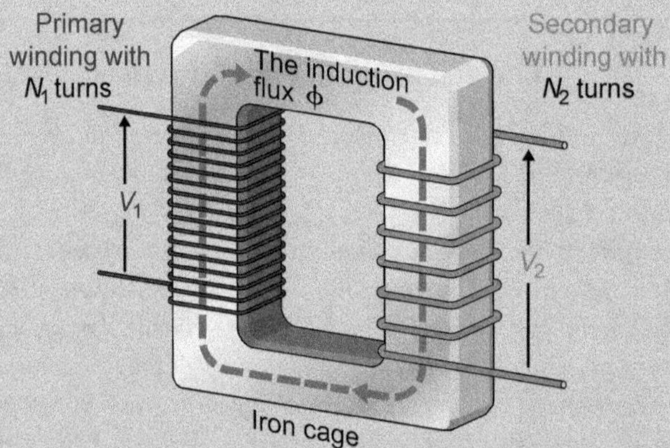

Principle of the transformer. Basically, there are two coil windings of conducting wire around an iron cage. The primary winding, for instance, is connected to a source of tension V and generates a magnetic field of strength proportional to the number of turns N_1. The iron cage leads the magnetic field and drives it to go through the secondary winding having N_2 turns; it produces a potential difference $V_2 = V_1N_2/N_1$ that, for example, is driving a mobile phone.

acquires a certain magnetisation contributing to the induction flux, this flux being multiplied by a considerable factor, of the order of 5,000!

Thus, in practise, a cage of ferromagnetic metal is required. For a similar reason, the pans used on the induction plates have to be of steel or cast iron or ferromagnetic alloys of iron and carbon.

The Microwave Oven

Another device for cooking that has been popular in the kitchens since the 1980s is the microwave oven. How does it work?

As implied by its name, this oven contains electromagnetic waves at frequency $\nu = 2.45$ GHz (2.45 billions of hertz). Since the speed of electromagnetic waves in vacuum is $c = 300{,}000$ km s^{-1}, the wavelength resulting from c/ν turns out about 12.2 cm. It is noted that they are centimetres and not microns as could be deduced from the word microwaves! Compare that length with the wavelength of the visible light (from 0.4 to 0.7 μm) or with the length of the FM radio waves reaching our houses, in general, of frequency around 100 MHz (and then at wavelength of the order of 3 m, see Chapter 3).

The structural element that generates microwave radiation in the oven is called a magnetron (Fig. 4). It is a magnet and a hollow metal cylinder with a heated wire

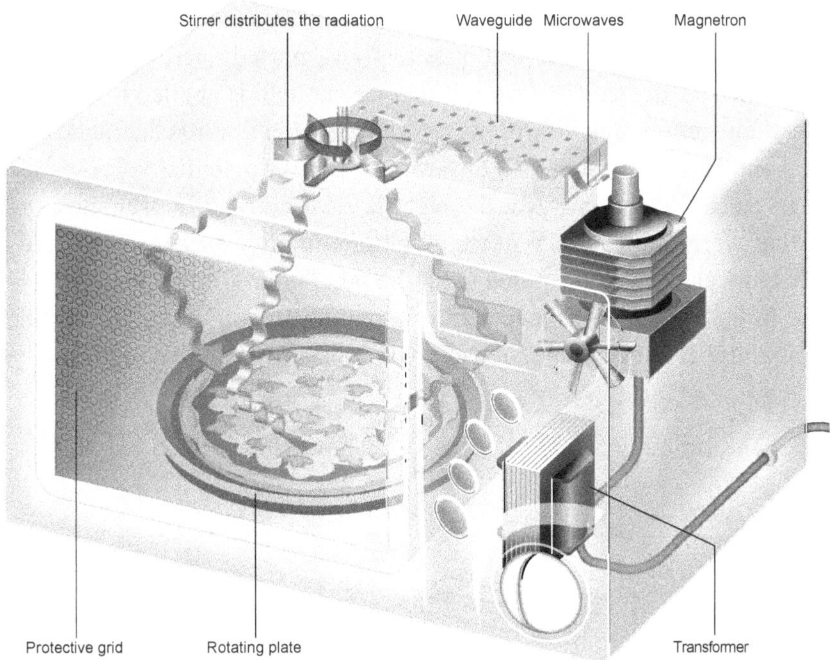

Figure 4. A microwave oven. Due to the effect of a high electric potential provided by a transformer, the magnetron emits an electromagnetic radiation that through a waveguide is sent inside the oven through a waveguide. A shaking device distributes the radiation in several directions in order to cook the foods in the most homogeneous way.

inside. Under the influence of the strong potential difference (several kilovolts) created by the transformer, the hot wire emits electrons, which, under the influence of a magnetic field, rotate around the wire and emit microwave waves. The structure is designed in such a way that standing electromagnetic waves of a given frequency are set up in it when the value of this frequency is maintained with an accuracy of 1%. The generated radiation is then directed to the inside of the microwave oven through a waveguide, which is similar to the devices discussed in Chapter 2.

How do the microwaves generated by the magnetron heat the food? To a large extent, this happens due to their interaction with the water molecules present in the food. To understand this, let us consider the structure of the water molecule, made of two atoms of hydrogen and one of oxygen, with covalent bonds sharing two electrons. The oxygen atom tends to attract the electron (one says that oxygen is more electronegative than hydrogen), and thus an excess of negative charge is set around the oxygen, while at the hydrogen site, there is an excess of positive charge. The molecule is not linear, the two O–H bonds forming an angle between them of about 100° (Fig. 5).

From an electrostatic point of view, things work as though the molecule were made by two electric charges of opposite signs, placed nearby: in other words, what is known as an *electric dipole*. If an electric field is applied to this dipole, it tends to align along the field, with its positive pole (+) towards the highest electric potential and its negative pole (–) towards the weaker potential.

In an electromagnetic wave, the electric field constantly oscillates, and this means that water molecules begin a vibrational motion in the microwave oven. Then this movement is transmitted to all the other atoms of the prepared food. An increase in the intensity of movement of atoms is nothing more than an increase in temperature! However, in reality, everything is a little more complicated: water

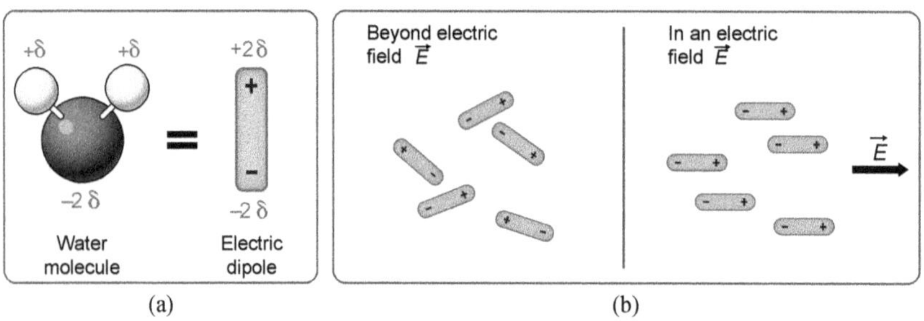

Figure 5. Schematic representation of the water molecule H_2O. The oxygen atom has a negative charge, while the two hydrogen atoms have positive charges (a). Due to the nonlinear geometry, the molecule corresponds to an electric dipole that is oriented in an electric field (b).

molecules are quantum objects and may not vibrate at any frequency. Therefore, in the microwave oven, the frequency of 2.45 GHz is selected, which corresponds to the resonant frequency of water molecules. The energy absorbed from electromagnetic wave in these resonant transitions then is redistributed by multiple relaxation processes, thereby heating the foodstuff.

The Microwave Oven, a Resonant Cavity

Better than the induction plate that heats only the bottom of the pan, the microwave oven radiates heat only where we intend it, since the heating process directly involves the foodstuff. So that the consumer does not share the fate of the chicken, the electromagnetic radiation in this case is limited to the working volume of the oven. The metallic walls of the oven (in general, metallic grids) are indeed very reflective and confine the microwaves that are present in the oven, even for a little while after the device is turned off.

For certain wavelengths, the microwave oven behaves as a resonant cavity, similar to the vibrating cord fixed at two extremities (see Chapter 11). For an appropriate wavelength of the excitation, the resonance occurs, and the cord takes the property of a stationary wave: in some points, called *loops*, the displacement of the cord is maximum, while, in other points called *nodes* (or antiloops), the cord does not move. The phenomenon of the stationary wave occurs when the length of the cord is equal to an integer number of half wavelengths: the distance between two successive loops or between two nodes is equal to half a wavelength.

This type of phenomenon occurs in a microwave oven. The radiation distribution is not homogeneous due to interferences (see Chapter 3) between incident electromagnetic waves and the same waves after reflections on the walls (see Panel on page 236). The constructors paid attention to this problem and found the solution: when the oven is operating, the plate where the food is set is put in rotation. We have experimentally verified that the radiation is strongly inhomogeneous when the plate is not rotating (Fig. 6). Even though the plate is rotating, some precautions are still necessary. For instance, it could be possible that part of the food to be cooked is at the centre of the rotating plate, and therefore it would receive the same radiation all the time. When the baby food is just taken out of the oven, it is better to gently mix.

A Big Haunch and the Skin Effect

The microwave oven can be used for other simple and peculiar experiments. Let us take from the refrigerator a rather big haunch of meat and put it on an

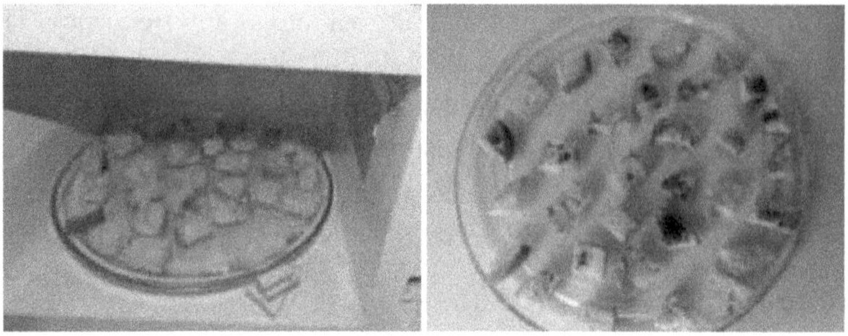

(a) (b)

Figure 6. The experiment shows the marked heterogeneity in a microwave oven during the heating. (a) A plate with bread chips is put on the oven plate which is prevented from rotating. (b) After turning on the radiation (for about a minute, at average power), some bread chips are carbonised, while others continue to be white, and others only slightly burned.

Stationary waves in a microwave oven

Lock the turntable in the microwave and scatter pieces of bread on it. After a short heating, they will be heated unevenly, which proves the presence of nodes and antinodes of the electromagnetic field in the furnace (Fig. 6). The reader may be tempted to draw an analogy with a one-dimensional vibrating string and come to the following conclusions: (a) One of the dimensions of the working chamber of the furnace is a multiple of half the wavelength $\lambda/2$ *radiation; (b) The distance between two successive antinodes or nodes (for example, two burnt areas) is equal to half the wavelength. Be careful with analogies! None of these conclusions about a microwave oven is true! In fact, the electromagnetic waves that cause resonance do not propagate perpendicularly to the walls of the oven. Let us denote the axes of the selected rectangular coordinate system* O_x, O_y *and* O_z. *When a propagating sinusoidal wave of maximum amplitude* E_0 *is travelling along the x-direction, the electric field is described by the formula* $E(x,t) = E_0 \cos(kx - \omega t)$, *where* ω *is* $2\pi c/\lambda$ *and* $k = 2\pi/\lambda$. *If the wave is propagating in an arbitrary direction, the electric field at the point* **r** *of coordinates x, y and z has to be written as* $E(r,t) = E_0 \cos(k \cdot r - \omega t)$, *where* **k** *is a vector directed along the propagation direction of the wave having norm* $k^2 = k_x^2 + k_y^2 + k_z^2 = 4\pi^2/\lambda^2$.

In order that stationary waves can occur in a parallelepiped box having parallel sides a_x, a_y *and* a_z, *the components of the* **k** *vector must satisfy the conditions*

(*Continued*)

(Continued)

$$k_x = \frac{\pi n_x}{a_x}, \quad k_y = \frac{\pi n_y}{a_y}, \quad k_z = \frac{\pi n_z}{a_z},$$

where $n_{x,y,z}$ *are integer numbers. The side lengths and the wavelength satisfy the condition*

$$\left(\frac{n_x}{a_x}\right)^2 + \left(\frac{n_y}{a_y}\right)^2 + \left(\frac{n_z}{a_z}\right)^2 = \frac{4}{\lambda^2}.$$

For an oven having sides $a_x = a_y = 29$ *cm and* $a_z = 19$ *cm, this equation is satisfied for* $(n_x, n_y, n_z) = (1, 1, 3), (0, 1, 3), (3, 2, 2)$ *and* $(4, 2, 1)$*. These different sets of values correspond to different types of standing waves or so-called modes. Many of these modes occur simultaneously in the oven (depending on the type of device) and, therefore, between the frequency and which pieces of bread burn, there is no simple connection.*

appropriate plate and then inside the microwave oven. Turn on and, after some time of irradiation, the meat starts to get brown, and after about 12 min, it appears to be perfectly cooked. When taken out from the oven and cut, we would find out that some part of the inner meat not only is not cooked but, even more, is also still frozen. What is happening? Only the zones near the surface have been irradiated! In fact, an electromagnetic field can penetrate a conducting medium only near the surface. The depth of penetration, also called the skin-depth, depends on the resistivity of the material and of the frequency of the wave incident on the conductor. The resistivity value indicates the ability of the material to conduct electrical current well (low resistivity) or poorly (high resistivity). This inability of the electromagnetic field to penetrate to a depth larger than δ is called *skin effect*. The effect increases by increasing the frequency since δ is inversely proportional to the square root of the frequency.

For the copper wires that are involved in most of our electric appliances and while the correspondent frequency is 50 Hz, the skin depth is about 1 cm. Therefore, the electric current is circulating all along the wire, and the skin effect can be disregarded. But at the microwave frequency, namely 2.45 GHz, the skin depth for copper is no more than a few micrometres.

Let us return to our badly cooked haunch: the skin depth is larger since the resistivity (ρ) of the meat is great, and the skin depth increases with the square root of that quantity ($\delta \sim \sqrt{\rho/\omega}$). The meat is not a good conductor, but, to a certain extent, it is a conductor; it contains water and inside it, there are ions that can carry charges. The resistivity of a muscle is of the order of ohm-metre, hundred million more than the one of copper, which is equal to $1.6 \times 10^{-8} \Omega \cdot \text{m}$. The skin depth for the meat inside the microwave oven is thus of the order of centimetre. Since the haunch is thicker, the inside can be heated only by conduction (see Chapter 18), as happens in a classical oven. In order to cook the haunch well inside, one must be patient! In a classical oven, the time required to cook a haunch is at least 1 h. By resorting to a microwave oven, it will not imply a faster time. Furthermore, the result could also be disappointing: certain chemical reactions that are produced with a grill and that provide a delicious aroma cannot be obtained with a microwave oven. They require high temperatures, certainly above 100°C, the temperature at which water boils (see Chapter 21). Thus in the microwave oven, you would not see in the haunch the appealing brown colour in the crust… If the microwave oven is not appropriate to cook the meat, at variance it could stimulate the search for novel attempts in the cooking field, along unexpected paths: for instance, the preparation of ice cream inside a warm *beignet.* Only the exterior paste will get hot, while the ice cream inside will be protected by the skin effect: furthermore, the chemical bonds are much stronger in the ice rather than in the liquid, and thus the microwaves will little affect the ice cream.

The reader being at this point familiar with the skin effect, we can return to the choice of 2.45 GHz for the frequency. This value is within the range of electromagnetic waves easily absorbed by the water molecules. In reality, the strongest absorption would occur around 20 GHz, and thus 2.45 GHz represents a compromise: at higher frequencies, the absorption would increase, but, in the meantime, the skin effect would increase as well, thus possibly making difficult an homogeneous cooking of the food.

Caprices of a Microwave Oven

It is known that metallic surfaces cannot be put inside the microwave oven. What happens if this prescription is violated? The authors of this book made the following experiment (the readers are strongly recommended to not repeat by themselves). The same amount of water was put in a metallic container and in a ceramic cup, and the microwave oven was turned on, under strict surveillance and for only 30 s. While at the end of the operation the water inside the ceramic cup was found pretty warm, the one inside the metallic container remained cold. The metal, where the penetration depth of the microwaves is very small, did not allow the penetration of the radiation. Although the water does not get hot, the metal can still absorb heat, for instance, through Foucault currents at the surface. Since the container had thick walls and the thermal conductivity of the metal is very high, the heat produced by the microwaves could be distributed over a rather large volume, and no serious damage occurred. On the contrary, in the case of very thin walls, as, for instance, gold-like covered plates, explosion or fire could occur.

During the experiment described above, a certain precaution was taken: the oven yields a certain power, and all the energy emitted in a given time must be dissipated. When one has water or foods on the plate, the energy is absorbed and they get hot. However, if the metallic container does not allow the penetration of the microwave inside where the foods are, then there is the risk that the energy finishes to damage other parts of the oven, for instance, the magnetron. For the same reason, an empty microwave oven should not be turned on. It should be remembered that metallic containers do disturb the propagation of the microwaves, and let us avoid placing them inside the oven.

It should be noted that not all earthenware or glassware is suitable for use in a microwave oven. The risk is that the plate gets as hot as the food it contains. A simple test to check if a given plate or any container can indeed be used is to place it in the oven together with a glass of water. If at the end of the irradiation the container is cold while the water is hot, that means that the container, or the plate, is appropriate.

Figure 7. Result of the attempt to cook an egg inside a microwave oven.

Let us proceed in listing the things for which a microwave oven cannot be used. It is not possible to cook an egg. During the irradiation, the material inside the egg rapidly gets hot and vaporises. The gas already inside the egg increases the pressure until the egg explodes (Fig. 7)! As with the badly cooked haunch, the explosion of the egg emphasises the risks of improper use of the microwave oven.

To conclude this chapter, let us address the last experiment. Is it really impossible to use the microwave oven to heat foods that do not contain water? We have attempted a suitable comparison by placing in the oven two cups: one containing water and the other peanut oil. After some minutes of irradiation, we found that the water was hot while the oil was still cold. This result could not be predicted *a priori.* In fact, materials that do not contain water but are formed by molecules implying an electric dipole (Fig. 5) can still be heated by means of microwave irradiation.

Chapter 17

Ab ovo

Is it possible to cook an egg on the top of Mount Everest? Knowing that a chicken egg boils in 3 min, how long would it take to cook an ostrich egg? How do you identify a raw egg from a hard-boiled egg? What is the secret to the tastiness of Japanese "onsen tamago"? The answers are all found in this chapter.

The Latin expression *Ab ovo* is taken from the *Ars Poetica* of the Latin poet Horace and means "from the egg". Namely *from the beginning*. This is an allusion to the egg laid by Leda, from which Helen would have emerged. Much later, Helen was seduced and kidnapped by Paris, and thus the Trojan War was initiated, which was narrated by Homer in the Iliad. As Horace observed, Homer, a skilled poet, did not begin his story *ab ovo*, but *in medias res*, in the heart of the matter. In this chapter, this expression simply means that, starting from the egg, we will offer to the reader some thoughts about the Physics involved in it.

The Egg Fight

A starting point for this chapter is suggested by another reminiscence in one of our famous stories, that of "Gulliver's Travels" by Jonathan Swift. This book tells the story of the war between the empires of Blefuscu and Lilliput following an edict from the Emperor of Lilliput who ordered his citizens to break their eggs by the small end. Gulliver believed this to be a private matter, and the authors of this book share that opinion. However, just for curiosity, the reader is invited to consider which end of the egg is easiest to break. Perhaps he has already taken part in a fight (appropriate for Easter Day) in which two players, each with a hard-boiled egg, try to break the opponent's egg (Fig. 1). What is the right tactic? Hitting on the small side or the big one? But also, should you choose a small egg or a large

Figure 1. Battles of decorated eggs are an Easter tradition. Each player strikes the opponent's egg: the winner is the one who manages to keep his egg intact.

one? Attack the opponent or wait for his assault? Some experienced practitioners claim that the attacker has an advantage. However, if the two eggs move at a uniform speed, this advantage does not occur under the Galilean principle of relativity.

On the other hand, attacking by the small end or by the big one does make a difference. Let us suppose that the two eggs are of equal size and that the axes are the same. Upon impact, the shells flatten out a bit at their ends because they have a certain elasticity. They then have a small common flat surface S, around which the shells form with the axis of the two eggs at two different angles $\alpha_1/2$ and $\alpha_2/2$. According to the principle of action and reaction, also called Newton's third law, the force F_1 exerted on the small end is opposite, therefore equal in absolute value, to the force F_2 exerted on the large end. To assess the effect of these forces on the shells, we must consider them as sums of forces distributed along the surface S and tangent to the respective surface of the eggs. For two-dimensional eggs, which do not exist but facilitate reasoning, there would be (Fig. 2) two forces on each side, f_1 and f_1' on the left, f_2 and f_2' on the right, with $F_1 = f_1 + f_1'$, $F_2 = f_2 + f_2'$,

$$f_1 = f_1' = F_1 \Big/ \left(2\cos\frac{\alpha_1}{2}\right),$$

and

$$f_2 = f_2' = F_2 \Big/ \left(2\cos\frac{\alpha_2}{2}\right).$$

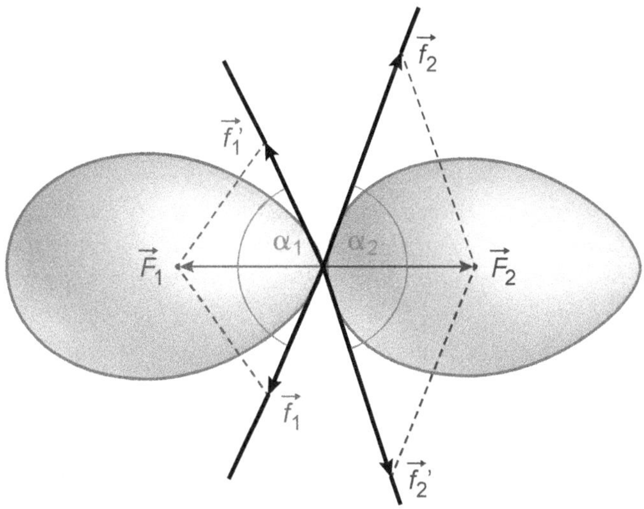

Figure 2. Egg fight; big end against small end. Who has the best chance of winning?

Since the angle α_1 of the small end is smaller than the angle α_2 of the large end, the force acting on the large end is greater, and it is the large end that breaks. For real three-dimensional eggs, the two forces f and f' are replaced by a continuous distribution of forces, and the reasoning is analogous: the forces acting on the large end and the small end are in the ratio $\cos\frac{\alpha_1}{2}/\cos\frac{\alpha_2}{2}$, therefore, higher on the big end. So, the big end breaks.

In the above reasoning, we have overlooked the role of the inside of the egg, which strengthens the resistance of the shell, especially for the small end, because there is an air pocket on the large side. This is a second reason for attacking from the short end. If you are having an egg fight, then you will attack from the small end. But what if your opponent also attacks from the small end? You can improve your chances by hitting the opponent's egg on the side, where the curvature is weakest.

The Spinning Egg

After the egg fight, another possible entertainment consists of spinning an egg around like a top. As is well known, this is a way of distinguishing a hard-boiled egg from a raw egg. The hard-boiled egg swirls for a long time, while the raw egg stops after a few turns due to the viscous friction of the material it contains.

The spinning top and the angular momentum

A top rotating rapidly around the axis does not fall if this axis is close to the vertical. How do we explain this phenomenon? For simplicity, let us study the motion of a top, which has the shape of a surface of revolution, and also assume that the contact point C between the top and the support is fixed (this is not the case for a Thomson's top or an egg). The nature of the motion of the top can be determined by the conservation law of its angular momentum also called the moment of rotation. This law for rotational motion plays the same role as the law of conservation of momentum for translational motion. The moment of rotation **L** *is a vector quantity. This vector is determined by the distribution of mass in the volume of the object, its angular velocity of rotation, as well as the axis about which it rotates. For example, the moment of rotation of a spinning bicycle wheel is a vector perpendicular to the plane of the wheel, the modulus of which is proportional to the angular speed of rotation. In particular, if the top is a set of material points with masses* m_i*, the position of which is determined relative to the point of contact by vectors* R_i*, and their linear velocities are* v_i*, then its angular momentum* **L** *is the sum of vector products* $m_i\mathbf{R}_i \times \mathbf{v}_i$ *(see Chapter 4).*

It can be shown that if the sum of the moments of external forces acting on the system is equal to zero, then its moment of rotation remains constant. In the case of a rotating top, and in the absence of friction, two vector components remain constant: the vertical one and the one directed along the axis of its rotation. If we launch the top in such a way that its axis of rotation coincides with the vertical, then the vector of its angular momentum **L** *will also be directed vertically. If the axis of rotation deviates from the vertical, then the projections of the vector* **L** *on these two directions remain constant, which implies an increase in the modulus of* **L***, and hence an increase in the rotation speed. Consequently, this causes an increase in kinetic energy. If the top rotates slowly, then this increase is small enough and can be compensated for by a decrease in potential energy resulting from a decrease in the vertical coordinate of the centre of gravity. In this case, the top falls. But if the top rotates fast enough, then such compensation cannot occur, and, since the total energy cannot increase, the top simply remains in the upright position.*

(*Continued*)

(*Continued*)

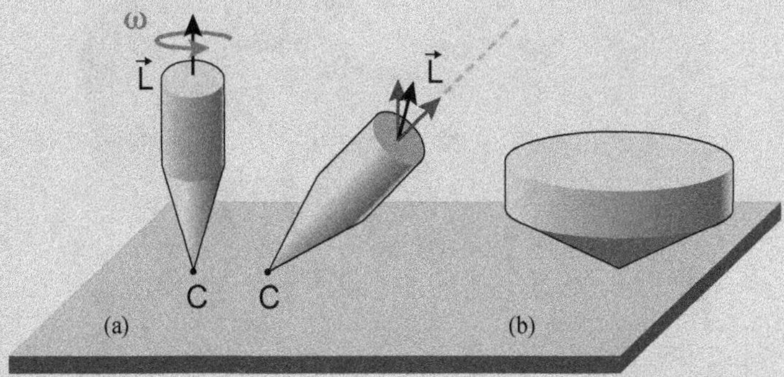

While the top is spinning fast enough, its axis of rotation remains vertical. The moment of momentum of the top relative to point C is marked with a black arrow. In contrast to the case of rotation of an ideal top without friction, in the case of rotation of an ideal top without friction, in the case under consideration, it is not completely preserved, but two of its projections still remain unchanged: the vertical component and the projection of the angular momentum directed along the axis of symmetry of the top. The right top (b), flattened, is more stable than the left, elongated (a).

What is the minimum rotation speed below which the rotating top becomes unstable and falls? It turns out that it is equal to $\omega_c = \alpha\sqrt{g/l}$, *where g is the acceleration of gravity and l is the distance from point C to the centre of gravity of the top. The coefficient* α *depends on its shape: it is less for a flattened top and more for an elongated top. Thus, the rotation of the flattened top is more stable.*

If, on the other hand, we give a hard-boiled egg a very rapid rotary movement on a very flat and not too smooth surface, we can witness an astonishing phenomenon: after a few turns, the egg rises and begins to turn on its tip, until it loses speed and then returns to the position where its centre of gravity is lowest. This is one of the entertaining wonders of mechanics, akin to spinning tops. Spinning tops are not only a fun spectacle for young and old alike, but also a classic problem topic for students (see Panel on page 244). One of the most spectacular spinning tops, which could be found in some stores before the era of video games, was shaped like a truncated sphere from which a cylindrical rod emerged. You can bring the top into a rapid rotation by twisting it with the rod with the thumb and forefinger, and after a few turns, the top, initially placed on its spherical part, overturns and sits on the rod. This strange behaviour was explained by

Figure 3. Wolfgang Pauli (Physics Nobel Prize winner 1945, left), Niels Bohr (Physics Nobel Prize winner 1922, right) and Thomson's top (bottom) in 1955.

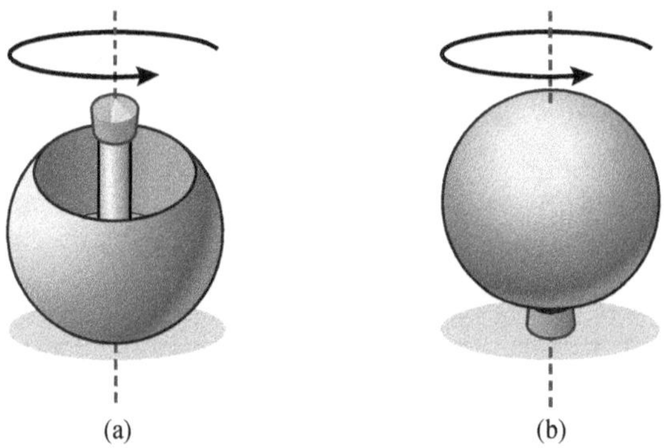

Figure 4. The amazing behaviour of Thomson's spinning top. If initially rotated in position (a), the top tilts after a few turns, then starts to rotate in position (b).

W. Thomson (later Lord Kelvin), so this toy is sometimes called Thomson's top (see Figs. 3 and 4).

The Chemistry of Egg Cooking

Of course, eggs are primarily used as food, not as toys. There are many ways to prepare them: hard-boiled eggs, omelettes, soft-boiled eggs, fried eggs, poached eggs, etc. This raises practical questions: for how many minutes should an egg be boiled to make it soft-boiled and not hard-boiled? Why are eggs boiled in

Proteins: Amino acid assemblages

Proteins are the result of the combination of large numbers of amino acids that form long chains. All proteins known in nature are combinations of 22 different amino acids. Their molecules are characterised by the presence of an NH_2 group (called amines) and a COOH group (carboxyl, with acidic properties), both of which are bonded to the same carbon atom. In addition, the latter is also associated with a group of atoms specific to each of the amino acids.

The reaction between two amino acids results in a water molecule and another molecule that is not yet a protein. But after at least 50 such reactions (usually there are much more of them), the resulting long molecule can finally be called a protein.

Under certain conditions and in the presence of water, the protein undergoes a reverse reaction — hydrolysis. This reaction takes place in the stomach, where the food is immersed in a very acidic environment: it provides the amino acids required to synthesise the proteins we need.

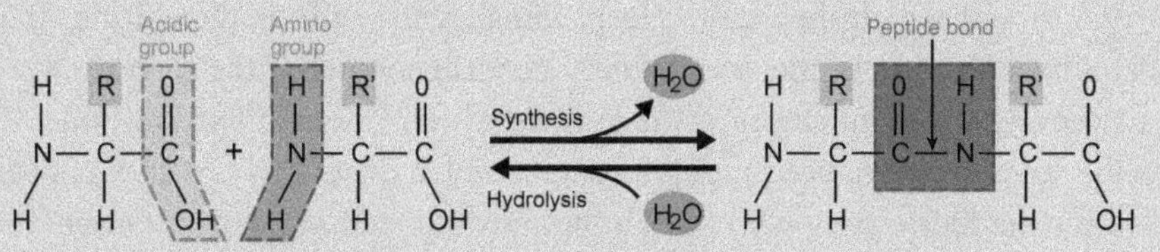

Two amino acids can be combined by a synthesis reaction: the amino group of one reacts with the acidic group of the other in order to release a water molecule to form a peptide bond. The reverse reaction is called hydrolysis. The groups of atoms R and R′ can be more or less complex (for example, a simple hydrogen atom for glycine or a CH_3 group for alanine).

saltwater? Why are eggs immediately immersed in cold water after boiling? Before answering these questions, let us get some idea of the physical–chemical processes that take place during the preparation of an egg.

The inside of an egg contains mostly water and proteins (see Panel above on this page). In the body, these long molecules provide many important functions, such as muscle work, catalysis of chemical reactions, or the transport of other molecules. To perform these tasks, they, being properly positioned in space, take one form or another. This configuration changes as a result of an increase in temperature or a change in the acidity of the environment: proteins fold and take on a shape that no longer allows them to perform their biological functions. This process is called denaturation.

From a gastronomic point of view, denatured proteins are often tastier and more acceptable for the digestive process than before they were prepared. Thanks to their ability to adhere, proteins can form networks, which, for example, can make soup thicker or make jelly. In boiled egg white, such a network connects water molecules. It is thanks to this that the protein is able to solidify: after all, almost 90% of egg white consists of water. In addition to hydrogen, oxygen and nitrogen, proteins often contain sulphur. These are the proteins contained in the egg, the decomposition of which produces a gas known for its unpleasant odour: hydrogen sulphide (H_2S). It gives the smell of rotten eggs to some thermal springs!

The Japanese thermal egg…

The yolk and white of the egg contain different proteins, which denature at different temperatures. Denaturation of white occurs between 58° and 80°C and that of yellow occurs in a narrower range, between 63° and 70°C. By letting an egg cook for half an hour in water at around 70°C, we obtain a food which Japanese gourmets are fond of and which they call Onsen Tamago: the yolk is then firmer than the white, which has a very pleasant creamy consistency under the tongue. We invite the reader to prepare an Onsen Tamago themselves. One will have to find a way to keep the water temperature roughly constant and at 70°C. It's not very convenient, except for the lucky guys who have a thermostatic bath. In Japan, Onsen Tamago is now sold already prepared. Traditionally, they were obtained by immersion in hot spring water: Tamago means "egg" and Onsen "thermal spring" (Fig. 5).

Figure 5. Cooking eggs in the traditional Japanese Onsen Tamago way by dipping them in a hot spring.

... and the Western boiled egg

The boiled egg prepared in the European way respects the opposite condition: the white must be denatured and the yolk must remain fluid. To this aim, the egg is immersed in boiling water for a few minutes. Getting a soft-boiled egg is not as easy as it seems. Moreover, if this is possible, it is mainly because the heat gradually penetrates the egg, and the temperature of the white increases before that of the yolk. If we want a soft-boiled egg, we should stop cooking when the final temperature of the yolk is $T_f = 63°C$. The time t it takes for this depends on the size of the egg, more precisely its small diameter d which can be determined with a calliper (Fig. 6). It also depends on the initial temperature T_0 of the egg and on the boiling point of water T_b (which varies, remember, with altitude).

English physicist Peter Barham has written a book called *The Science of Cooking* in which he proposes, for the cooking time t of a soft-boiled egg, the formula obtained by the professor of University of Exeter Charles D. H. Williams (Fig. 7)

$$t = 0.15\,Kd^2 \tag{1}$$

where t is in minutes, d is in centimetres, and

$$K = \ln \frac{2(T_b - T_0)}{T_b - T_f}$$

Figure 6. Determining the small diameter of an egg allows the preparation of an ideally cooked soft-boiled egg.

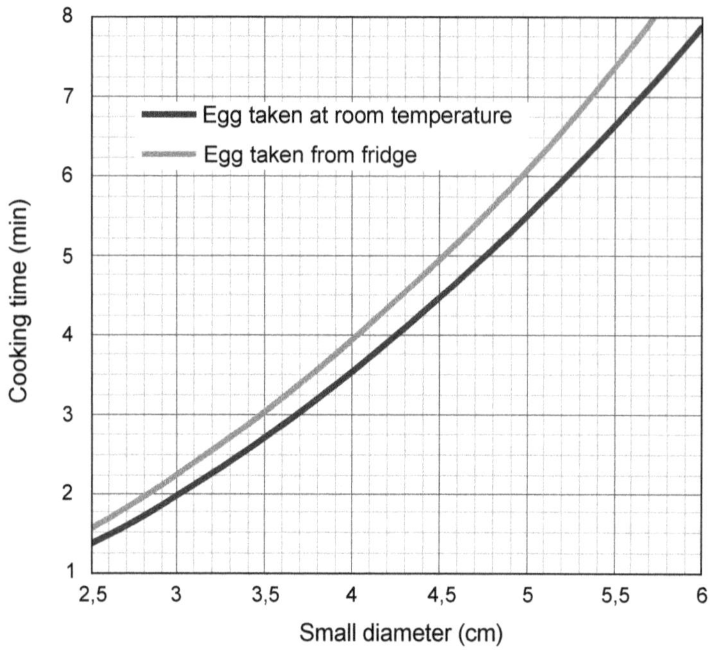

Figure 7. Time needed (in minutes) to cook a hard-boiled egg, depending on its small diameter (in centimetres).

where ln denotes the natural logarithm. For $T_b = 100°C$, $T_f = 63°C$ and $T_0 = 20°C$, $K = 1.46$. But if your egg comes out from the fridge and is only at 5°C, then $K = 1.64$.

Note that the cooking time (Formula (1)) increases very quickly with the diameter d of the egg. Under the standard conditions (usually that means at normal atmospheric pressure at sea level), the boiling point of water is $T_b = 100°C$ (212°F). Therefore, according to the Williams formula, the time required to boil soft ($T_f = 63°C$) a typical egg with $d = 4$ cm, just from the fridge ($T_0 = 5°C$), must be $t = 3.50$ min. For a jumbo egg with $d = 6$ cm, the time is almost twice longer: $t = 7.88$ min.

Note that the formula also indicates a necessity to enhance cooking times at higher altitudes. We have already mentioned that the boiling point of water notably drops with altitude. Therefore, the times recommended in books written at the sea level should be appropriately extended for Alpine cooking. For instance, at the altitude of 5,000 m (16.5 thousand feet), water boils at 88°C (190°F). Thus making a soft-boiled egg will take 5.8 min instead of 4 min that would suffice at sea level.

Formula (1) is quite general. If an object of a given shape, at an initial temperature T_0, is put into water at temperature T_b, the time necessary to reach a temperature close to T_b in the whole object is proportional to the square of the smallest diameter, with a coefficient which depends on T_0, T_b and the material which constitutes the object.

The search for the unbreakable egg

Now let us discuss the value of salting the water in which we boil eggs. Hull fracture is a common occurrence. The white can then escape through the breach and forms flakes in the water (Fig. 8). Fracture may have two causes: it can result from the turbulent movement of boiling water, which throws the egg upwards, and the egg breaks when it falls back; or it can be due to the expansion of the material inside the egg, including the air. It is often said that fracture can be avoided by salting the water. Indeed, this is a way to increase Archimedes' force so that the falling egg hits the bottom of the pan at a lower velocity. A salt concentration of about 15% is necessary to compensate for the egg's weight, but much lower concentrations also seem to be effective. Another advantage of salting the water is that if the egg breaks and if the egg white flows outside, the flow is stopped early because salt stimulates the denaturation of the protein in the egg white. Concerning the second cause of shell breaking, namely the expansion of the material inside the egg, it is recommended to pierce both ends of the egg with a pin before immersing it in water. This allows hot air to escape without breaking the shell.

Figure 8. When immersed in boiling water without precaution, the eggs may crack. A thread of coagulated white announces the disaster.

Trial by fire

Whether broken or unharmed, our eggs are finally cooked! You can pull one out of the water with a spoon and, to wow people, drop it in your hand. It's hot, but you can take it. You then invite your friend (not your best friend, preferably) to grab that egg in turn. He will not stand the test. Why? The still wet egg was cooled a bit on the surface by the spray of water, but once it is dry, it is much hotter. Jules Verne may have had the same experience; he who tells in Michael Strogoff the adventure of his hero, prisoner of a wicked emir who condemns him to be blinded by the blade of a red-hot saber. But when the burning blade is placed in front of his eyes, Michael Strogoff's eyes fill with tears at the thought of his mother being mistreated by the emir, and these tears save the sight of the hero, who miraculously finds her at the last chapter after pretending to be actually blind.

We now come back to our egg, assuming it is a hard-boiled egg. You let it cool slowly, and, a few hours later, you try to peel it. You can only do it with some difficulty, losing many shreds of white that stick to the shell. It is because you made the mistake of not throwing the egg, immediately after taking it out of hot water, into a pot of cold water. The sudden cooling would have loosened the egg white from the shell, which has a different coefficient of thermal expansion.

Last recommendation for inexperienced egg lovers: they now know how to distinguish between a hard-boiled egg and a raw egg. But how do you distinguish a fresh egg from an egg that has passed the expiration date? All you need is a glass of water in which you just drop your egg. The fresh egg sinks and sets; the cooler egg, carrying a larger air pocket, stands up; and the non-edible egg floats (Fig. 9). Some of the protein has broken down, giving rise to hydrogen sulphide which

Figure 9. The left egg can be eaten soft-boiled, the middle one can be eaten hard-boiled, and the right one should be discarded.

escapes through the pores of the shell. This causes the egg to lose weight, and that weight eventually becomes less than Archimedes' force so that the egg floats. Since you don't always have a glass of water with you when shopping, there is another way to recognise fresh eggs. A trained observer only needs to look through the brightly lit egg. The fresh egg is translucent to some extent, allowing a trained eye to distinguish it from a rotting egg, which hydrogen sulphide makes opaque. This was the principle of an instrument called *ovoscope* that conscientious grocers once used.

The Empty Eggshell

While the raw egg, hard-boiled or soft-boiled is fertile in teaching and entertainment, the empty eggshell is not without its attractions. To empty an egg, you obviously need to pierce two holes at both ends with a pin: one large enough to extract the white and the yolk, and one to let air in. A hole of 1 mm^2 is sufficient to suck up the not denatured protein (which is excellent), but you may need to poke the yolk with the pin. Since the hull is empty, it should be washed and allowed to dry.

What to do with this shell? A first possible experiment is the observation of its electrostatic properties, already noted by Michael Faraday in the 19th century. Let us bring a plastic object, previously rubbed against a woollen fabric and therefore electrically charged, to the side of our eggshell. This one is attracted and follows the comb, as a dog follows its master.

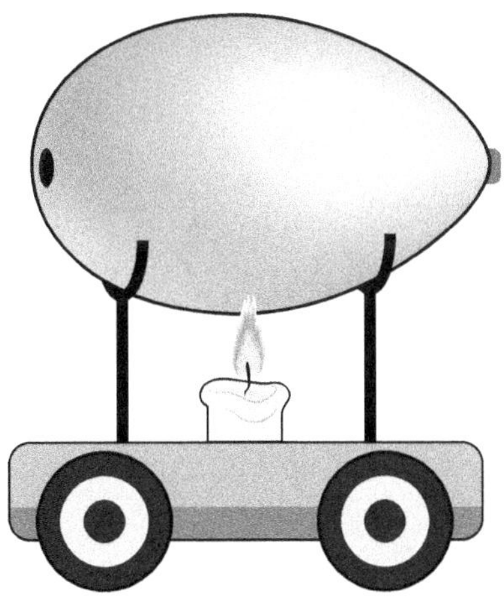

Figure 10. Jet egg.

How to explain this behaviour? By rubbing the comb against the woollen material, electrons are torn from the wool and transferred to the comb, which charges negatively. The comb then attracts light objects like pieces of paper… or an eggshell. This method of electrifying by friction was already known to the ancient Greeks in the 6th century BC. Of course, they had not observed it for plastics but for amber; fossilised coniferous resin. The word *electricity* comes from the Greek ελεχτρον (electron), which means *amber*.

The last experience we offer is for jet propulsion and may delight your children or your grandchildren, but also perhaps the child that you yourself have secretly remained. Fill the shell halfway with water using a syringe, seal one of the holes (e.g., using chewing gum), and find a cart like the one in Fig. 10. Install the shell there, put a candle underneath, and light it. After a few moments, the expelled water vapour will ensure the propulsion of the machine, by virtue of the conservation of the Newtonian momentum or amount of movement, like a jet plane.

Chapter 18

The Secrets of Baking Pizza

Authentic Neapolitan pizza has few ingredients but requires a very hot wood-fired oven to prepare. Therefore, real pizzerias proudly display their ovens and bake pizza right in front of the guests. In just one and a half minutes, the pizza swells, inviting the gourmet spectator to taste it immediately... Let us tell you more about the heat exchange that occurs when baking in a wood-fired oven.

A Brief History of Pizza

Pizza in its usual form — a filling baked on yeast dough — appeared in the alleys of Naples in the first half of the 18th century. The first, very simple fillings included only tomatoes and mozzarella. Tomatoes were brought to Europe by conquistador caravels at the beginning of the 16th century from Peru. As for mozzarella, a cheese traditionally made from buffalo milk, it is the legacy of the Lombards, who imported their buffaloes to Campania after the fall of the Roman Empire. Important: mozzarella does not need to be stored in the fridge, it cannot withstand the cold! Note that by adding basil leaves to the recipe (Fig. 1), we reproduce the three colours of the Italian flag which became the symbol of the country after unification in 1861. This is the traditional Margherita pizza, named after Queen Margherita, wife of the second king of Italy, Umberto I — however, this is sometimes disputed.

At the beginning of the 20th century, many Italians emigrated, and with their resettlement around the world, pizza became known everywhere. Today, it is no longer just an Italian national dish — many Americans even believe that pizza was invented in their country! Outside of Naples, pizza began to take on a wide variety of forms: for example, the dry and thin Roman pizza is distinguished from the

Figure 1. Pizza Margherita, stuffed with basil, mozzarella and tomatoes, reproduces the three colours of the Italian flag.

fluffy Neapolitan pizza with a thick crust. We got acquainted with the process of its preparation thanks to Antonio — *a pizzaiolo* or pizza maker.

Making Pizza Dough

Antonio starts preparing the pizza dough one day in advance. The pizza dough is similar to that one for bread: it is also made from wheat flour, water and yeast (if you wish, you can add a pinch of salt and a little olive oil). The amount of water should be slightly more than half the amount of flour, and about 20 g of yeast should be added per kilogram of flour.

After the dough is well kneaded, the pizzaiolo leaves it to mature, during which time the dough increases in volume. What is happening? Under the influence of yeast, fermentation occurs, similar to alcoholic fermentation (see Chapter 14). When fermented, various complex processes take place during the test: microbiological, physical and biochemical. Wheat flour is 70% starch (see Panel on page 257), which the yeast breaks down. In this case, the fermentation product of interest to us is not alcohol, but gaseous carbon dioxide. It forms tiny bubbles that remain in the dough and thus cause it to "rise".

Now that the dough can be baked, let us look at how heat spreads in the pizza oven and in the pizza itself. Recall that there are three heat transfer mechanisms (Fig. 2):

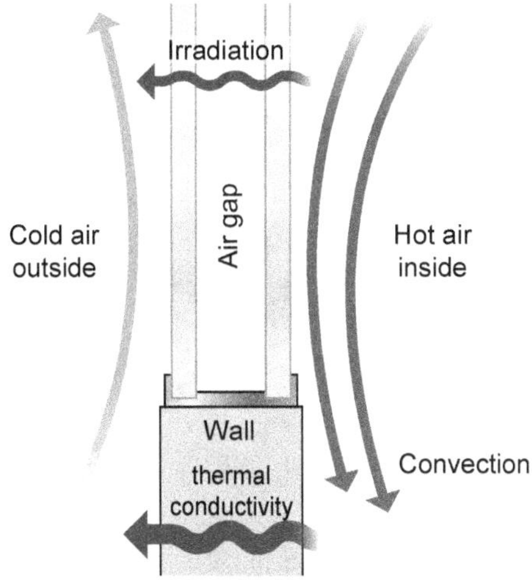

Figure 2. Three types of heat transfer using the example of a double-glazed wall. An air gap between the two panes prevents heat exchange between the outside and inside air.

- *Thermal radiation*: Heat transfer is carried out by absorption or emission of electromagnetic radiation (for example, the sun heats the earth through radiation, see Chapter 7);
- *Convection*: In liquids and gases, it represents the general movement of matter (for example, heat emitted by the Earth's crust spreads in the atmosphere by air currents, see Chapter 7);
- *Heat conduction*: Atoms and electrons from hot regions sequentially transfer kinetic energy to cold regions (for example, the yolk of an egg immersed in boiling water is cooked due to thermal conductivity, see previous chapter).

What is the distribution of roles between heat conduction, radiation and convection in the process of pizza preparation? Using simple calculations, let us try to find answers to these questions, employing approximate data instead of unknown values.

Sugars, starch and carbohydrates

Our great-grandfathers were familiar with white starch powder. By adding water to it, a special paste was obtained with which collars were impregnated. When dry, starched collars held their shape and gave those who wore them an elegant and

(*Continued*)

(*Continued*)

prim look. Remember this when you eat noodles or rice, which become sticky if cooked for too long: this is also the effect of starch!

Irish chemist Thomas Andrews (1813–1885), who studied liquid and gaseous states of matter, like his contemporaries, wore a stiff starched collar.

The molecules that makeup starch (polysaccharides) can be thought of as a combination of a large number of glucose molecules (monosaccharides). Starch, like other sugars, is composed exclusively of carbon, oxygen and hydrogen. So, the formula of sugar that we add to coffee is $C_{12}H_{22}O_{11}$, *and glucose in honey consists of* $C_6H_{12}O_6$ *molecules. Fruits contain fructose, which is an isomer of glucose (and therefore has the same formula* $C_6H_{12}O_6$*). Formulas for sugars and starch can be written as* $C_n(H_2O)_p$, *where n and p are integers. This is why these substances are called "carbon hydrates". However, today this expression is no longer used by chemists: they prefer the term "carbohydrates". The carbohydrates that plants produce for us are the main source of energy for the body. This energy allows us not only to control the muscles, especially the heart, but also to fuel all the chemical processes that support life. Simple sugars such as sucrose and glucose, which are directly absorbed by the body, are quickly absorbed, while starch slowly releases energy, being converted into glucose by enzymes found in saliva and in the stomach.*

How Much Energy Does It Take to Make a Pizza?

In Antonio's pizzeria, the oven, the base of which is lined with refractory bricks, is fired with wood. The temperature in it, according to Antonio, reaches 325–330°C.

What does the supplied heat do in a pizza? On the one hand, it heats up, and on the other, it induces physical transformations and chemical reactions that will give it texture, taste and smell. The temperature of the dough T_0 at the outlet of the refrigerator is about 5°C. The pizzaiolo gives it a disc shape and adds the topping. The pizza, when baked, reaches an average temperature of T_1 and then it is removed from the oven. It seems reasonable to estimate T_1 at 100°C: the oven is hot enough to quickly bring the dough to this temperature, and the water in the dough evaporates and keeps the temperature at this value. Indeed, at normal pressure, water is known to be in equilibrium with vapour at a fixed temperature of 100°C. Of course, the water contained in pizza is not pure (in particular, it contains salt), but the temperature will not deviate significantly from 100°C.

The amount of heat required to heat the pizza from T_0 to T_1 is

$$Q_1 = \rho_d \ell C_d (T_1 - T_0) S,$$

where S is the area of the pizza, ℓ is its thickness, ρ_d is the density, and C_d is the specific heat capacity of the dough. Let us assume that the latter is equal to the heat capacity of the yeast dough, that is, $C_d = 2{,}800$ J kg · K^{-1}. The dough left for one day fermentation is full of bubbles and has a low density ρ_d = 800 kg m^{-3}. Regarding its thickness, the official manuals of Roman pizza makers recommend a thickness of 4 mm after baking, with a tolerance of 20%. Taking into account the loss of volume during cooking and a deviation of 20%, let us accept ℓ = 5 mm. Thus, from the above formula, we obtain Q_1/S = 1,000,000 J m^{-2}, or 100 J cm^{-2}. In the calculations, we neglected the filling (cheese, vegetables), the thickness of which is not uniform.

Now let us estimate the energy spent on physicochemical reactions that occur during baking. Some chemical reactions, called exothermic, release energy; others, endothermic, absorb it, but this amount of energy is insignificant in comparison with the energy Q_2 required for the evaporation of water. Without a wood-fired pizza oven, we baked bread in an electric one and found that the evaporated water was about α = 10% of the original weight. Suppose that for pizza the percentage is the same — therefore, during baking, it loses the mass $\alpha \rho_d \ell S$. Evaporation of pure water at normal pressure consumes energy W = 2,257 J g^{-1}, or 2,257 kJ kg^{-1}. Assuming that the evaporation of liquid from the pizza consumes the same energy, we get

$$Q_2 / S = \alpha W \rho_p \ell = 225 \times 800 \times 5 = 900 \text{ kJ m}^{-2}, \text{ or } 90\text{J cm}^{-2}.$$

Ultimately, the total heat Q_0 transferred to the pizza is $Q_0 = Q_1 + Q_2$, that is, per unit area $Q_0/S = 190$ J cm^{-2}. About half of this heat is used to evaporate the water, and the other half is used to heat the pizza from the fridge temperature to the water boiling point.

Where Does the Heat Come from?

The heat Q_0 is supplied by the oven in a short time τ, just 2 min, explains Antonio. Remember: it takes twice as long to boil a soft-boiled egg (see Chapter 17). This short time is due to the thinness of the pizza and the high oven temperature. And also due to the fact that the pizza heats up both from the bottom and from the top! The brick bottom heats pizza up using thermal conduction, and in addition, the pizza is exposed to infrared radiation from the cupola of the oven and in general from all sides.

First, consider the heat flow entering the pizza from below through the working surface of the oven by means of a heat conduction mechanism. Suppose that its inner surface heated by a flame (Fig. 3) has a temperature $T_2 = 330$°C. In the place where it fits pizza, the top of the brick is constantly chilled by cold dough having a temperature $T_0 = 5$°C, and, therefore, its temperature T_3 must be lower than T_2. The amount of heat transferred by thermal conductivity in time t through the pizza's surface S, is proportional to the temperature difference $(T_2 - T_3)$ between its lower and upper boundaries: $t(T_2 - T_3)/d$, where Λ is the thermal conductivity of the refractory brick and d is the thickness (see Panel on page 261). Let us assume that this heat is only used to heat the pizza. Then the amount of heat received per unit area of pizza is

$$\frac{Q}{S} = \Lambda t' \frac{T_2 - T_3}{d},$$

Figure 3. Temperature distribution in the brick bottom of a traditional wood-oven depending on the height z. The temperature in the pizza and on the part of the bricks adjacent to the surface (dashed line) depends on time. Conversely, the temperature in the brick remains relatively stable while placing the pizza in the oven, baking it, and loading the next pizza.

where t' is the pizza's baking time. According to Antonio, it is about 120 s. Now we must admit that we do not know exactly what temperature T_3 is equal to. Let us trust Antonio again: in his opinion, it is about 200°C. Knowing that the thickness of the working surface is $d = 2$ cm, and the thermal conductivity of the brick is $\Lambda = 0.86$ W m^{-1} K^{-1}, we get:

$$Q/S = 0.86 \times 120 \times 130/0.02 \text{ J m}^{-2}, \text{ or } 70 \text{ J cm}^{-2}$$

Thermal conductivity of materials

Thermal conductivity is a relatively slow process, especially in solids. The mechanism of this phenomenon was analysed at the beginning of the 19th century by Joseph Fourier, whom we mentioned in Chapter 7.

If there is a temperature difference ΔT between two regions of space separated by a material of thickness Δx, the heat flux ϕ that flows between these two regions through the surface of area S is:

$$\phi = \Lambda S \frac{\Delta T}{\Delta x}.$$

Here the heat flow is measured in W, area S in m^2, ΔT in K, and Δx in m, and where Λ is the thermal conductivity of the material, measured in W m^{-1} K^{-1}. This heat flux is the amount of heat which is transferred between two surfaces per unit of time. The higher the thermal conductivity, the greater the heat flux, and the more efficiently the material transfers heat.

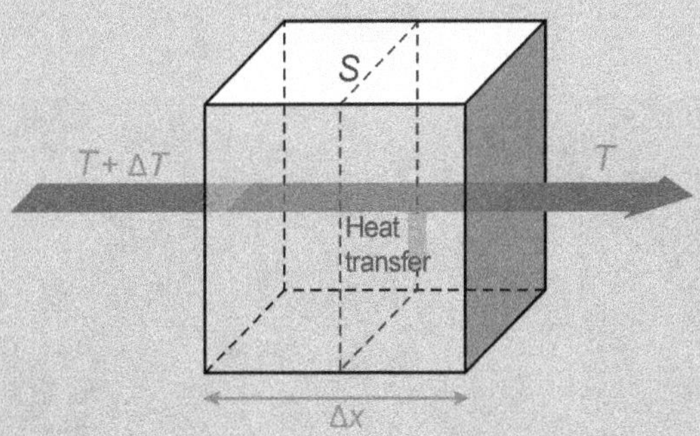

Thermal conductivity through a wall of width Δx between two domains with temperatures $T + \Delta T$ and T. The amount of heat transferred during the time Δt is equal to

$$\Lambda S \Delta t \frac{\Delta T}{\Delta x}.$$

This amount of heat obtained due to thermal conductivity is much less than the total heat $Q_0/S = 190$ J cm^{-2}, evaluated above. Thus, the role of radiation in the process of pizza baking cannot be disregarded.

Role of Radiation

Above we addressed the transfer of heat by means of the thermal conduction. However, when the chicken is skewered in an electric oven, the meat does not touch hot walls at all. And yet it is baked! In this case, the heat is transferred to the chicken partly by the radiation that is emitted by heating element and walls, and partly by means of convection of heated air. Returning to our object of interest — pizza, we turn to the discussion of the role of radiation.

Antonio's brick oven (Fig. 4) has a double heat-insulated cupola: its temperature is slightly higher than that of the working surface of the oven, but for the assessment, we will also take it equal to 330°C (that is, 603 K). Being heated to this temperature, the cupola, like the side walls and working surface, emits electromagnetic, mainly infrared, radiation. Assuming that the radiated power per unit area is determined by the Stefan–Boltzmann law, and that the pizza has two surfaces of area S, we get

$$\frac{Q'}{St} = 5.67 \cdot 10^{-8} \cdot 603^4 \ \mathrm{Wm^{-2}} \approx 7{,}500\,\mathrm{Wm^{-2}}.$$

Figure 4. The pizzas are baked in a wood-fired oven. The pizzaiolo masterfully kneads the dough, puts the filling on it, and places the pizza in the oven.

Thus, within 120 s until the pizza is in the oven, it gets to unit surface the heat $\frac{Q'}{S} = 7{,}500 \cdot 120 = 900\ \text{kJ m}^{-2}$, or 90 J cm^{-2}.

Here, one should notice that, in its turn, the pizza also irradiates out a "flow" of the intensity $I_{pizza} = \sigma(T_{pizza})$.[4] Since a major part of the baking time is required for the evaporation of water contained in the dough and toppings, we can assume T_{pizza} = 100°C = 373 K: the toppings will boil at this temperature until they (and the whole pizza) are well cooked. This results in a pizza's radiation intensity

$$\frac{Q''}{St} \approx 1{,}100\ \text{W m}^{-2},$$

i.e., 15% of the obtained radiation, the pizza returns back to the oven.

To summarise briefly: total heat Q_0 supplied to the pizza, from one side is provided by thermal conductivity, and from the other — by radiation, that is, $Q_0 = Q + Q' - Q''$ (Fig. 5). Adding 70 J cm^{-2} supplied by the heat conduction from below and $0.85 \cdot 90\,\text{J cm}^{\ 2} \approx 75\,\text{J cm}^{\ 2}$ received by radiation, we actually get 145 J cm^{-2}, that is, somewhat less than the value Q_0/S = 190 J cm^{-2}. Recall that our calculations are approximate, for example, we neglected the heating due to convection, as well as the role of pizza toppings, and, probably, by the radiation arriving from the overheating with respect to the pizza's body working surface.

So, almost 70% of the heat received by a pizza comes from radiation. The important role of radiation is also evidenced by the presence of burning spots on the edges of the pizza and on the top of the filling (Fig. 6). If it baked only due to

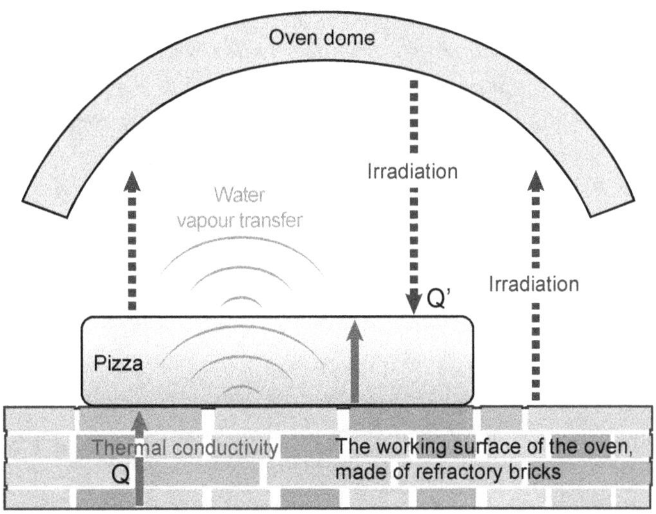

Figure 5. Processes of heat transfer during the cooking of a pizza.

Figure 6. The pizza is ready, as evidenced by the presence of slightly burnt parts on the crust and filling. This charring is probably due to the fact that the evaporation of water from the surface was not compensated by the inflow of water from the inside.

the heat coming from the working surface of the oven, the top of the pizza would be the least hot, and therefore it would not have burned.

Infrared radiation provides approximately equal heating of the pizza both from the top and the bottom. In addition, the homogeneity of the temperature distribution is partly ensured by heat transfer due to the convection of water, which

Stefan–Boltzmann law

The power of thermal radiation per unit area at the same temperature for different sources is different, but it is always less than or equal to the power emitted by an absolutely black body (see Chapter 7). Recall that this is the name for a body that absorbs all the radiation it receives.

The power that such a body emits from the unit surface at a temperature T, according to the law, derived by physicist Joseph Stefan (1835–1893), is equal to σT^4*, where the constant* σ *is equal to* 5.67 10^{-8} $\mathrm{Wm^{-2}\,K^{-4}}$*. When Stefan presented this law to his colleagues in 1879, its justification was based on the analysis of experiments. Only in 1884, Austrian physicist Ludwig Boltzmann mathematically proved that the power emitted by a black body should be proportional to* T^4*. His demonstration, based on thermodynamics, was brilliantly elegant. Yet, it is of little interest today. Indeed, the formula determined by Planck in 1900 (see Chapter 7) contains much more information than the Stefan–Boltzmann law, since it details the*

(*Continued*)

(*Continued*)

power emitted by the black body for each value of the frequency, instead of just providing overall power. By integration over the frequency, the Stefan–Boltzmann law can be deduced together with the expression for the constant σ, *that is*

$$\sigma = 2\pi^2 k_B^4/(15c^2h^3),$$

where c is the speed of light in vacuum, k_B *is the Boltzmann constant, and h the Planck constant. We have already introduced these fundamental constants in Chapter 7 and will come back to them in Chapter 22.*

While the numerical value of the constant of proportionality σ *had been experimentally determined by Stefan, it is clear that Boltzmann could not determine it theoretically. After all, the expression for* σ *contains Planck's constant, which, in 1884, 16 years before Planck's work, had not yet been derived!*

evaporates in the volume of the pizza. This steam quickly escapes through the top surface of the pizza, partially condensing it and thus warming it up. The rest of the steam goes into the oven.

Which Oven Is Preferable: Electric or Wood-Burning?

Today some pizzerias are moving away from the traditional wood brick oven in favour of an electric one, which is usually equipped with a steel working surface. Steel has a high thermal conductivity; as a result, cold dough, which is placed on steel, barely cools the working surface, since the heat coming from it is almost instantly absorbed by pizza. Therefore, if we want the pizza's lower surface to be at temperature T_3 = 200°C, as recommended by Antonio, the temperature at the base of the steel plate should not be much higher; a reasonable estimate is 230°C, or 500 K, instead of 600 K in a wood-burning stove.

It follows from this observation that the walls and cupola of the steel oven also have a much lower temperature than in the case of a brick oven. As a result, the power of irradiated heat, according to the Stefan–Boltzmann law (see Panel on page 264), is about two times $[(\frac{600}{500})^4 \approx 2]$ less than in the case of the wood oven. Thus baking in an electric stove will require more time and will be less uniform than in a wood oven. In addition, the traditional brick oven provides a very strong flow of heat during the first seconds of baking, but as the top side of the brick surface cools down, this flow decreases, which avoids excessive heating of the pizza base. The steel surface does not have this property. If you add the delicious smell of wood, you understand why we prefer Antonio's pizza made in the wooden oven rather than one made in a modern, even luxurious pizzeria that uses a non-wooden oven.

In this chapter, we have asked more questions than the ones we have answered to. Our goal was not to teach the reader how to bake pizza correctly, but to undertake a quantitative analysis of a fairly mundane action. We will leave the reader with the opportunity to experiment, to measure those values that we have estimated approximately. For example, the weight loss of a pizza during baking is easily measurable, and the role of radiation can be determined by hanging in the oven an object to be heated (not necessarily edible), not in contact with a working surface.

Chapter 19

Noodles, Spaghetti and Physics

Many of our readers presumably are spaghetti lovers, and a few of them will sometimes cook this dish. The latter are aware of the importance of the correct cooking time. Perhaps, however, they are not familiar with the physicochemical processes that correspond to cooking and wonder why the cooking time is different for different types of spaghetti. Moreover, the cooking time for the pasta of the same type must be increased if one is at high altitudes. In the extreme case of cooking at the top of Mount Everest, as we will see, the noodles will be too "al dente".

In this chapter we hope to satisfy the curiosity of our gourmet readers. But we will also see that the study of spaghetti can lead us to unexpected apertures in various fields of physics. After noting that cooking of the hollow types of pasta such as bucatini meets with the capillary phenomena, we will open a window to the science of polymers and the strength of materials. So searching for the answers to seemingly trivial questions about the common pasta leads us to much more serious problems in modern science.

An Overview of the History of Pasta

Contrary to some claims, it does not seem correct that pasta was imported by Marco Polo in 1295 upon his return from China. Its story actually begins much earlier on the Mediterranean coast, when prehistoric man gave up nomadic life and began to sow seeds for his food. In the first millennium BC, the Greeks were already making pasta in the form of thin pancakes they called λαγανον. The Romans used the same term (laganum) which may well be the origin of the current word *lasagna*. The poet Horace was delighted (Satires, I, VI, 115) to find when he returns home a dish of leeks, chickpeas and pancakes:

Inde domum me ad porri et ciceris
refero laganique catinum.

The growth of the Roman Empire brought the expansion of pasta to Western Europe. Pasta was a means of preserving grain, and was particularly suitable for tribal migrations. When the Arabs conquered Sicily in the 10th century, they introduced a kind of pasta that can be considered the ancestor of spaghetti. It was named *itryah*, which for Sicilians became *trie* and which can be considered the ancestor of spaghetti. In Sicily, pasta began to be made at the start of the second millennium. Based on a will written by a notary named Ugolino Scarpa in Genova, we know that macaroni were already consumed in Liguria as early as 1,280. And in Boccaccio's Decameron, macaroni became a symbol of gastronomy.

The first corporations of pasta makers ("Pastai") appeared in the 16th century in Italy, becoming part of the political landscape. At that time, macaroni was considered aristocratic food. We can indeed imagine that making tubes from flour is not easy without the appropriate tools. The invention of the mechanical press brought down production costs so that in the 17th century pasta was consumed by all social classes. Naples became a centre of production and export. In northern Italy, pasta became popular at the end of the 18th century, mainly thanks to Pietro Barilla who, after founding a small factory in Parma, became one of the main producers of the Italian food industry.

Pasta Making

When you buy a package of pasta, it is usually made from durum wheat semolina (Fig. 1). The seed of this cereal is yellow and hard, whereas the soft wheat, used to make bread, has a white, crumbly seed. Durum wheat is ground in the form of semolina, which is mixed with water and kneaded, so as to obtain a malleable dough. This gives a dough similar to bread dough or pizza dough (see Chapter 18), except that it does not contain yeast.

Modern pasta production processes are based primarily on extrusion (which involves pressure, forcing fresh dough through a hole or die) and stretching. Extrusion was invented by metallurgists and used for the first time to make long metal rods with particular profiles (Fig. 2). It can be performed hot or cold. Stretching is analogous to extrusion, but the die is located at the exit of a duct. In metallurgy, drawing produces wires up to 0.025 mm in diameter. Extrusion can also be applied to polymers, ceramics and food. Figure 2 shows some examples of spaghetti dies.

Figure 1. Ears of durum wheat which are used in particular for making pasta.

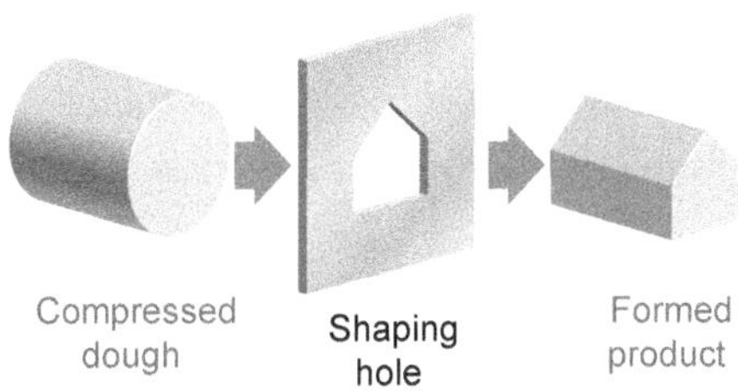

Figure 2. The principle of extrusion. By compression, a material having some fluidity is forced to pass through a rigid die, which gives it the desired shape. Extrusion can be compared to stretching, a process used in metallurgy, in which the material also passes through a die, but is pulled from the front instead of being compressed from the rear.

In addition to pressing, there is another moulding method: it is rolling of dough. During the rolling process, the dough passes between two cylinders which compress it. More or less thick plates are formed, which will then be cut off — for example, to obtain lasagna sheets (Fig. 3).

These production steps play an important role in the preparation of the product. In wheat groats, starch molecules form granules with a diameter of 10–30 microns, which are surrounded by various proteins. Due to the addition of water and mechanical stress arising when kneading dough and shaping, two of them, gliadin and glutenin, combine with water and create a continuous network,

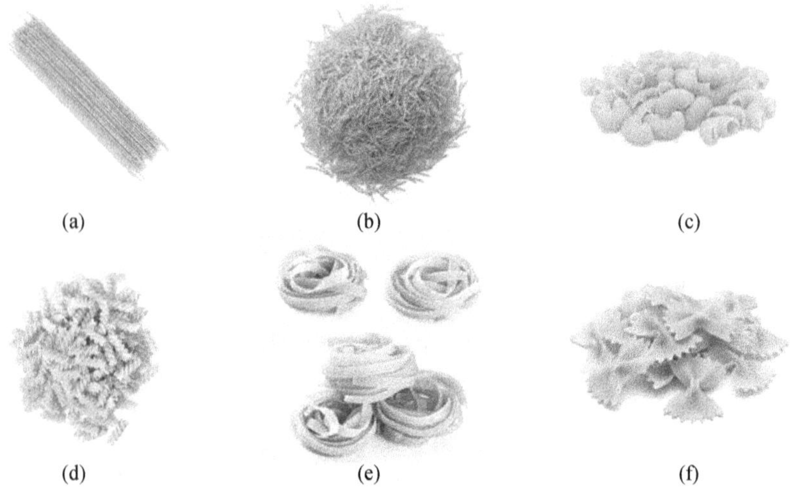

Figure 3. The spaghetti (a) and vermicelli (b) are obtained by extrusion; blade just cuts off the length wanted at the output of the press. By varying the speed flow of the dough in various parts of the press, one obtains the curved shapes, such as shells (c) or spiral pasta, e.g., fusilli (d). The tagliatelle (e) are obtained by rolling, just like the farfalle (f) which are then die cuts.

which is a substance called gluten. This net envelops the starch granules. The quality of such a network is of paramount importance for ensuring the integrity of the pasta during the cooking process, during which these granules gradually swell.

After shaping the pasta, the excess water added while kneading the dough should be removed. The point is that water may cause unwanted reactions between starch and gluten. To prevent this, drying is needed, which also allows the products to be stored for a long time. A long time ago, the pasta was dried in the sun, but as technology developed, the drying temperature gradually increased: from 50–55°C in the 1970s to over 100°C nowadays.

The Science of Cooking Noodles

Cooking time is directly related to the ability of starch molecules to absorb water. Indeed, a cooked noodle is twice as heavy as a dry noodle. Water enters the gluten network and diffuses towards the centre of the pasta which is thrown into a pot of boiling water. When the temperature reaches around 70°C, the starch molecules form a kind of gel. Noodles are considered to be cooked “al dente” when the starch gel absorbs the quantity of water just sufficient to make it soft enough to eat. Cooking pasta therefore involves penetrating hot water into the initially dry noodles.

Overcooked pasta is the shame of unlucky cooks. When gluten cannot hold starch and part of it comes to the surface, pasta becomes sticky. The degree of "stickiness" depends not only on the time of cooking but also on the method of making pasta: if dried with a hot method, then the gel mesh turns out to be resistant, its cells are smaller, so such pasta usually does not stick.

In practise, the cooking time depends on the diameter of the pasta, but also somewhat on atmospheric pressure, therefore on the altitude and the weather. To take an extreme case, anyone with the crazy idea of going to cook noodles on top of Mount Everest would be sorely disappointed. At this height of 8848 m, the atmospheric pressure is in fact $3.5 \cdot 10^4$ Pa, which corresponds to a boiling point of water of 73°C, a temperature at which the starch gelation is very slow. It is therefore very hard to cook noodles there, except with a pressure cooker.

Cooking Time

Let us return to a moderate altitude and ask ourselves why the cooking time of pasta depends on the diameter of the spaghetti. The material will be, for simplicity, assumed to be cylindrical (spaghetti, vermicelli) and of diameter d. What is the time it takes for the water to penetrate to the centre of the cylinder? The penetration of water is a phenomenon of diffusion analogous to the diffusion of heat in an egg immersed in boiling water, and we can therefore think that the time necessary for the penetration of water to the centre of the cylinder is proportional to d^2, as seen in Chapter 17. However, the cooking time τ also includes a subjective term, which depends on the consumer's taste. So we will write the cooking time of the spaghetti in the form

$$\tau = ad^2 + b. \tag{1}$$

Constant $\boldsymbol{b}$ for Germans, who like well-cooked pasta, turns out to be larger than for Italians, who prefer undercooked pasta (sometimes the coefficient $\boldsymbol{b}$, as will be shown below, may even become negative). The first term in Eq. (1) is the time it takes for the water to penetrate to the centre of the spaghetti. It is independent of the nationality of the consumer, but depends on the physical properties of the dough, including the ease with which boiling water diffuses. It also depends, as we have said, on the temperature of the boiling water and, therefore, on the atmospheric pressure.

We went to a supermarket and got a variety of cylindrical pasta of different diameters but hopefully of the same composition. We note in Table 1 for each type of subject the recommended cooking time, as well as the diameter d

Table 1. Diameters and cooking times for different types of the cylindrical pasta.

Type of pasta	Diameter d (mm)	Recommended cooking time (min)	d^2 (mm^2)	Cooking time calculated according to Eq. (1) (min)
Capellini no. 1	1.15	3	1.32	2.2
Spaghettini no. 3	1.45	5	2.10	5
Spaghetti no. 5	1.75	8	3.06	8.1
Vermicelli no. 7	1.90	11	3.61	10
Vermicelli no. 8	2.10	13	4.41	13
Bucatini	2.70	8	7.29	22.5

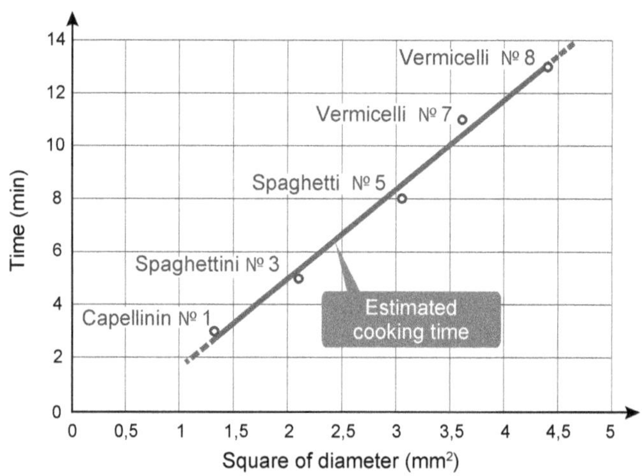

Figure 4. The cooking time of cylindrical pasta is a linear function of the square of the diameter.

(in millimetres) measured with a precise calliper. We also calculated the square of the diameter. Then we have in a diagram (Fig. 4) plotted for each type of pasta a point whose abscissa is the square of the diameter and the ordinate is the recommended cooking time. We see that the points are on a straight line according to Eq. (1).

Figure 4 makes it possible to determine the parameters a and b. We have chosen to adjust them so that Eq. (1) gives the recommended cooking time for spaghetti no. 3 and vermicelli no. 8. We find $\boldsymbol{a} = 3.4$ min/mm^2, while $\boldsymbol{b} = -2.3$ min. This negative value is characteristic of the Italian culinary culture, which is fond of pasta "al dente". However, it does pose a problem: it seems that pasta below a certain diameter would not even need to be cooked! This diameter is 0.82 mm, and the capellini in Table 1 is not far from this value, as the reader can easily verify. This result is manifestly absurd which demonstrates that Eq. (1) is an approximation which would not be acceptable for very thin noodles.

The Case of Hollow Pasta

How to estimate the cooking time of bucatini, a type of cylindrical pasta with a hole in the centre? Their diameter is about 2.7 mm, and the estimated cooking time according to Eq. (1) is about 23 min, which is very different from the recommended one, namely 8 min. Obviously, the above formula turns out to be inapplicable for hollow products. Indeed, the water in this case does not cover the distance $d/2$ (where d is the outer diameter) to reach the core of the bucatini but only the distance $(d - d_i)/2$, where d_i is the diameter of the hole, is equal to about 1 mm. Thus, in the formula, the diameter d should be replaced by the difference between the outer and inner diameters, which is $d - d_i = 1.7$ mm. The formula now gives a cooking time close to the recommended one. However, to generalise this reasoning to all tubular pasta products, such as bucatini or *penne*, should be followed with care: if the "tube" is too narrow, water may not penetrate inside, and if it is too wide, then heat will enter into the pasta both from the inside and the outside!

Do Spaghetti Form Knots?

When you cook spaghetti, they tend to get tangled with each other. Yet the authors of this book have never observed that one single spaghetto tangles with itself and forms a knot. The reason is simple: the length of a spaghetto is *too short*. For any long and flexible object (spaghetto, string, rope, cable, et alia), the critical length L_c exists, beyond which the formation of a knot is almost inevitable, whereas below this length the probability to tie is weak.

The probability to find unknotted polymer chain exponentially decreases with the growth of its length L

$$w \sim exp\left(-\frac{L}{\gamma \cdot \xi}\right),$$

where ξ is the characteristic length at which the polymer can change its direction at a right angle, and $\gamma \approx 300$ is a large factor, obtained as a result of numerical and theoretical modelling. Applying this formula to spaghetti, where $\xi \approx 3\,\text{cm}$, and assuming that the probability of self-knotting becomes noticeable at $w \sim 0.9$

Polymers

Carbon, together with hydrogen, easily forms long-chain molecules called polymers.

There are many biological polymers, such as starch. And above all, the plastics that have invaded our daily lives are made of polymers to which are added various additives. Expanded polystyrene is widely used as impact protection. Polyethylene makes up plastic packaging.

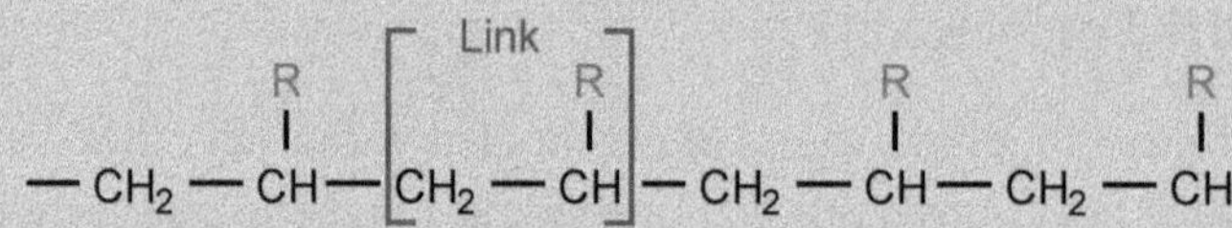

Polymers are soluble in some solvents. They can take many forms there, possibly forming knots. The study of these forms has been the subject of numerous works which won the 1974 Nobel Prize in Chemistry for the American Paul-John Flory (1910–1985), then the Nobel Prize in Physics 1991 for the French Pierre-Gilles de Gennes (1932–2007). Flory had to be content with imagining these molecules unfolding in their solvent. Nowadays, thanks to the atomic force microscope, which in a way "feels" the atoms with a point of nanometric size, we can see such molecules deployed on a flat support.

(*Continued*)

(*Continued*)

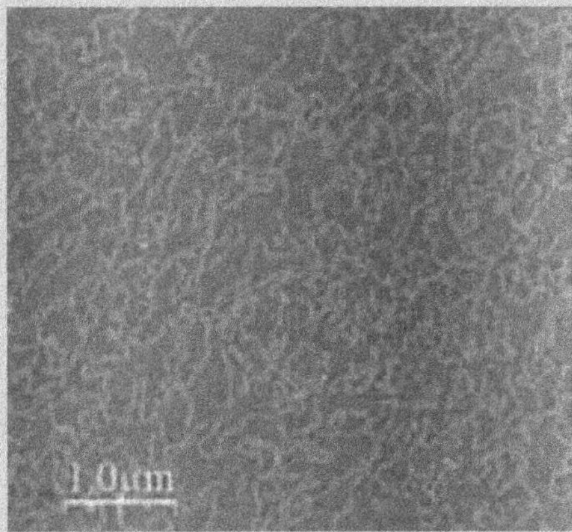

DNA molecules on a nickel surface, observed by atomic force microscopy. Courtesy of A. D. Chepelianski.

Figure 5. Since the rope is long and flexible, it most likely gets entangled.

(10% of spaghetti becomes knotted), one can estimate the value of critical length as $L_c \approx -\gamma\xi \ln(0.9) \approx 0.9$ m. The length of a standard spaghetto is 23 cm, and this is not long enough to form knots.

The reason why physicists are interested in that question is not because they are particularly fond of pasta, and if they are, then they do not care if it is knotted, but among the long and flexible objects, there are polymers, objects of interest to physicists as well as to biologists and chemists (see Panel on page 274).

We all know that strings, wires, etc., tend to get tangled. Jerome K. Jerome, in *Three Men in a Boat*, deplored this peculiarity of towropes.

The model studied is not necessarily applicable to a rope that is wound in such a way that it occupies a minimum volume (see Fig. 5), i.e., by forming loops which tend to become entangled as soon as they are handled (and not, of course, spontaneously as the author of *Three Men in a Boat* pretends to believe, see Panel below on this page).

The Breaking of Spaghetti Strands: Material Resistance and Flexure Waves

The breaking of the spaghetti is easy to observe, but the result of the observation is unexpected. Take a dry spaghetto by its two ends (one for each hand) and bend it, causing it to bend more and more. It will obviously break pretty quickly, and you would expect it to shatter into two pieces. However, the break often produces three or more pieces. This unexpected behaviour caught the attention of several physicists, among whom one was Richard Feynman. Indeed, if the breakage of a

Naughty ropes

There is something very strange and unaccountable about a tow-line. You roll it up with as much patience and care as you would take to fold up a new pair of trousers, and five minutes afterwards, when you pick it up, it is one ghastly, soul-revolting tangle.

I do not wish to be insulting, but I firmly believe that if you took an average tow-line, and stretched it out straight across the middle of a field, and then turned your back on it for thirty seconds, that, when you looked around again, you would find that it had got itself altogether in a heap in the middle of the field, and had twisted itself up, and tied itself into knots, and lost its two ends, and become all loops; and it would take you a good half-hour, sitting down there on the grass and swearing all the while, to disentangle it again.

Jerome K. Jerome, *Three Men in a Boat*

spaghetto is not of the slightest practical interest, it gives an easily achievable image of a break similar to that of a beam, for example.

It was not until 2005 that the research of two French physicists, Audoly and Neukirch, made it possible to understand the rupture of spaghetti. They studied the behaviour of a thin elastic rod under the effect of bending deformation. Their study was purely numerical but gave a good understanding of the process of rupture. When applying mechanical stress, the first fracture occurs at the weakest point of the rod. We can think that the two pieces return to their form of equilibrium (the straight one). In fact, they do it in a complicated way. Initial fracture produces flexure waves in the two fragments of the rod. These waves obviously end up decaying, but if the elastic constants and the length of the rod meet certain conditions, the propagation of the waves can lead to a new fracture. It should be noted that these waves are superimposed on the initial deformation which existed before the rupture, and which relaxes slowly, in a time much longer than the period of the flexure oscillations. If we add the two deformations, the one that oscillates quickly and the one that relaxes slowly, we find that their sum can be large enough at certain points, to cause new fractures. Audoly and Neukirch verified their theoretical calculations by recording the rupture of a spaghetto with a high-speed camera (Fig. 6).

It should be noted that these authors are specialists in applied physics and that the study of spaghetti is not only for them an entertainment or a means of attracting the attention of the general public, but above all a test, on inexpensive

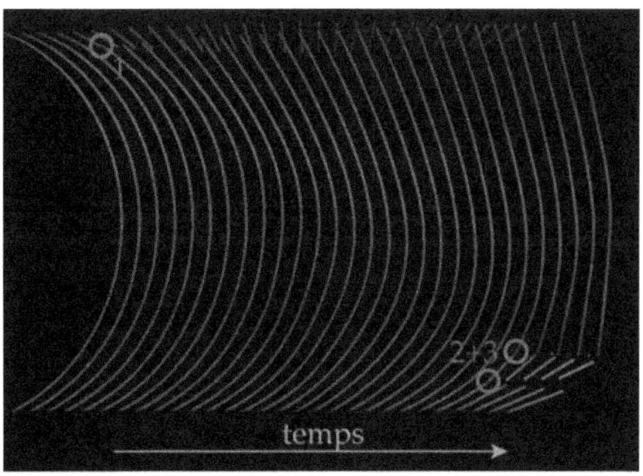

Figure 6. Breaking a dry spaghetto into several pieces. From B. Audoly and S. Neukirch, Fragmentation of rods by cascading cracks: Why spaghetti does not break in half, *Physical Review Letters* **95**, p. 095505 (2005).

equipment of the experimental and theoretical concepts and techniques whose applications are quite practical.

We invite the reader to keep some spaghetti dry, break them up while the others cook, and count the number of pieces. Be careful, the spaghetti must be very dry!

Chapter 20

The Physics of Good (and Bad) Coffee

Why is one coffee delicious and another not? And more generally, why do coffees made in different ways have different tastes? There are good and bad manufacturers, good and bad roasters, and most importantly, different ways to make coffee! But even when making coffee alone and in the same way, say espresso, using one and the same apparatus, the same temperature and filtration pressure, the resulting drink can differ significantly depending on the skill of the barista: the degree of grinding of coffee beans, other tricks that are only known to him. And how to compare drinks obtained by different methods — Turkish coffee with Scandinavian? Thus, coffee of different tastes is obtained more or less successful depending on the skill of the user.

In our globalised world, the 21st century traveller will find the same drinks in every corner of the globe, be it Paris, New York or Kathmandu. But there is one exception — coffee. If you order coffee at a bar in Naples, you will be offered a cup slightly larger than a thimble, which contains a viscous black liquid under an appetising foam. Ask for a coffee in Chicago and you get a huge glass containing half a pint of hot brown water. In this chapter, we will not discuss the merits and demerits of different ways of brewing coffee; instead, we will address the various processes and physical phenomena that take place during its preparation.

Coffee in Brief

Coffee drink is made from the beans of the coffee tree (Fig. 1), which grows in tropical and equatorial regions (Arabian Peninsula, Latin America, Africa...). Coffee beans contain sugars, fats, proteins, flavours and the alkaloid caffeine, which appears to affect the particular areas of the brain responsible for memory and concentration, providing a boost to short-term memory. After harvesting and drying, the coffee beans are roasted at a high temperature of about 240°C. At the same time, as a result of chemical reactions between proteins and sugars (Maillard reactions, see Chapter 21), the grains acquire a beautiful brown colour. In addition, roasting releases many volatile aroma molecules: over 800 different compounds! The beans are then ground, and the coffee powder thus obtained is brought into contact with hot water to extract valuable aromas. As we will see in the following, various preparation methods differ, the temperature of the water, the pressure at which the process takes place, and the duration of contact between the coffee powder and the water (Fig. 1).

Boiled Coffee

The "boiled coffee" method is very old and is still used in Finland and northern Scandinavia. The roasted coffee is coarsely ground and then poured into the water

Figure 1. A branch of a coffee tree. Each berry contains two coffee beans. Sorting ripe (red) berries which must be separated from rotten (black) or not yet ripe (green or yellow) berries is one of the difficulties of harvesting.

(10 g of coffee in 150 to 190 ml of water). The whole is then brought to a boil for 10 min. Pour the coffee into the cups without filtering, and let stand for a few minutes, while the suspended ground particles sink to the bottom under the effect of gravity. We will not recommend this process because the aromas of coffee, embodied in fairly volatile molecules, are carried away by the water vapour that escapes during boiling. These aromas are therefore lost for the taster.

Drip Coffee Maker

Drip coffee makers are common in America, Northern Europe and France. The conical paper filter filled with coarsely ground coffee is placed in a sort of funnel above the coffee maker, usually made of glass, and very hot water is slowly furnished into it. The water bathes the coffee by dissolving the soluble substances present in it, then passes through the filter, and falls into the coffee maker. This method has the advantage of requiring little handling, as the hot water is usually poured continuously from an electric device, and the cleaning of the coffee maker is reduced since the coffee is confined in the disposable filter. The process gives, in 5 or 6 min, a coffee that is relatively poor in aromas: indeed only a few essences are able to pass through the paper filter in the absence of pressure. The typical dose is around 6 g of coffee per 140 ml cup of water.

Turkish Coffee

When making Turkish coffee (the method which is used, for example, in Greece and Serbia), very fine ground coffee is mixed with sugar and then filled into a metal conical container, usually made of copper or brass, called *ibrik* (Fig. 2).

Figure 2. The ibrik, a suitable utensil for brewing Turkish coffee.

Cold water is then added, and the ibrik is heated, traditionally by placing it in hot sand or, more modernly, on a gas or electric stove. You can also throw coffee powder directly into the ibrik filled with boiling water. Heating the container creates convective currents (see Chapter 7) that carry some of the coffee powder to the surface, where it forms a kind of crust. When water is approaching its boiling temperature, the bubbles begin to turn the crust into foam. At this point, the ibrik must then be removed from the heat or sand. The process is repeated twice to form a thick foam layer. The coffee is then poured into small cups. Before drinking it, wait a few seconds for the coffee to cool and most of the coffee sediment falls to the bottom of the cup. Turkish coffee has a very distinctive flavour and is very popular with those who like to eat a little of the coffee residuals.

The Moka (Geyser) Coffee Maker

The coffee maker that we are going to describe can be found in almost all Italian kitchens! It is made of three parts: a lower container where water is poured, a metal funnel equipped with a filter constituted by a perforated metal plate between which the ground coffee is placed, and an upper part which collects the beverage (Fig. 3). A coffee lover will hardly be able to be creative in its preparation in such a coffee maker: unlike other methods, it is imperative to strictly follow the rules dictated by the design of the device.

This is a rather complex invention. The filter is obviously located in the very centre of the apparatus. Cold water is poured into the lower compartment up to the safety valve, leaving a small volume filled with air. Ground coffee is filled into the filter, practically without tamping. The coffee maker is then closed by screwing the top onto the base, and the filter is located in the middle. Now the lower part of the funnel, made in the form of a tube, is submerged in the water, and almost touches the bottom of the container. The coffee maker equipped in this way begins to heat up over low heat. After a few minutes, the water that has reached the boiling point enters the filter through the funnel. It should be noted that in this design, the funnel operates in the opposite mode to the usual one used for overflowing liquids: water enters it through a narrow part and exits through a wide one. Passing through a filter filled with ground coffee, the water is saturated with its aromas and transforms into coffee beverage. The resulting drink, which rises under pressure from the bottom, passes through another narrow tube and finally ends up at the top — you just have to pour it into a cup!

What happens in a moka from a physics point of view? The lower part of the device is tightly connected to the upper and sealed with a rubber gasket (Fig. 4). When water is heated, the pressure of saturated water vapour increases

Figure 3. The Italian coffee maker, traditionally called moka. This model was patented by Alfonso Bialetti in the 1930s. The coffee maker is fitted with a safety valve which prevents an explosion in the event of excessive overpressure (due, for example, to a too fine and too compacted grind).

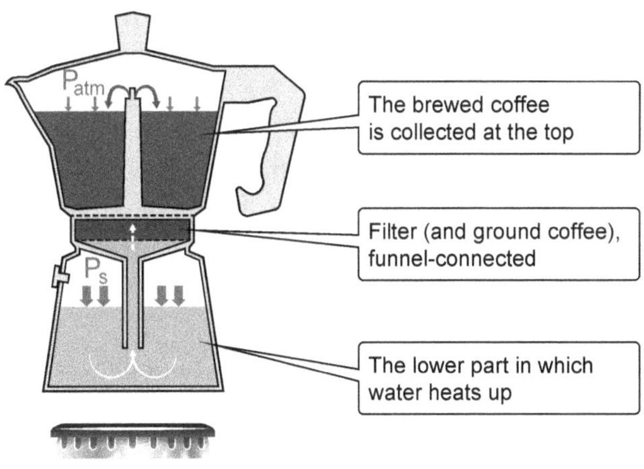

Figure 4. Principle of the Italian coffee maker. The saturated vapour pressure increases with temperature. When it exceeds atmospheric pressure, water is pushed into the funnel and through the coffee grounds, then spurts out into the upper compartment.

(see Chapter 15), and the liquid evaporates into the space not filled with water. The water temperature quickly reaches 100°C: the pressure of saturated water vapour becomes equal to atmospheric pressure. Thus, the steam, like a spring, begins to push the water through the funnel, forcing it to pass through the ground coffee in the filter. As the heating continues, the temperature and pressure continue to increase. The water displaced by the steam and already converted into coffee ends up at the top of the appliance. To prevent the filtration time from being too short, you can intervene in the process when the noise indicates that the finished drink begins to pass through the top: the heating should be stopped, and this will slow down the passage of water through the coffee.

Temperatures of about 100°C and pressures slightly higher than 1 atm ensure that the coffee is saturated with all the aromas, but high temperatures, alas, can lead to the disappearance of some of them. Thus, the geyser coffee maker produces strong, aromatic coffee, which, however, does not achieve the quality of a good espresso. To avoid overheating the water, you can go and brew coffee high in the mountains, in an alpine refuge, where the atmospheric pressure is lower. It may not be possible to cook pasta at the top of Everest (see Chapter 19), but the coffee there is probably better than at its foot!

The Physics of Filtration

It is obvious that the taste of the coffee produced by any coffee maker depends on the quality of the powder and the temperature of the water. Furthermore, it depends on how long this water remains in contact with the coffee, which is also the time it takes for the water to pass through that coffee. The coffee particles form a veritable labyrinth of conduits through which water will make its way: the extraction time will be shorter the larger these conduits are and the greater the pressure in the lower compartment. The overall water flow follows a physical law: Darcy's law (see Panel on page 285) in which the permeability κ of ground coffee occurs. What is the latter worth? An estimate was made by Italian physicist Concetto Gianino (2007) who carried out the experiment in an Italian coffee maker. The quantities involved in the problem are as follows:

- the height of ground coffee, equal to $L = 0.014$ m in the coffee maker used;
- the filtering area, equal to $S = 14\ \text{cm}^2$;
- the viscosity of water which, at 100°C, is of the order of $\eta = 0.3 \cdot 10^{-3}$ Pa·s;
- the density of the water, $\rho = 1{,}000\ \text{kg m}^{-3}$.

In the experiment, the mass of prepared coffee was about 0.07 kg, for an extraction time a little less than 1 min. According to Darcy's law, the permeability is then: $\kappa = 3.5 \times 10^{-9}/\Delta P$, where ΔP is the pressure difference between the two sides of the filter. It is therefore sufficient to know ΔP to deduce κ. The measurement is not easy, since you must have access to the lower compartment of the coffee maker. Ingeniously, Gianino took advantage of the presence of the safety valve to introduce a temperature probe. Knowing the temperature of liquid water in equilibrium with the vapour, he deduced the pressure (see Chapter 15). The result of the measurement is that ΔP is of the order of 3 kPa. With this data, it turns out that the permeability of ground coffee is $\kappa \approx 10^{-12}$ m^2. This is not very fine grinding, yet the found value is remarkably high, close to that of clean, highly-permeable sand. Obviously, the permeability depends on how the coffee has been ground and how tightly it is tamped into the filter. Experts recommend not to tamp it at all, so as not to drag out the filtration time and to carry it out at a relatively low temperature.

Darcy's law

In the mid-19th century, two French engineers, Henry Darcy and Jules Dupuit, made the first experiments on the movement of water in tubes filled with sand. This was the starting point of the science of filtration, currently applied to the movement of liquids through solids with interconnected pores or cracks. Darcy enunciated the so-called linear filtration law, which bears his name. This law determines the volume Q of a fluid which flows per second through a porous material of section S and thickness L, hereafter called filter under the effect of the pressure difference between the ends of the filter ΔP

$$Q = \kappa S \Delta P/(\eta L). \quad (1)$$

Here, η *is the viscosity of the fluid and the permeability* κ *is a characteristic coefficient of the porous material. It has the dimensionality of the area and therefore is measured in m*^{2} *in the international units system. In fact, the permeability of common materials, such as sand, is on the order of* $1\,\mu m^2$*, i.e.,* 10^{-12} m^2*, a unit which is sometimes called darcy when it measures permeability. In the case of the "conventional" coffee maker (Figs. 5 and 6),* ΔP *is related to the weight of the water above the coffee.*

(Continued)

(Continued)

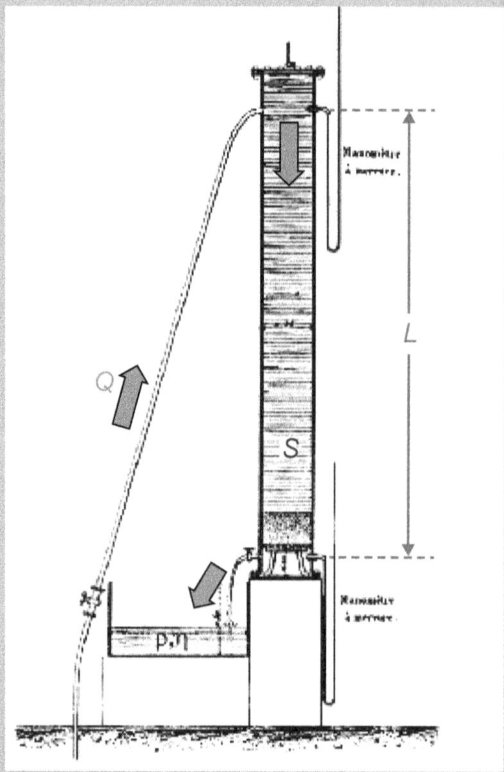

Device for measuring the permeability of a filtering material such as sand. A flow of water with flow rate Q passes through in a cylinder of section S and length L filled with sand. The pressure difference $\Delta P = \rho g \Delta h$, measured by two tubes (see Chapter 10) positioned at the top and bottom of the cylinder, is linked to the flow by Darcy's law.

Old-Fashioned Coffee Makers: "Napoletana", French "Filter Coffee", *et alia*

The ancestor of the Italian coffee maker moka is the Neapolitan one ("Napolitana", see Fig. 5) somewhat reminiscent of the latter, since it has two separable compartments and, between the two, the metal filter where the coffee is filled. The essential difference is in the "motor" which pushes the water through the filter: in a Napolitana it is simply gravity. As soon as the water begins to boil, the heat is turned off and the appliance turned over. The overpressure ΔP, defined as above

Figure 5. Neapolitan coffee maker. When the water in the bottom boils, the coffee maker is turned over, and the coffee passes through the filter by gravity.

(pressure difference between the two sides of the filter), is due to the weight of a water column of a few centimetres. It is therefore of the order of kPa.

Other coffee makers used in the first half of the 20th century (Fig. 6) also took advantage of the overpressure produced by the weight of the water column. However, they were not turned: hot water was poured directly over the coffee powder. An advantage of this device is that it allows one to achieve easy determination of the permeability κ, all other quantities in Darcy's formula (1) being known (see Panel on page 285).

In short, the principle of these coffee makers (which we will call *conventional*) is the same as that of the paper filter coffee maker. Only the material that constitutes the filter changes! However, the same grind cannot be used with both types of coffee makers: an earthenware or metal filter is perforated with holes much larger than the pores of the paper filter, which prohibits the use of too fine coffee powder. On the other hand, filtering through a paper filter is faster, because some

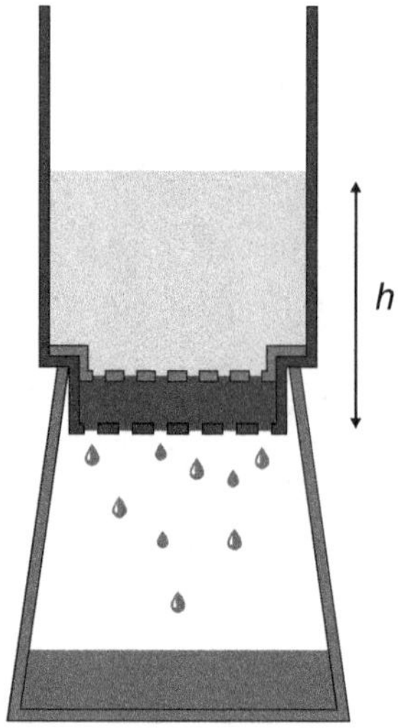

Figure 6. Principle of the "conventional" coffee maker. These coffee makers were often made of earthenware. The boiling water is poured into the upper part and passes through the coffee powder contained between the two perforated plates. The resulting coffee flows into the lower part.

of the hot water passes through the paper above the coffee or only passes through a small thickness of the coffee powder. Some *aficionados* claim that old-fashioned coffee tastes better than the one from an Italian coffee maker, because the slower filtering helps the aromas to recover, and the lower temperature prevents them from evaporating.

In the mid-20th century, the best coffee served in French bars was indeed a "filter coffee". The device was individual and included a cup surmounted by a receptacle comprising at its base a metal filter in two parts, which could be brought closer or removed from one another using a central screw by compressing thus more or less the coffee (Fig. 7). Insufficient squeezing produced tasteless juice. Overtightening produced nothing at all and was a fatal error. In fact, the pressure due to the weight of the water column in the container is relatively low: around 500 Pa, which quickly turns out to be insufficient for the water to overcome the surface tension forces and pass through the grind when the latter is too compact.

Finally, let us mention the "French press", which is a popular coffee brewing device consisting of a cylindrical glass with a lid and a piston with a fine wire mesh filter. The coffee powder is deposited into the glass which is then filled with water. Applying force to the piston in the French press, one pushes a growing, permeable layer of ground coffee through the hot water (Wadsworth, 2021).

Figure 7. The French press is a popular device for brewing coffee, comprising a cylindrical beaker fitted with a lid and plunger with a fine wire mesh filter. The plunger is used to drive the solid coffee particles to the bottom of the beaker, separating the hot liquid above.

Nowadays, being the busy people we are, we no longer use old fashioned coffee makers.

Experimental measurement of the permeability of ground coffee

It is known that the permeability of ground coffee for a geyser (nondrip) coffee maker is about 10^{-12} *m*^{2}*. How to measure it yourself for a general case?*

The most convenient way to experiment is with a "conventional" coffee maker (Fig. 6). The excess pressure ΔP *in this case directly depends on the height of the water level h:* $\Delta P = \rho g h$*, where g is the acceleration of gravity. To apply Darcy's law, all you need to do is measure the surface area of the filter, weigh the amount of water that will be poured into the coffee maker, and note the filtration time. A difficulty arises from the fact that h depends on time. The reader is invited to find a solution to this problem.*

You can even test the effect of viscosity η *on filtration time using water of different temperatures. At 100°C, the viscosity is 0.0003 Pa·s; at 30°C – 0.001 Pa·s. This is due to the fact that the viscosity of liquid decreases with increasing temperature since it determines the degree of rigidity of intermolecular bonds. On the contrary, in gases, the interaction of molecules is very small, and the viscosity at a given pressure increases with increasing temperature.*

Espresso

The impatience which is characteristic of the modern human was already common in the 19th century. Legend says that at this time, a subject of the Kingdom of the Two Sicilies, refusing to wait for his coffee *alla Napoletana* to be ready, succeeded in convincing one of his friends, an engineer in Milan, to invent a machine to make a good, aromatic coffee in less than a minute. The engineer accepted the challenge... and created the espresso machine. Several inventors have contributed to its improvement. One of the first espresso machines was presented at the Universal Exhibition in Paris in 1855 by Edouard Loysel de Santais; the result was a bit unstable. Around 1900, the Milan engineer Luigi Bezzera developed a commercial version, initially reserved for bars and restaurants; it then spread to individuals.

What is the principle of espresso? The passage of water through the ground coffee is done under high pressure (up to about 15 atm or even a little more) and at moderate temperature (88–92°C) so that certain aromas, unstable at high temperature, are not broken down here. The cup of coffee is obtained in about 30 s so that espresso is less rich in caffeine than conventional coffee. Also, an important distinguishing feature of espresso is the delicate brown foam. It consists of tiny gas bubbles trapped in a liquid film. In this way, the aromas that shape the coffee taste are bound and do not evaporate. In addition, the foam limits heat exchange with the surrounding air, and the drink cools more slowly.

In modern machines, water is brought to the necessary pressure by means of an electric pump (see Fig. 8). Previously, it was produced by a lever: in the raised position, the necessary amount of water was introduced and then the lever was

Figure 8. A family-use espresso machine. In modern models, the lever has been replaced by an electric compressor.

lowered to let the water penetrate the powder. The pressure was therefore exerted by the action of the arm, multiplied by the effect of the lever. The reader will easily verify that it is not necessary to be Goliath to gain the required pressure.

Variations on the Espresso Theme

With an espresso maker and a good coffee mix, it is possible to be inventive. For example, in Italian bars, you can drink: a concentrated "caffè ristretto" ("restricted" coffee) which is prepared with a standard quantity of coffee but with less water; a "caffè lungo" ("long" coffee) prepared with a normal quantity of coffee but with more water; a "caffè macchiato" ("dappled" coffee) which is espresso with a dash of milk; a "caffè corretto" ("corrected" coffee) which is espresso with a liquor, and 50–60 other variations on the espresso theme.

Special discussion requires the "cappuccino" which is espresso in a medium-sized cup to which is added milk "beaten" with the vapour in order to obtain a light, frothy foam. A good barista can pour the milk on the coffee so as to write the first letter of your name on the surface, while the Barista Championship winner can paint the scene from "Swan Lake" (see Fig. 9). Finally, one can simply add a bit of cacao to the foam.

Figure 9. The scene from "Swan Lake" painted at the surface of cappuccino by the winner of the World Barista Championship Pietro Vanelli.

They say that nowadays in Naples, a few bars still serve the "caffè prepagato" ("prepaid coffee"). A well-dressed gentleman accompanied by a lady enters a bar and orders three coffees, two for themselves, and one "prepaid". After a short time, a poor man enters the same bar and asks, "Is there a caffè prepagato?" And then the barista pours him a cup of free coffee.

Naples always remains Naples… and not only in the Toto movies.

Air humidity and degree of coffee beans grinding for espresso

Let us now address the matter of the "espresso" brewing not from a technological point of view but as an art, ultimately based on the laws of Physics. The final intermediary between the coffee machine and the consumer is the "barista". A lot depends on a barista's art and devotion to the profession. And it is not just because they regularly cleans the coffee machine of sediment, warms it up early in the morning and pours out the top 10 cups without offering them to the first customers. It is also necessary (but not enough) to take care of the cleanliness of the grinder, grind 7–9 g of a good mixture of coffee beans per serving immediately before brewing, and much more. Among the secrets of the mastery, the important element is the attentive observation of changes in the air temperature and humidity, with the related appropriate adjustment of the grinding degree of the grains. At first glance, their recipe for quality coffee making seems paradoxical: with increasing humidity or air temperature, coffee beans should be grounded more roughly. Let us try to understand the physical reasons for this advice. To do this, we rewrite Darcy's law (1) in explicit form for the filtration time ($\tau = V/Q$)

$$\tau = \frac{1}{\kappa}\frac{\eta VL}{S\Delta P} = \frac{1}{\kappa}\frac{\eta L^2}{\Delta P}, \tag{2}$$

i.e., relating the time of preparation of a cup of coffee τ *to the volume of the obtained beverage* (V)*, the viscosity of water* η*, the sizes of the filter* (S, L)*, and the difference of pressure on it* (ΔP)*. The optimum pressure and temperature for the espresso brewing are set in the coffee machine in advance, and the barista does not change them in the process of work. The viscosity of water also remains unchanged in Eq. (2): it depends only on the temperature, which we assume to be constant. The mass of the drink in the cup can vary depending on the customer's request — from 25 ml for the ristretto, up to twice as much.*

Namely by the filtration time, the barista judges the quality of the drink: for the espresso, this time should be 18–25 s, while it may be twice as much. If coffee

(*Continued*)

(*Continued*)

"falls" into the cup in a shorter time, then such an underexposed drink turns sour, with a light loose "foam". Overexposed coffee, on the contrary, has a dark-coloured cream and turns bitter.

As one can see from Eq. (2), the only parameter by which the barista can affect the brewing time τ is the porosity coefficient κ of the coffee powder in the filter. Varying it, the proper filtration time is achieved, despite the change in temperature and humidity of the ambient air.

Let us imagine that the coffee shop is located in the open air at the "South Pole Scientific Station" in Antarctica.[1] *At a temperature of –40°C to –50°C and in polar conditions, the vapour density in the ambient air (namely the absolute humidity) is negligible. The smallest particles of coffee (of the size r_1) formed during the grinding process in the coffee grinder are electrified (free electrons from the grinder's knives transfer to them).*[2] *The fact that the ground coffee is electrified is evident since this powder sticks, say, to a coffee grinder.*

The electrostatic forces that arise between them have the character of attraction and lead to the formation of effective agglomerates of these particles, whose sizes are noticeably larger $R_a \gg r_1$. In the subsequent filtration, it is these composite agglomerates that represent obstacles to water leaking through the filter: water becomes a coffee drink when washing them. Thus it is the size of the agglomerate particle R_a that determines the value of the porosity coefficient κ, and hence the brewing time of the coffee portion.[3] *Thus this selects the size of the grind so that the time τ would belong in the range of 18–25 s.*

---

[1] The Southernmost habitation on Earth.

[2] The reason for this process is simple, it is exactly the same as the electrification of hair when combing a plastic comb or synthetic clothing when it is worn. This is friction, in which a certain amount of charge is transferred from one body to another. As early as 1733, the French scientist Charles François de Cisternay du Fay (14 September 1698–16 July 1739), after carrying out numerous experiments, proved that all the types of electricity known at that time, that is, of electricity of different origins — the celestial (lightning), the animal (obtained from creatures, for example, from electric acne), are reduced to two types. These are "the glass one", obtained by rubbing glass on silk and "the resin one", formed when rubbing the resin on the wool. After the experiments of Benjamin Franklin, 15 years later, it became clear that this division corresponds to our today's positive ("glass") and negative ("resin") electric charges.

[3] Provided that the *barista* compacts the powder in the filter always with the same mechanical force.

(*Continued*)

(Continued)

We will now transfer our coffee shop to the equator, for example, to Singapore. Here, the air is saturated with water vapour, the relative humidity reaches 90%, the temperature is the same 40°C but here with the sign "+". As a result, absolute humidity increases hundreds of times in comparison with the icy desert of Antarctica. Therefore, if we choose the characteristic grinding size r_1 also here, the agglomerates that make up the coffee powder in the filter will turn out to be smaller than R_a. Indeed, the tropical air is saturated with moisture, and the electrostatic forces are substantially weakened.[4] Therefore, the coefficient of porosity of the coffee powder ground in a warm and humid environment will be substantially less than that in cold and dry air. The filtration time [Eq. (2)] will increase, and the coffee will be overexposed. To avoid this, a good barista changes the grind size of coffee beans almost every half hour, depending on changes in humidity and air temperature.

Another important factor is that water is not a bad electrical conductor, and so the charges are redistributed much faster in slightly damp coffee than in the dry one. Hence in humid air, the coffee beans should be grounded more roughly than in dry air.

[4] Recall the dielectric permittivity ε of the medium in the denominator of the Coulomb law. For water, $\varepsilon = 81$.

Table 1. Various methods of preparing coffee.

Method	Pressure	Temperature	Time
Classical method	A little more than 1 atm	A little less than 100°C	5–15 min
Turkish method	1 atm	100°C	5 min
Italian coffee maker	More than 1 atm	A little more than 100°C	1 min
Old Neapolitan coffee maker	A little more than 1 atm	100°C	5–15 min
Paper filter	A little more than 1 atm	Less than 100°C	3 min
Espresso	10 to 18 atm	88–92°C	15–25 s

Instant Coffee… and a Closing Word

Let us finish our review of preparation methods with instant coffee, a soluble powder that is simply poured into hot water. It is obtained by evaporating, at high temperature and low pressure, a coffee prepared with finely ground beans. The powder obtained is stored under vacuum, which prevents oxidation of the aromas and allows it to be kept for a long time. The drink is prepared by simply throwing the powder into hot water. The characteristics of the main methods of brewing coffee are summarised in Table 1. Everyone will choose their preferred method!

Let us leave the last word to a character born under the pen of Eduardo De Filippo, an Italian playwright of the 20th century: *Io, per esempio, a tutto rinuncierei tranne a questa tazzina di caffè, presa tranquillamente qua*. In English: *I, for example, would give up everything except this little cup of coffee that I take here quietly.*

Chapter 21

Science, Cooking and Liquid Nitrogen Ice Cream

Phenomena belonging to the realm of Physics are very frequently involved in cooking activities. But, do not forget the role of chemical transformations! The preparation of good dishes implies a variety of physical and chemical mechanisms, and their study can open a new perspective for the art of cooking. It can also help in understanding the empirical recipes that are transmitted generation by generation. Furthermore, it stimulates the elaboration of quite novel and original cooking dishes.

Cooking: Physical Modifications…

Culinary techniques consist in the combination of some ingredients and their transformation in order to obtain the food we like. What does transformation mean? Let us refer, for instance, to a sauce called mayonnaise, made of egg yolk, vinegar, salt, pepper and oil. From the point of view of physics, it consists of an emulsion of oil in water, stabilised by some compounds from the egg yolk. An emulsion is nothing other than a dispersion of small droplets (here, the droplets of oil) within another liquid (here, the water) which is not miscible with the former (Fig. 1).

In recipes for the mayonnaise, water never appears in the list of the ingredients: it is already present in the vinegar and the egg, possibly in other ingredients that the cooker adds in order to improve the taste according to his personal feeling.

The mechanism that makes the oil droplets stable (more precisely, thermodynamically metastable) is based on the presence of molecules (supplied by the egg) in the mayonnaise, which have hydrophilic and hydrophobic parts. The first forms hydrogen bonds with water molecules and is thus attracted to the water, and the

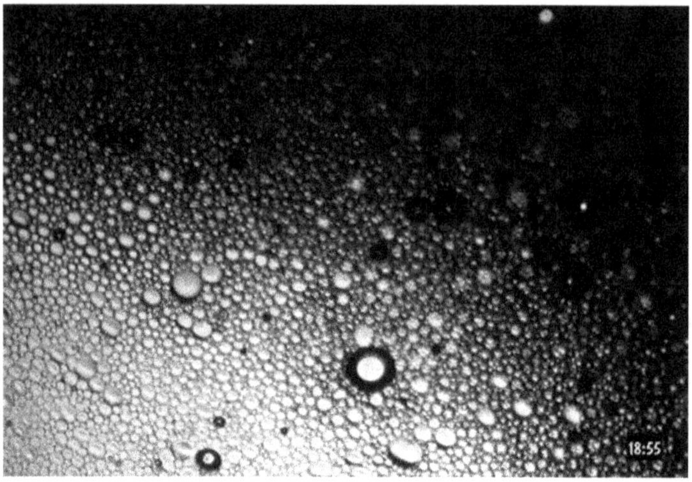

Figure 1. Micrometric view of the structure of a mayonnaise sauce at the beginning of its production when there is enough water (from the yolk and the vinegar) to accommodate oil droplets. The size of the oil droplets depends on the particular process and the oil quantity used: from 0.1 μm up to 0.1 mm.

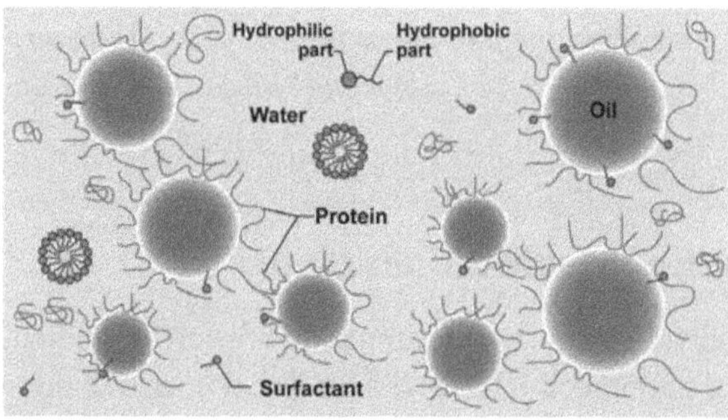

Figure 2. The mayonnaise is formed by oil droplets dispersed in water and stabilised by proteins and small surfactant molecules, the phospholipids.

second is repelled by the water molecules. Such molecules envelop the oil droplets, directing their hydrophobic part to it, and prevent them from mixing with water due to their hydrophilic part (Fig. 2).

In a well-made mayonnaise, the size of the oil droplets is rather small (in the range 0.1–100 μm). To obtain them, one has gently to beat, for a rather long time, the egg, while adding the oil drop by drop to allow an easy fractionation. By following this advice, it should not be difficult to achieve the success of a good mayonnaise. That procedure is basically physics; having promised to the reader the chemistry as well, now we move to it!

... and Chemical Reactions

What could possibly be the cuisine if food could not be cooked (see Chapter 16)? For our pleasure, the cooking processes change their consistency, colour, texture, odours and taste as well. It essentially involves a sequence of chemical reactions that change the ingredients of the recipe at the molecular level.

Let us describe what happens during the cooking of meat, already performed by the prehistoric man after the discovery of fire. The muscles of the meat are mainly formed by water and long organic molecules: the proteins (see Panel on page 247).

In raw meat, proteins form twisted chains. When cooked, these chains are denatured: they unfold and then join together, forming a gel. Above, we discussed the coagulation of egg white proteins during cooking (see Chapter 17). The denaturation of meat proteins begins at relatively low temperatures — from 55 to 80°C. However, cooking meat dishes takes time because the heat penetrates in meat, which is largely water based, slowly. In fact, every pot-au-feu lover knows that until cooked, the meat should boil in broth for several hours over very low heat; this time is required to separate the proteins into parts with the release of amino acids that give the dish flavour. This process tenderises the meat, disrupting the collagenic tissue and dissolving the proteins.

If the matter is to prepare a roast in the oven, then things are different. A good roast is pink on the inside while the external surface must take a brown crust. To obtain such a result, the cooking temperature must be above 100°C. In this case, chemical reactions occur, namely thermal decomposition of organic and inorganic compounds (pyrolysis), as well as reactions between proteins and carbohydrates, so-called the Maillard reactions (glycation reactions) (see Panel on page 300). These reactions occur at any temperature (for example, they are responsible for the opacification of the crystalline lens in people affected by diabetes), however they occur more quickly when the temperature is around 140°C, this implying that the surface of the roast is dry; in fact, the water, though not pure, can hardly go over 100°C. In order to grill the exterior of a roast without it drying out, the thermostat of the oven is set in the range 160–170°C (see Fig. 3).

A Novel Discipline Is Emerging

Following Antoine Lavoisier, Étienne François Geoffroy, Henri Braconnot, Michel Eugène Chevreul, Louis-Camille Maillard and others, step by step, researchers have revealed the mysteries of recipes that culinary experts have passed down for about 3,000 years without really asking themselves what physical–chemical

Figure 3. When cooking the meat, one causes the progressive denaturisation of the proteins with modifications in the colour and in the tissue. The exterior part of the roast acquires a temperature well above 100°C producing the pyrolysis reaction (decomposition of the molecules due to heat), as well as the Maillard reaction which is responsible for the development of the brown colour in the crust and also many other processes such as hexose dehydration, oxidations. Depending on the temperature reached inside the roast, the cooked meat is called "saignante (with blood)" — for red meat (the temperature inside is below 60°C), or "à point (medium rare)" — for pale pink meat (internal temperature about 70°C). At temperatures above 80°C, the cell walls of the muscles break, and the meat turns grey. (From H. This, "Elementary Treat of Cooking", 2002.)

Maillard's reactions

Maillard reactions are chemical reactions that provide colours, tastes and aroma to foods when they are cooked at high temperatures. They were discovered in the year 1912 by chemist Louis-Camille Maillard (1878–1936).

These reactions involve one protein and one sugar (see figure) that form a bond, yielding a novel compound. This compound immediately reacts with other compounds present in the foods in a complex way. Finally, through three different compounds, one obtains molecules with a pleasant aroma, very good taste and brown colour. In our roast, the proteins simply react with the glucose already present in the muscle tissue, which provides the energy required for its functioning.

The Maillard reaction plays a relevant role not only for the roast meat but also in the roasting process of the coffee grains (see Chapter 20), in fried cooking,

(*Continued*)

(*Continued*)

for the crust formation on the bread, and for the occurrence of various colours in beer!

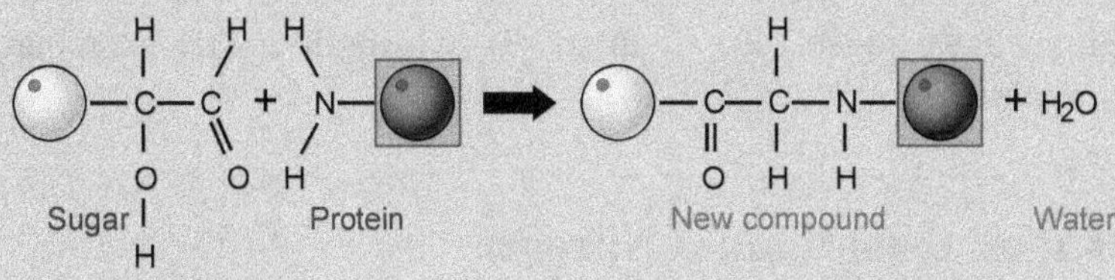

First step of the Maillard reaction involving one protein and one sugar molecule as the glucose. Only the atoms really involved in the reaction have been explicitly reported. The remaining part of the protein has been represented as a dark circle, while the light circle represents the remaining part of the sugar.

mechanisms were driving their art. Why not follow the inverse way, trying to find novel recipes by starting from our knowledge of physics and chemistry? In this way, you can revolutionise the foundations of culinary art by creating innovative dishes with new organoleptic properties. This approach has been taken by the so-called molecular cuisine. Its leaders are eminent scientists, in France, physical chemist Hervé This. He is the co-inventor of molecular and physical gastronomy (molecular gastronomy for short) alongside names such as the British physicists Nicholas Kurti and Peter Barham, the French physicist Jean Matricon, and Italian physicists Ugo and Beatrice Palma, and Davide Cassi. Molecular gastronomy is a scientific activity, but it is natural that science has found applications and inspired some distinguished cookers.

Among other innovations promoted by molecular gastronomy, we shall suggest how to prepare a "soft" egg by cooking it in alcohol: alcohol coagulates the egg proteins without altering the taste (while the taste of the ethanol is evidently added). One can also fry a fish in a mixture of molten sugars rather than in the oil (to avoid the taste of sugar, one can simply envelop the fish with a leaf of leek). The readers will find other amazing recipes in books dealing with molecular gastronomy or molecular cooking style, the art of cooking is now resorting to new techniques transferred from laboratories.

Besides the potentialities opened by the discoveries of new associations of tastes, molecular gastronomy yields some hope for people suffering from allergies or diabetes or alimentary intolerances. The ingredients that the doctors indicate as forbidden can simply be substituted by others, and still preserve the flavor! For example, the yolk of the egg is currently used to bind sauces since it contains proteins having emulsion properties. Unfortunately, it is also high in cholesterol. Why not replace it with soy lecithin, which has similar properties? And make chocolate mousse or mayonnaise without eggs in the same way! In addition, in case of gluten intolerance (see Chapter 19), wheat flour can be replaced with potato flour, which contains starch.

Ice Cream by Using Liquid Nitrogen

Finally, we address the way to produce ice cream in a particular way, as inspired by molecular gastronomy.

First, let us recall what an ice cream is. The cream is based on milk, sugar and various flavourings; sorbets are prepared starting from fruits, sugar and water. The soft structure is related to the presence of microcrystals of ice mixed with air bubbles in a supersaturated sucrose solution. Usually, the preparation of the ice cream requires the appropriate machine: the mixture is progressively cooled down, while the rotation of a proper device keeps it in motion. In this way, one avoids the growth of large ice crystals and as well of too compact a mixture. The preparation takes about 1 h and resorting to the usual machines does not always provide satisfactory results.

Figure 4. The production of ice cream by means of liquid nitrogen is made on demand. When in contact with the cold vapors of nitrogen, the water vapour is condensed thus causing the formation of white clouds, an appealing effect!

Recipe of the ice cream by liquid nitrogen

To get liquid nitrogen, one can simply ask one of the companies furnishing it in containers of 10 or 15 l, and it can last several days.

Warning: *Handling liquid nitrogen implies some risks analogous to the ones when dealing with boiling water: you will need gloves and protective goggles.*

Prepare the cream in a metallic container (one that can resist strong temperature differences). For instance, mix 100 g of sugar, 25 cl of milk, 25 cl of sour cream and add a flavouring, and possibly salt according to your personal taste.

With the eyes protected by the goggles, pour the liquid nitrogen on the cream, about a volume for two volumes of cream, and mix with a wooden spoon; then repeat the operation.

Serve and taste!

Let us leave the ice cream machine and return to the scientific approach. We can easily obtain an excellent ice cream in some tens of seconds by resorting to liquid nitrogen. The transition of the nitrogen from the liquid to the gas state occurs at about –196°C at ordinary pressure. When pouring liquid nitrogen into a workpiece (see Panel above on this page), the very rapid evaporation of nitrogen (which can be compared with the behaviour of water poured into smoking oil) simultaneously provokes both instant freezing of cream into microcrystals and the formation of nitrogen bubbles (Fig. 4). The bubbles inside ice cream does not cause any danger: in fact, this gas is present in 78 percent of the air we breathe.

Part 4

The Strange Quantum World

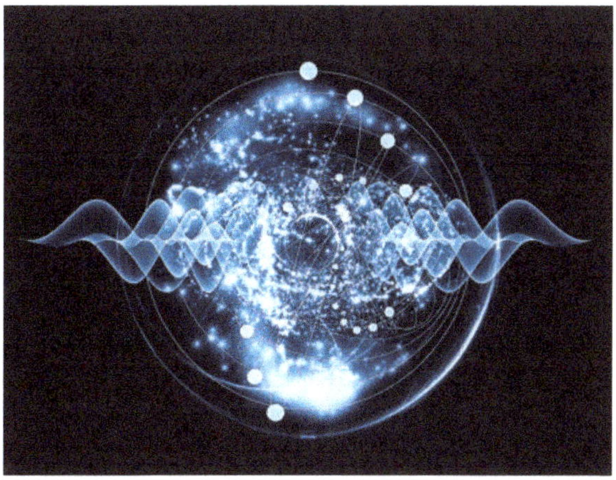

In the final part, dear reader, we will have a journey into an amazing world inaccessible to the uninitiated: a world usually open only to those who are not afraid of the most complex equations and the most unusual mathematical methods.

We ask you to be patient and trust us, as Dante did during the journey through Hell. Do not try to understand everything at once, because the secrets of the quantum world, like the nooks and crannies of Dante's Inferno, are countless and still await their researchers. Most likely, some of them are not fully understood by your guides either...

Chapter 22

Uncertainty, the Real Base of Quantum Physics

It is not possible to locate a particle and measure its velocity at the same time. This "uncertainty principle" is contrary to our intuition. It is the real basis of quantum physics that controls the behaviour of the world at the nanometric scale.

The year 1900, which marked the beginning of the 20th century, is also the date of the emergence of quantum mechanics. It was then that Max Planck found a solution to the problem posed by Gustav Kirchhoff four decades earlier (see Chapter 7). Planck's solution was based on the assumption that the energy of the physical system is quantised — that is, for example, if monochromatic light of frequency υ is confined in a mirror chamber, then its energy will necessarily be a multiple of one "quantum" of energy equal to $h\upsilon$, where h is Planck's constant. At first, this hypothesis seemed relatively innocent. However, 30 years later, it turned out that it challenges the deterministic understanding of physics.

The Uncertainty Principle

In the year 1927, German physicist Werner Heisenberg formulated the following principle, known as the uncertainty principle. Let us refer to a particle of mass m moving along the 0x axis at velocity v. If one can arrive to estimate its velocity with the precision Δv, then it is impossible to evaluate its position x with a precision Δx better than . $/m\Delta v$, where $\hbar = h/(2\pi) = 1.054 \times 10^{-34}$ J × s (joules times seconds).

In other words, $m\Delta x\Delta v \geq \hbar$. This property can be extended to a particle, defined by the three coordinates x, y and z, moving along the three spatial directions. Instead of dealing with the velocity v, the particle momentum $\boldsymbol{p} = m\boldsymbol{v}$ is more frequently used. The Heisenberg relation is then referred as

$$\Delta x \Delta p_x \geq \hbar \tag{1}$$

and the analogous relations for the other two components.

This inequality is indeed surprising. The Newton laws (see Chapter 4) in principle let one obtain the position and the velocity of a particle, for all the future time once the initial conditions are known. In Newtonian physics, known as classical mechanics, there is no room for uncertainty. However, this determinism, valid in the macroscopic world, no longer applies at the atomic scale. This inconsistency is what we are going to explain, by addressing an illustration of the Heisenberg principle.

Let us imagine sending some particles (say electrons or neutrons) towards a wall having a hole of diameter Δx (see Fig. 1).

Some particles can pass through the hole, and, when that occurs, their position in the plane of the wall is known with a precision Δx. Then the components of their velocity parallel to that plane have to be known with an uncertainty inversely proportional to Δx. Even in the case that the velocity of the particles before their arrival at the wall was strictly perpendicular to it, after the exit from the hole, the beam of particles is distributed within a cone. In other words, the particles are experiencing the same diffraction phenomenon that affects the light rays when passing through one hole (see Chapter 3).

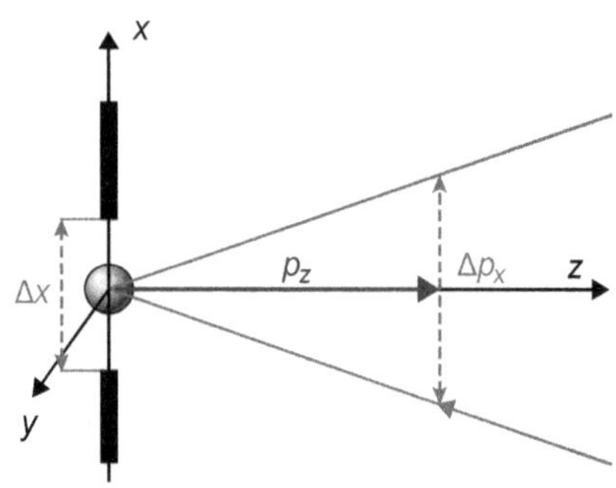

Figure 1. If a particle is passing through a hole or a window having size Δx, its position along the x direction is evidently known with the precision Δx. In light of the Heisenberg inequality, the momentum along this direction cannot be estimated with a precision better than Δp_x. If the particle belongs to a beam parallel to the direction z of the momentum p_z, the passage through the window causes a spread of the beam by an angle $\Delta p_x/p_z$.

Uncertainty and Measurement

According to Heisenberg (Fig. 2(a)), the quantum uncertainty is related to the interference of the particles being detected and the experimental apparatus used to measure them. Let us address this issue.

Let us assume that we wish to study the motion of one electron. How to do it? Our eye is evidently inappropriate, having too poor a resolution. What happens if we use a microscope? The resolution of the microscope is related to the wavelength of the radiation used in the observation. For the visible light, it is of the order of 100 nm, and the particles having a size less than that value are not observable. Thus the atoms are not observable since their size is of the order of 0.1 nm; the electrons are also not observable.

We can imagine using a microscope that uses a radiation of very short wavelength, as, for instance, the X-rays or the γ rays, having a wavelength of the order of 0.01 nm. Do we have a device that can measure the position and the velocity of the electron with the desired precision?

We shall analyse more closely our imaginary experiment. To detect the position of the particle, we have to use at least one quantum of the electromagnetic radiation. The energy E of this quantum is hc/λ (c being the light speed in the vacuum). The energy transported by the quantum is considerable when the wavelength λ is small. The momentum of the quantum is proportional to this energy,

Figure 2. Werner Heisenberg (1901–1976) (left) and Niels Bohr (1885–1962) (right). These two theorists of quantum indeterminism were good friends until the Second World War. In 1941, Heisenberg visited Bohr, who escaped to the United States shortly soon after. British writer Michael Frayn imagined the interview in his famous play "Copenhagen", performed in London in 1998.

and, when colliding with the electron part of this momentum, is evidently transferred to the particle. Therefore, using our estimate of the position of the electron by means of X-rays or γ rays, it turns out that the momentum of the electron *a fortiori* is uncertain. An exact analysis of the collision process points out that the product of the two uncertainties cannot be smaller than the Planck constant, thus again returning to the Heisenberg uncertainty principle.

One could think that our description is valid only in that particular case, or that the method devised for the measurement is not suitable. It is not so. The most distinguished scientists (in particular Albert Einstein, as we shall address) have tried to imagine some experiment that *in principle* could measure the position and the momentum of the electron with a precision better than the one expected according to the Heisenberg principle. All the attempts failed. The uncertainty principle is a deep law of the nature of the microscopic world. So deep a law that one cannot think that the uncertainty is anyway related to the perturbation of the measurement: several experiments prove that the law is intrinsic and that the Heisenberg relation is valid also in the absence of any perturbation that the measurement could possibly imply.

Deterministic World and Quantum World

The uncertainty principle is evidently in contrast with our intuition. Does the Heisenberg relation contradict the basic aspects of determinism? For one object of mass m, one would have $\Delta x \Delta v \geq \hbar/m$. If we apply this condition to a ball of typical mass of around 1 kg, the limit of that product is about 10^{-34} m^2 s^{-1}, namely practically zero. In the case that the position of the ball is known with the precision Δx around 10^{-10} m (namely about the size of an atom!), then the uncertainty in the velocity turns out to be extremely small, say around 0.03 nm h^{-1}. Therefore, we can conclude that *in practice* the macroscopic world is deterministic, in agreement with our intuition.

For what sizes do quantum effects have to be taken into account? Let us refer to the motion of very small material particles in a liquid: Brownian motion (see Panel on page 311).

Let us consider a particle of Brownian character with a mass of the order of 10^{-13} kg and a diameter of around 1 micron. The uncertainty relation implies that $\Delta x \Delta v_x$ has to be larger than $\hbar/m$, namely about 10^{-21} m^2 s^{-1}. If we wish to evaluate the position by an accuracy of 1% of the size, then the uncertainty Δv_x cannot be less than 10^{-13} m s^{-1} which is still very small. In fact, the speed of a Brownian particle is more than a million times larger, say of the order of 10^{-6} m s^{-1}. Therefore, one can conclude that also particles as small as Brownian ones are correctly described by classical mechanics. Thus the uncertainty relation becomes essential

Brownian motion

A particle immersed in a liquid experiences a chaotic motion.

When small particles are in suspension in a liquid, they are experiencing an irregular motion: Brownian motion. This is a phenomenon observed in the year 1827 by Scottish botanist Robert Brown. When watching pollen grains by means of a microscope, he noticed that for the smallest particles (diameter about 1 or 2 microns), a random motion was occurring (see figure). This motion is due to collisions of the particles and the molecules of the liquid. Thus Brownian motion is a kind of message sent from molecules to human beings in the 19th century: "you cannot see us but we are there!". The message was decoded in the 19th century by French physicists and in particular by Louis Georges Gouy (1854–1926). From him, nowadays, one knows that the motion of the molecules is strictly related to the temperature: the greater the motion, the higher the temperature. In fact, temperature is a measure of the kinetic energy of the molecules. In Brownian motion, a part of that energy is transferred to the small particles.

only for particles significantly smaller than a Brownian particle. So, it becomes extremely important for the electron. It is so important that, as will be shown below, it makes it possible to estimate the size of an atom.

From the Uncertainty Principle to the Atomic Radius

Let us refer to the simplest atom, hydrogen, namely just one proton and one electron. The first essentially correct description of this atom was given by New Zeland physicist Ernest Rutherford (1871–1937). In that description, the electron of charge –e and the proton of opposite charge[1] are kept bonded by the electrostatic interaction: the electron rotates around the proton similar to the

[1] The elementary charge is $1.6 \cdot 10^{-19}$ C (coulombs).

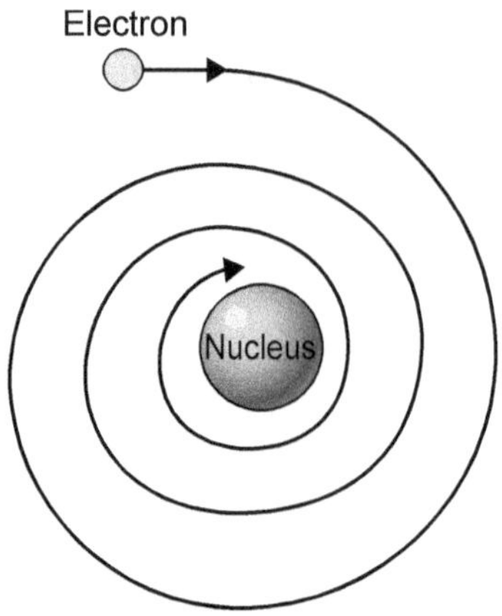

Figure 3. According to classical mechanics, the Rutherford atom is unstable: the electron should fall into the nucleus.

rotation of the Earth around the Sun. Note that such a description implies a problem: according to classical electromagnetism addressed by Maxwell in the 19th century, an electric charge in circular motion emits electromagnetic radiation. Consequently, the energy of the electron must decrease, and thus it must fall into the nucleus (Fig. 3). Instead, it does not fall! To explain this fact, it was necessary to introduce a new physical principle that would go beyond the framework of Newtonian physics. It was Heisenberg's uncertainty principle.

According to that principle, the poor electron must be affected by a kind of continuous oscillation within a certain room, with a speed badly determined, but certainly non-zero. From this consideration, we are going to derive an order of magnitude of the atomic size! Let v and $2R$ be the order of magnitudes of the velocity and of the diameter of the sphere within which the electron has to "oscillate". The uncertainty equation implies an order of magnitude $2mRv > \hbar$. Therefore, the kinetic energy of the electron, given by $mv^2/2$ must be greater than $\hbar^2/(8mR^2)$. By adding the electrostatic energy due to the electron–proton interaction, one finds that the total energy W of the electron is given by

$$W \geq \frac{-e^2}{4\pi\varepsilon_0 R} + \frac{\hbar^2}{8mR^2} \tag{2}$$

ε_0 being the vacuum permittivity, a constant equal to 8.85×10^{-12} F m^{-1}. The energy of the atom cannot be smaller than the minimum of Eq. (2) which occurs for the value $R = R_0$ where

$$R_0 = \frac{\pi\varepsilon_0\hbar^2}{me^2} \approx 0.013 \text{ nm}. \tag{3}$$

The equilibrium state of a mechanical system corresponds to the minimum of the potential energy (see Chapter 11). In the fundamental state, having the smallest energy, the radius of the atom cannot be much greater than R_0, otherwise the potential energy of the electron would be too great. On the other hand, it cannot be smaller than R_0, otherwise the kinetic energy would be too great. This is why the electron cannot fall into the nucleus! Equation (3) gives an idea of the size of the hydrogen atom, which is of the order of the angstrom (namely a factor of 10 smaller than 1 nanometre).

The Emission Spectra of Atoms: The Key to Atomic Structure

Being in its ground state (minimum of total energy), an atom cannot lose energy. However, an atom can receive energy, passing at the same time into one or another "excited" state. Yet, it does not remain excited for an infinitely long time — after a while, emitting light, an atom returns to its ground state. This light corresponds to the emission of precisely defined frequencies, that is, the emission spectrum of the atom presents itself the "spectral line series" (see Chapter 7).

These frequencies form a so-called discrete set, that is, they can be numbered. To explain the origin of such a line spectrum, it is reasonable to assume that the values that the energy of a given atom can take also constitute a discrete set. Since light can be emitted only in the form of photons (see Chapter 7), the energy conservation law requires that the energy $h\upsilon$ of each photon be equal to the difference between two allowed values of the energy of the atom (Fig. 4).

Thus, the discrete form of the radiation spectrum is explained, at least qualitatively. It remains to find out why the values of the energy of the atom constitute a discrete set.

At the beginning of the 20th century, the question of the nature of the atom — the smallest particle of a substance that is the carrier of its properties — was one of the central issues in physics. The proposed models, being internally contradictory or inconsistent with the experiment, were refuted one after another. And so, in 1913, Danish physicist Niels Bohr (Fig. 2, right) proposed a mathematically simple theory of the atom, explaining the existing experimental data, but based on such unusual assumptions, which he himself called postulates.

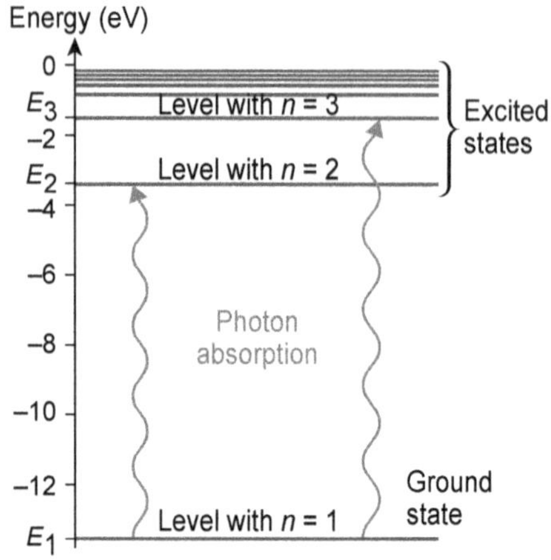

Figure 4. Energy diagram for the hydrogen atom. The atom transits from the ground to an excited state by absorbing the photon of energy $\Delta E = h\upsilon$ corresponding to the energy difference between two levels of an atomic state. The energy here is expressed in electron-volt (1 eV = 1.6×10^{-19} J).

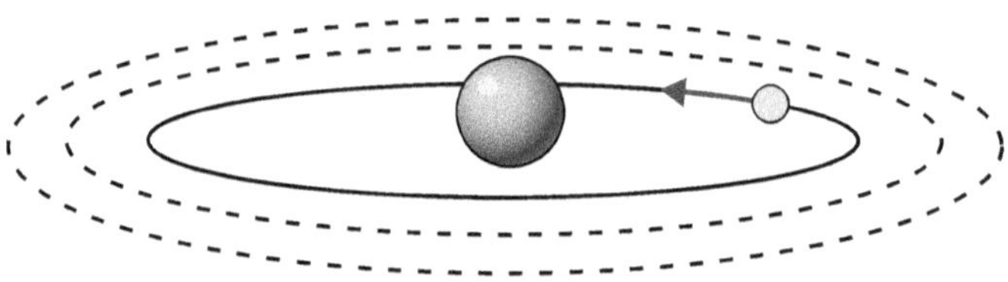

Figure 5. The hydrogen atom as imagined by Rutherford and Bohr at the beginning of the 20th century.

The Atom According to Niels Bohr

The atom imagined by Niels Bohr did not take into account the uncertainty principle. As in the Rutherford model, the electron rotates around the nucleus, as the Earth rotates around the Sun: however, it can rotate only around particular orbits (Fig. 5). For example, the circular orbits are allowed only when the product of the momentum mv times the radius R of the orbit is an integer multiple of the Planck constant

$$mvR = n\hbar. \tag{4}$$

The momentum and the radius of the orbits are related since the centrifugal force (see Chapter 4) mv^2/R must compensate for the electrostatic attraction. In the case of the hydrogen atom, where the nucleus is just a proton, the attraction force

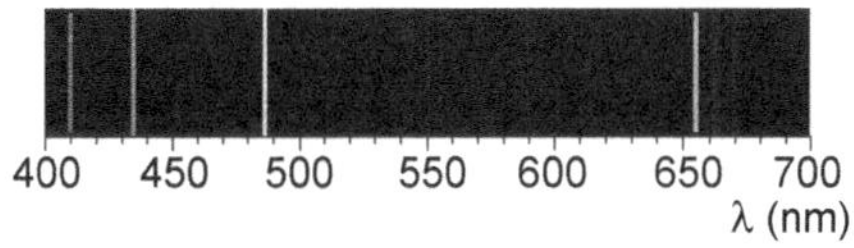

Figure 6. The Bohr model provides an explanation of the emission spectrum of the hydrogen atom in the range of visible light. The spectral lines at 410, 434, 486 and 656 nm correspond to transitions from the excited states $n = 6, 5, 4$ and 3 towards the state $n = 2$ (see Figure 4).

is given by $e^2/(4\pi\varepsilon_0 R^2)$. Thus one can obtain the values of the radius R_{n-1} of the orbits that are allowed as a function of the number n. For $n = 1$, the corresponding value exceeds that one calculated by formula (3) by four times, and it corresponds to the ground state. The reader can easily derive by themselves the general formula corresponding to excited states.

The Bohr model, which goes back to the year 1913, explains rather well the main characteristics of the emission spectra of atoms (Fig. 6); however, its limits were soon realised. It did not allow any explanation of phenomena that later on were discovered, for instance, the experiment performed by Davisson and Germer (1927) that pointed out that electrons are experiencing the same diffraction and interference phenomena of light (Chapter 3). Those experiments are totally incompatible with a deterministic physics. The value R_0 for the distance of the electron from the nucleus has to be an average value: the uncertainty principle does not allow a precise value for the distance nucleus–electron. The Bohr theory was completed about 10 years later by the introduction of a revolutionary concept: the probability of the presence of an electron.

The Probability of Presence

Let us imagine that we have been able to locate the position of the electron at a given time. Could the position a second later be predicted? No, since the measurement of the position of the electron necessarily induced an uncertainty in the velocity. No theory and no experiment can predict what is going to happen to the electron. So, what to do?

Let us change the strategy and put a mark on the place where the electron has been found; then another mark in correspondence to a second measurement of the position; then let us repeat many times the same operation. Although it is not possible to predict where the electron will be after a given measurement, still we will discover that the distribution of the marks indicating the various positions does follow a precise rule. The local density of the marking points, which is a function of the position in the space, indicates the probability to find the electron in correspondence to a given measurement. We have to give up the idea to describe the motion of the electron, but still we can provide the probability to find it in any given position. In the nano-world, the behaviour of the electron is characterised by a probability!

A reader not familiar with this concept cannot appreciate the role of chance in the laws of nature. Einstein, despite the fact that he was at the origins of quantum mechanics (Fig. 7), was shocked by the proposed concept of quantum indeterminism. A convinced determinist, he once told Niels Bohr: "God does not play dice."[2] Nevertheless, as you will see, this probabilistic theory is supported by strong experimental evidence.

Therefore, in the nano-world, the state of an electron is defined by a probabilistic laws. The marks we have placed in correspondence to the various positions form a kind of cloud, as the water droplets in the sky form clouds of variable density. This "electronic cloud" represents a description of the electron better than the small planets in orbits around the nucleus as was previously described by Rutherford.

Figure 7. The famous Solway meeting in 1927 that gathered nearly all of the builders of the quantum mechanics. Seventeen of the attendants would win Nobel prize! From left to right in the first line: I. Langmuir, M. Planck, M. Curie, H. A. Lorentz, A. Einstein, P. Langevin, E. Guye, C. T. R. Wilson and O. W. Richardson. In the second line: P. Debye, M. Knudsen, W. I. Bragg, H. A. Kramers, P. Dirac, H. Compton, I. De Broglie, M. Born and N. Bohr. In the third line: A. Piccard, E. Henriot, P. Ehrenfest, E. Herzen T. de Donder, E. Schrödinger, E. Verschaffelt, W. Pauli, W. Heisenberg, R. H. Fowler and L. Brillouin.

[2] Bohr answered: "Einstein, stop telling God what to do".

De Broglie Wave and the Schrödinger Equation

What is controlling the structure of the probability clouds? Does there exist an equation that similar to the Newton equations in classical mechanics (see Chapter 4) describes quantum mechanics? Yes, this equation does exist! It was devised by Austrian physicist Erwin Schrödinger (1887–1961) in the year 1925, and it represents the basis for atomic physics and theoretical chemistry.

Schrödinger's theory generalised the revolutionary idea proposed a year earlier by French physicist Louis de Broglie (1892–1987), which was that a wave of length $\lambda = h/p$ can be associated with any particle with momentum p.

Thus, any particle can exhibit a wave-like or a corpuscular-like behaviour, depending on the particular situation, just as well as light (see Chapter 7). As for the electromagnetic wave theory developed by James C. Maxwell (1831–1879) where an electric field $E(x, y, z, t)$ (which is a function of the three spatial coordinates and of the time) is introduced, the Schrödinger equation describes the state of a particle by means of a "wave function" $\psi(x, y, z, t)$: its modulus square gives the probability of the presence of the particle in a given point at the time t. This rule is inspired by an analogy with optics, where the modulus squared of the electric field yields the probability to find the photon at a given point. The difference is that the electric field is revealed by other effects, as, for instance, the force on a charged object, while the wave function devised by de Broglie does not have a clear physical meaning.

Using the Schrödinger equation, it turned out to be possible to find the spatial distribution of the probability density of the electron for its possible states in the hydrogen atom. By plotting these probability density distributions on a plane in different colours, one gets an image of various atomic orbitals (regions in which the probability of finding an electron is highest). Such images replace the electron orbits of Bohr's model of the atom (Fig. 5) and graphically represent the behaviour of electrons in an atom. Calculations based on the Schrödinger equation explain the existence of discrete energy levels, which are the reason for the line spectra observed during the emission and absorption of light. Similar but more complex calculations allow us to understand how chemical bonds are formed between atoms. Note that the works of de Broglie and Schrödinger preceded Heisenberg's discovery of the uncertainty principle. The latter is simple, concise and elegant but contains less information than the Schrödinger equation.

The experiment by Davisson and Germer

The concept of the relationship between waves and particles, the so-called wave–particle dualism, proposed by de Broglie, led to the idea of using optical research methods with particle streams replacing light. So, in 1927, American physicists Clinton Davisson and Lester Germer bombarded a nickel crystal with electrons. As a result, they obtained diffractograms similar to those arising when crystals are irradiated with X-rays (see Panel below on this page). To interpret the obtained diffraction patterns, electrons had to be assigned a certain wavelength, and it coincided with the value predicted by de Broglie. Thus, the experiment brilliantly confirmed his hypothesis.

Studying matter by means of diffraction experiments

The reactor at the Laue–Langevin Institute in Grenoble. The neutrons are produced by nuclear reactions and are used for spectrometric studies of condensed matter. The container of heavy water is immersed in a pool that can absorb the radiation. The blue light is related to the Cerenkov effect. The reactor is controlled by means of special bars absorbing the neutrons, and that can be extracted in relation to the amount of uranium still present.

Electron diffraction is rarely used to study crystals because electrons are absorbed by matter much more strongly than X-rays (see Chapter 9). Of much greater interest is another, more elementary particle — the neutron! When it comes to observing light atoms or studying atomic magnetic properties, neutron diffraction is preferable to X-rays. The latter makes it possible to draw up maps of electron density, while polarised neutrons make it possible to investigate not all, but only electrons located on the outer shells of an atom — precisely those that determine its chemical and magnetic properties. The disadvantage of this method is that expensive and bulky nuclear reactors are required to produce neutrons (see figure), while an X-ray facility is easy to equip even a modest laboratory.

Zero-Point Motion of Atoms

The uncertainty principle yields interesting information on the motions of atoms in solids. By the word "solids", we mean crystals (see Chapter 9) since the crystalline form is the one stable for most bodies at low temperatures. The atoms are not fixed in the crystal: they oscillate around their equilibrium positions. The motions have small amplitude, and the distance between two neighbouring atoms remains the same as the one corresponding to their average positions, namely of the order of a few angstrom. In general, these oscillations are due to temperature. An increase in temperature implies an increase in the oscillation amplitudes. What happens when the temperature approaches 0 K, namely minus 273.15°C? One could think that the oscillations stop and the atoms become fixed. Then their positions would be known, and their velocity would be zero which would violate the Heisenberg uncertainty principle (Eq. (1)). Therefore, the atomic motions cannot stop even at zero temperature, and a zero-point motion remains.

Let us refer to the case of a simple monoatomic system, for instance, to the crystal of equivalent hydrogen atoms or oxygen or iron. We will achieve a simplified estimate, still qualitatively suited, for the motion of one atom with respect to its neighbours by assuming that this reference atom is attracted towards the equilibrium position by a force proportional to the distance as if it was held in place by an elastic force. The motion of the atom from its equilibrium position along the O*x* axis is described by the equation

$$x(t) = x_0 \cos(\omega t - \alpha),$$

where x_0 indicates the amplitude of the oscillation (analogous equation holds for the other coordinates). Along this axis, the velocity of the atom is

$$v(t) = \omega x_0 \, sin(\omega t - \alpha).$$

The Heisenberg relation implies $\Delta x \Delta v_x \gtrsim \hbar/m$ and therefore ωx_0^2 must be at least of the order of $\hbar/m$, m being the mass of the atom in consideration. On the other hand, ω is of the same order of magnitude for most elements, say around 10^{13}–10^{14} Hz (this typical value for the vibrational frequency in solids is known as the Debye frequency). By substituting the mass of the atom with Am_n, A being the atomic number and m_n the average mass of the nuclei (about 1.67×10^{-27} kg), one deduces that x_0 should be of the order of $10^{-11} m / \sqrt{A}$. This condition implies an upper limit on the amplitude of the zero-point motion, which, in general, is rather small in comparison to the average distance between atoms, which is around 0.1 nm or more. Therefore, one should not think that the zero-point motion can affect the stability of the crystal. A doubt can occur for a small atom such as hydrogen or helium (A = 1 and A = 4). In reality, the helium atom represents an exception: the zero-point motion makes the solid unstable at all temperatures, provided the pressure is <2.5 MPa. All other solids, including hydrogen (H_2), can indeed become solids when the temperature goes towards zero, regardless of pressure.

Quantisation of the magnetic moment

We have already seen that, according to quantum mechanics, at no time is it possible to establish the exact values of the position $\boldsymbol{r}$ and velocity $\boldsymbol{v}$ of an electron rotating around a nucleus. The properties of its magnetic moment are even more unusual.

The magnetic moment is a vector quantity that characterises the property of an object to orient in a magnetic field. For instance, the needle of a compass is reorienting in the terrestrial magnetic field and indicates the Northern direction. In the group of elementary particles and the objects belonging to the atomic scale, a large number of them possess a magnetic moment: the electron, the proton, the neutron and many nuclei of atoms or molecules.

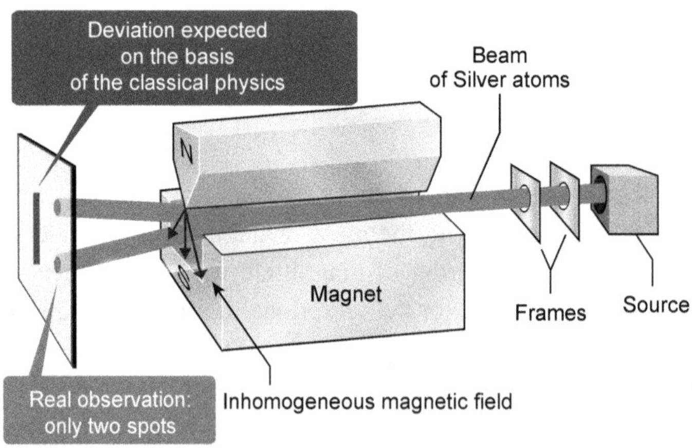

Figure 8. Principle of the experiment performed by Stern and Gerlach. Silver atoms go through an inhomogeneous magnetic field, vertically oriented. Classical physics would predict that the beam is simply homogeneously spread. Instead, it is observed that the beam splits up into only two components.

The components of magnetic moments in space are usually indicated by μ_x, μ_y and μ_z. When the compass needle is oriented in a certain direction, then all three components of its magnetic moment are clearly defined. Unlike a compass, an electron or a neutron are objects belonging to the quantum world. For them, only one of the three components of the magnetic moment can be measured, and it can only take on two opposite values: $-\mu$ or $+\mu$. This seemingly paradoxical statement was confirmed experimentally: as early as 1922, Otto Stern and Walter Gerlach were the first to obtain experimental data in favour of quantising the magnetic moment of representatives of the quantum world. In their experiments, they directed a beam of silver atoms (which, thanks to the electrons of the outer shell, have a magnetic moment) through an inhomogeneous magnetic field. As a result, it was found that this beam is split strictly in half, which proves the quantisation of the magnetic moment into only two discrete values (Fig. 8). Indeed, if the magnetic moment could take at least three values, then the beam would be divided by three, and if the magnetic moment of silver atoms could change continuously, then the beam would simply disperse into a cone.

A few more comments about the silver beam. By labelling the direction of the magnetic field as *x*, then one concludes that there is a state of the magnetic moment having $\mu_x = -\mu$ and a second one with $\mu_x = +\mu$. If the particle is in the state with $\mu_y = \mu$, the magnetic moment in the *x*-direction will turn out $\pm\mu$ with the same probability, and thus the average value of a large number of measurements will be zero. This is the same as the average measurements that one could perform with the magnetic moment in the state $\mu_y = -\mu$. To realise the meaning of this property in the quantum mechanical framework, it is stated that the state $\mu_y = \mu$ in reality is a mixture of the state having $\mu_x = +\mu$ and $\mu_x = -\mu$.

Schrödinger's Cat

The "mixture of states" concept properly describes the situation occurring at the atomic scale. It is amusing to imagine its extension to the macroscopic world. Similar to the magnetic moment that can take two states, Schrödinger pointed out that in the framework of quantum mechanics one cat could simultaneously take two states: alive and dead (Fig. 9). The extension of quantum mechanics to this macroscopic scale implies that a state can exist which is a sort of mixture of an alive cat and a dead cat. When we open the box, there is the same probability (namely 50/50) to find the cat dead or alive, but before the box is opened, the cat is in a superposition of state dead and state alive (see Panel on page 322).

"This is absurd!", the reader may decide. In fact, the cat is either alive or dead, and before the door is opened, it does not matter if we know its condition. This understanding of the situation under consideration is based on a different, once

Figure 9. In the framework of the hypothetical experiment, Schrödinger imagined that a cat was kept inside a hermetic box. A device based on random disintegration of one radioactive atom could open a bottle containing a poison gas. After a certain time, the probability that the atom is disintegrated and then the bottle is releasing the poison is ½. Until the experimenter opens the box, quantum mechanics states that the atom is simultaneously disintegrated and undecayed, and therefore that the cat is simultaneously alive and dead.

Schrödinger's cats of today

Doubts about the determinism of physics of the microworld began to appear since 1924 when Louis de Broglie proposed the idea of wave–particle dualism, and, three years later, Clinton Davisson and Lester Germer proved it empirically. As a result of these discoveries, Niels Bohr and Werner Heisenberg came to conclusions that subvert the classical concepts of determinism when applied to the quantum world, and Erwin Schrödinger came up with a joke about a cat alive and dead at the same time. Schrödinger's cat turned 80 in 2020! However, as he grows old, he becomes more and more alive. More recently, thanks to the efforts of scientists, it materialised from the field of abstract reasoning and became a reality. Of course, this is not a real cat, but a tiny object, which is only jokingly called "Schrödinger's cat". This name today means any relatively macroscopic object brought into a state of quantum superposition. This kitten (which is just a few atoms) indirectly became one of the 2012 Nobel Prize winners in physics awarded to Serge Arosh and David Wineland.

existing, interpretation of quantum mechanics, based on the concept of a hidden parameter. According to this concept, the description of the world is deterministic, but some parameters required for its implementation are not available to us. Modern science refutes this concept. Yet, the indeterminism of quantum physics gives rise to paradoxes that are intuitively difficult to accept. Let us describe one of them.

Einstein, Podolski and Rosen (EPR) Paradox

In the year 1935, Einstein and his collaborators Boris Podolski and Nathan Rosen addressed one paradox that gave origin to several scientific studies, some of them even rather recently. The EPR paradox involved a situation that presently we indicate as "entanglement". This approach no longer involves a single object, the Schrödinger cat, but two objects. Let us imagine taking a cat and a dog, even though this was not the case considered by Einstein. We assume that one of the two is dead, without knowing which. The state where the cat is alive and the dog is dead will be indicated as $|+ -\rangle$, while the state with the cat dead and the dog alive will be $|- +\rangle$. When the two states are mixed, one says that "entanglement" occurs. The entangled state will be represented by the symbol

$$\left(|+-\rangle + |-+\rangle \right)/\sqrt{2}.$$

Until the cat and the dog are kept in two well-separated boxes, one does not know if the cat is alive or if the dog is alive, the other being dead. But if by opening the box of the cat we find it dead, we know that in the other box the dog is alive and vice versa, if the cat is found alive, we know that the dog is dead. The two observations are correlated. If this correlation is preserved even when the two boxes are brought 1,000 mi apart, in order to know what is happening to the dog, it is necessary to open the box of the cat. Thus one gets instantaneous information while we know that no signal can be transmitted faster than light! We could also say that by opening the box where we find the cat alive, instantaneously the death of the dog a long distance away is caused, while before it was just semi dead. Certainly, there is no way to predict the result of the box opening since we could find with the same probability, ½, the cat is alive or dead and the same for the dog. Thus, we can understand while Einstein, Podolski and Rosen were perplexed. At the end of the article, they suggested efforts for the formulation of a novel quantum mechanics that could overcome the problems they had addressed. That novel theory could be based on the presence of hidden parameters, non-accessible to the measurements and not yet included in the theory by Schrödinger and Heisenberg.

Bell Inequalities and the Experiments by Aspect

The EPR paradox was disputed by many distinguished researchers, including Bohr. Other eminent scientists, including Louis de Broglie and David Bohm, like Einstein, would have preferred to restore determinism. The discussion lasted a long time and bore a philosophical connotation. In 1964, John Bell was able to make it more concrete and showed that deterministic physics, even with latent variables, must include some measurable inequalities that contradict the usual form of quantum mechanics.

The Bell inequalities have been tested by Alan Aspect and his collaborators in Paris in 1982. A situation analogous to the one addressed in the previous section was explored. The objects they used were not dogs or cats since quantum mechanics does not imply the application to macroscopic objects. The objects they used were photons. Their polarisation (namely the plane of vibration of the electric field) can take two perpendicular directions, similar to the two states for the cat and the dog (dead and alive). The study involved the correlation among the polarisation of the photons. This is just one of the several difficulties of the experiment, another relevant one being that the photons propagate very fast, and therefore there are many things to do before they lose their correlation. Summarising the experiments by Aspect proved that the Bell inequalities were not satisfied. Thus the quantum mechanics as it is presented in a variety of books is basically correct and cannot be implemented by any theory including hidden variables. The experiment by Aspect transformed the *Gedankenexperiment* of Einstein, Podolski and Rosen into a real experiment, including the possibility of quantitative evaluation.

Thus, the EPR paradox is real: quantum physics is paradoxical. The microscopic world is paradoxical.

Chapter 23

Physics, Geometry and Beauty

In the previous chapter, we paid attention to the absence of deterministic laws implied by quantum mechanics, and we emphasised how our vision of the world is affected by that. Let us now disregard any philosophical trouble and simply observe Nature as it is. How does one not wonder when seeing these special consequences? Our emotional feeling was already triggered by the beautiful symmetries of the crystals (Chapter 9) or by the variety of the Chladni patterns (Chapter 11).

Scientists are not insensible to the beauty of Nature. Louis de Broglie spoke of "the mysterious beauty that the electric flash displays" (in a speech in honour of Jean Perrin, 1962). Heisenberg wrote a paper about "The relevancy of the beauty in the exact sciences" (Die Bedentung des Schonen in der exakten Naturwissenscaft) by reporting the following: "The internal relationships [of the atomic quantum theory] in their mathematical abstraction display an incredible degree of simplicity and beauty, that we can only accept with humility. Not even Plato could imagine such beauty. It cannot be invented, it existed after the creation of the world." Einstein pointed out the following: "the simplicity and the beauty [of Nature] for me is a real aesthetic question... I greatly admire the mathematical models that are offered by Nature".

By following those words of these great physicists, let us admire some artistic realisations that Nature provides at the microscopic scale. A major role in this area is played by a particularly abundant element: carbon (Fig. 1).

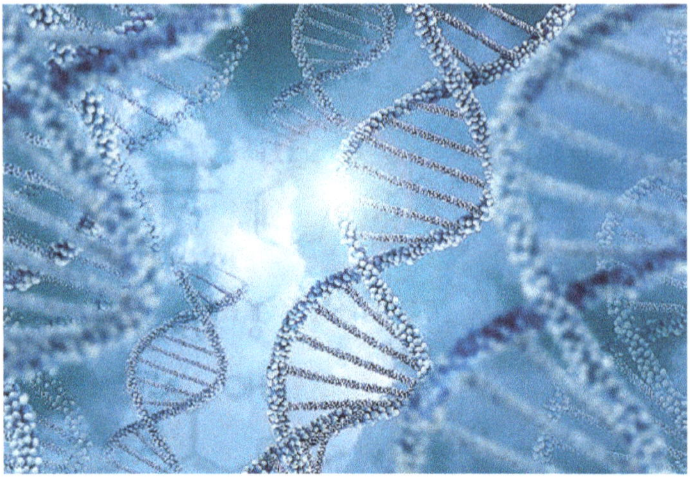

Figure 1. The double screw of the DNA molecule that codes genetic information has been derived thanks to X-ray diffraction (Chapter 9). The molecule is formed by four types of nucleotides assembled in a complementary way on each wire that gives it possibility of duplication.

Metamorphoses of Carbon

The Italian chemist and writer Primo Levi (1915–1987) (particularly known for his imprisonment in Auschwitz and his related book *If This Is a Man*) wrote about carbon: "Carbon is a peculiar element: it is the only one capable to form bonds with itself along long and stable chains without requiring high energy; for life on Earth (the only one known to us until now), these chains are indeed required. That is the reason for assigning carbon the name "element of the living world". These long chains of carbon atoms can fix hydrogen or oxygen atoms or nitrogen or phosphorus and are the basis for the molecules granting life on Earth." Let us recall, for example, the proteins and the sugars (Chapters 17 and 18) or the very perfect molecular structure of DNA (Fig. 1).

Furthermore, carbon has other astonishing properties that Primo Levi could not be aware of in 1970. Until rather recently (around 1990), institutional books reported only two crystalline variants of carbon: graphite and diamond.

Diamond is a rare crystal, and Nature can provide big samples only under extreme conditions of temperature and pressure. On the other hand, diamond is the simplest form of carbon (Fig. 2) where each atom has four nearest neighbours with strong chemical bonds, called covalent bonds (see Chapter 16). The fact that the carbon atom wants to have four nearest neigbours is the consequence of simple laws of chemistry. In fact, the carbon atom, the sixth atom in the periodic table of the elements, has six electrons. In a simplified model, the electrons are placed in shells around the nucleus. Two of those six electrons are close to the nucleus and do not have any role in chemical processes, and they form what is called a closed shell. The other four electrons are placed more apart from the nucleus. This second shell has the possibility to host eight electrons, and the atoms tend to acquire an electronic structure where the external shell hosts the maximum allowed number so that it becomes complete. This is the so-called rule of the octet for the early chemists, somewhat equivalent to the rule of the duet for the light elements. This rule has been satisfactorily explained by modern chemists by resorting to the rules of quantum mechanics. To comply with this rule, the atoms can gain or lose electrons in order to form ions or to put in common the electrons with other atoms to obtain covalent bonds.

In diamond, each of the four electrons of this second shell of a given atom is coupled to another electron of the second shell belonging to another atom so forming four C–C bonds, thus satisfying the octet rule. Therefore, each atom has four nearest neighbours (Fig. 2). The structure derived in this way is very stable, and hence diamond is very hard.

As was already mentioned, the diamond form of carbon is very rare. Due to a kind of caprice of chemistry, the carbon atoms have the tendency to form a two-dimensional structure in which each atom is bounded to only three other

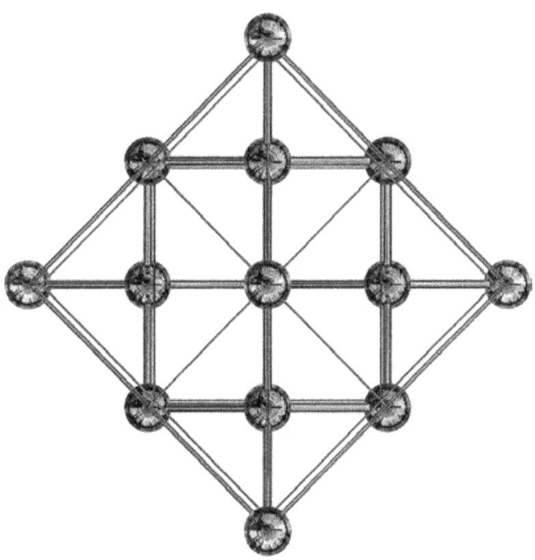

Figure 2. The crystalline structure of diamond. The crystal lattice is face-centred cubic (Chapter 9) where four further carbon atoms are set. Each atom has four nearest neighbours yielding a regular tetrahedron disposition.

Figure 3. Graphene is composed of carbon atoms forming a two-dimensional honeycomb lattice.

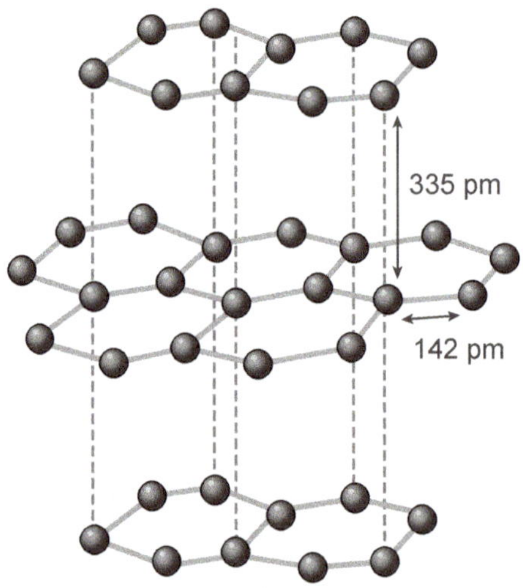

Figure 4. Crystal structure of graphite. Graphite is a sequence of layers of graphene with weak bonds between two of them. The weak bonds are represented by the dotted lines.

atoms. This is a two-dimensional crystal: graphene (Fig. 3). The remaining unused fourth electron of the outer shell (one per atom) is ready to participate in the formation of a weak bond, which connects one graphene layer to another, another to the third, etc. As a result of this packing, graphite is formed: the most common form of crystalline carbon (Fig. 4). This solid dark substance can serve, for example, as a lead in a regular pencil.

The weak bonds connecting the graphite layers turn out to be fragile, and it is quite easy to break them. For example, by sticking adhesive tape on graphite and tearing it off, it is easy to separate several layers; repeating this procedure several times, in the end, it is possible to obtain a single graphene layer. This simple and successful method, which has found widespread application, won the Nobel Prize in Physics for André Geim and Konstantin Novoselov in 2004.

Often the electron of the outer shell, which does not find a covalent bond, slightly strengthens the three bonds of its fellows with the electrons of neighbouring atoms. Instead of being packed with other layers into three-dimensional graphite, the graphene layer deforms and becomes flat and spontaneously forms unusual structures. Let us describe some of them.

When Carbon Plays Soccer

As a result of observations and speculative analysis, researchers in different parts of the world have come to the conclusion that a small amount of a specific substance appears in the soot and candle flame, the molecules of which consist of 60 carbon atoms (C_{60}).

The carbon atoms and their mutual bonds form 20 hexagons and 12 pentagons, thus reproducing the shape of a soccer ball (Fig. 5(a)).

Following the discovery of the C_{60} molecule, other bigger molecules, still made by hexagons and pentagons, have been discovered and created. Such is the molecule C_{70} which rather resembles the ball for playing rugby with 25 hexagons and 12 pentagons (Fig. 5(b)). Another one worth mentioning is the C_{540} molecule (Fig. 6) which has been obtained by graphite vaporisation under irradiation by laser light or by resorting to the electric arc. These molecules have also been found in stardust.

Family history

At the end of the 19th century, the Russian chemist Dmitri Mendeleev came up with a system for the classification of chemical elements. He compiled a table in which he arranged them according to the degree of increase in atomic mass so that the elements in each column have similar chemical properties. So, in the column corresponding to carbon, below it, there are silicon (Si) and germanium (Ge). Like the carbon atom, the atoms of these elements have four electrons on the outer shell, which are ready to participate in chemical bonds. Thus, silicon and germanium also form crystals with a diamond structure (Fig. 2). These crystals are widely used in electronics: when certain impurities are added, they become semiconductors (see Chapter 28).

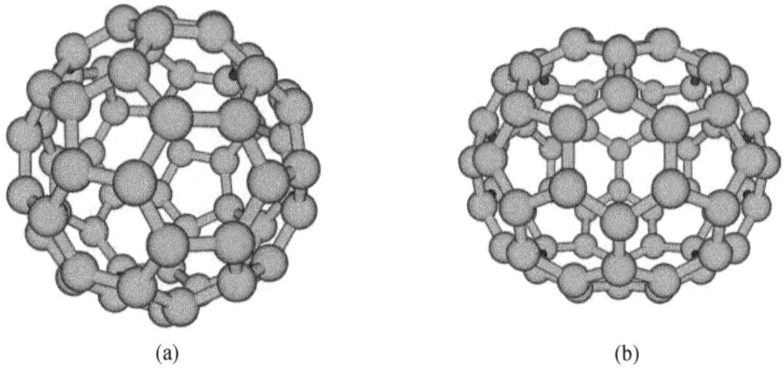

Figure 5. (a) The C_{60} molecule reproduces the shape of a soccer ball, which is obtained by assembling pieces of leather with pentagons and hexagons. (b) The C_{70} molecule resembles a rugby ball.

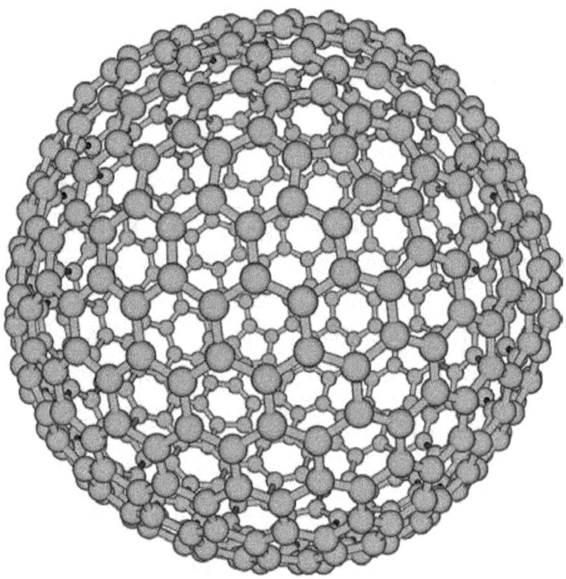

Figure 6. C_{540} molecule.

All these molecules take the form of a convex polyhedron and have the formula C_{2n}, with n being the integer and a variable number of hexagons, but always 12 pentagons. How do we explain that? We are going to prove that the faces are hexagons or pentagons, and then the number of pentagons is necessarily 12. Let h be the number of hexagons and p the number of pentagons. According to a geometry theorem by Euler, if a convex polyhedron has f faces, s vertices and a corners, then these three numbers must obey the equation

$$f + s = a + 2.$$

Now f, s and a can easily be written as functions of the number h of hexagons and the number p of pentagons, under the condition that $f = h + p$. Since each hexagon has six sides while each pentagon has five of them and each side is common to two faces, then

$$a = (6h + 5p)/2.$$

For the s vertices, one can observe that each vertex is common to three faces, since there is no way to get a larger number of faces while two would be absurd. Since each hexagon has six vertices while each pentagon has five of them, each vertex being shared by three faces, one must have

$$s = (6h + 5p)/3.$$

By returning to the Euler equation and by taking into account the numbers we have found, one derives that p must be 12. One should remark that a polyhedron with $p = 12$ and $h = 0$ does exist, it is the regular dodecahedron corresponding to the molecule C_{20}, the smallest one in the family of the molecules we are addressing. As regards the fact that all the faces are either pentagons or hexagons, this is not so astonishing: a square face, for instance, would imply that two C–C bonds define an angle of 90°, while, on the contrary, the outer electrons tend to distribute themselves in the space in a way to equilibrate the group of atoms around them.

The molecules we have described are called fullerenes from the name of an American architect, Buckminster Fuller, who designed geodesic structures resembling the C_{60} molecule (the first one discovered, sometimes also known as fullerene).

Physicists have discovered and described hidden beauties of structures created by Nature, that, in reality, some artists had already displayed. The great Italian Piero della Francesca (from Tuscany, who died at almost 80 years of age in 1492) was probably the first to describe the geometric structure corresponding to a soccer ball, resulting from the truncation of regular icosahedron by five planes. Piero della Francesca was most interested in mathematics. One does not find that geometric structure in the "Libellus de quinque corporibus regolaribus" but in the book published in the year 1510 by one of his pupils, Luca Pacioli. The illustration of the book by the title *De Divina Proportione* was by Leonardo da Vinci (Fig. 7)!

Carbon Nanotubes

Carbon is also capable of forming particular tubes. One can find a variety of forms and sizes, with the diameter from one to several tens of nanometres (Fig. 8).

Figure 7. The structure figured out by Piero della Francesca and drawn by Leonardo da Vinci for the book published by Luca Pacioli.

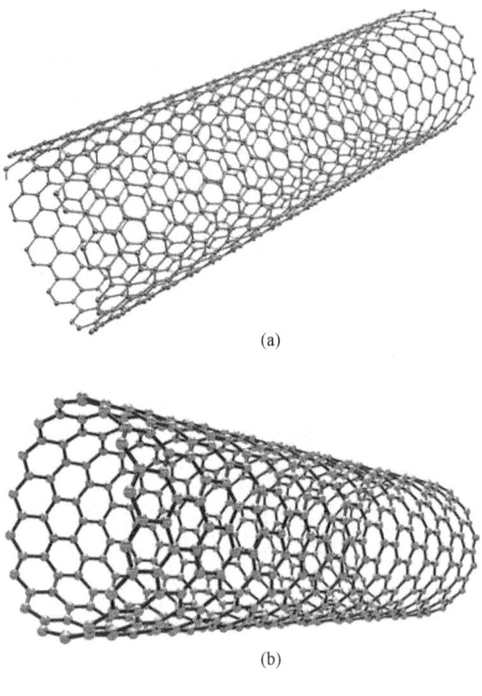

(a)

(b)

Figure 8. Two forms of carbon nanotubes. (a) The armchair structure has bonds perpendicular to the axis of the tube. (b) The zigzag structure has bonds parallel to the axis of the tube.

Carbon nanotubes are obtained in laboratories in different ways: by tearing them off from a block of carbon by means of laser light, by electric arc as well as by depositing appropriate carbon vapours. In fact, we, without knowing it, have been producing them for centuries, but only with the advent of improved tools are we able to detect these nano-objects (see Chapter 28).

What can be done with these objects? The nanotubes, the fullerene and other graphene derivates have electric and optical properties that have excited the curiosity of physicists and that promise an appealing future, for instance, as photovoltaic cell elements (see Chapter 28). Graphene-based electronics could possibly let the Moore law increase its life expectancy, allowing one to create transistors even smaller than the ones based on silicon (see Chapter 28).

Nanotubes also have remarkable mechanical properties: they are extremely resistant to stretching. Therefore, we will soon be able to see light and durable bicycles and tennis rackets made of composite materials reinforced with carbon nanotubes with a diameter of several microns. Another metamorphosis of this "special element" celebrated by Primo Levi!

Chapter 24

Perpetual Motion in Superconductors

In some materials, below a certain characteristic temperature, an amazing phenomenon occurs: their electrical resistance completely disappears. This phenomenon was first discovered in 1911 in samples made of mercury, the critical temperature of which is only Tc = 4.15 K. In the next few years, superconductivity was found in other materials, but always at very low temperatures. The phenomenon seemed inexplicable. Several decades passed before the first theory was developed to explain it satisfactorily. As we will see, even today, over a century later, more and more superconductors are throwing up amazing mysteries for researchers.

The Discovery of Superconductivity

In 1908, Dutch physicist Kamerlingh Onnes (1853–1926) succeeded in liquefying helium, which at ordinary pressure has a boiling point at a temperature of 4.2 K. This remarkable technical achievement opened up the opportunity for scientists to study the resistance of metals at very low temperatures (see Panel on page 337). The result was not long in coming: on April 28, 1911, at a meeting of the Royal Netherlands Academy of Arts and Sciences in Amsterdam, a fundamental discovery was announced: at temperatures below 4.15 K, the electrical resistance of mercury completely disappears (Fig. 1).

The absence of electrical resistance means that, once set in motion, the charges in a closed circuit will move forever (Fig. 2). Indeed, researchers in England were able to make the current circulate in the superconductor for several years without any slightest damping; the experiment was interrupted only when the cooling of the device was disrupted due to a strike at the power plant.

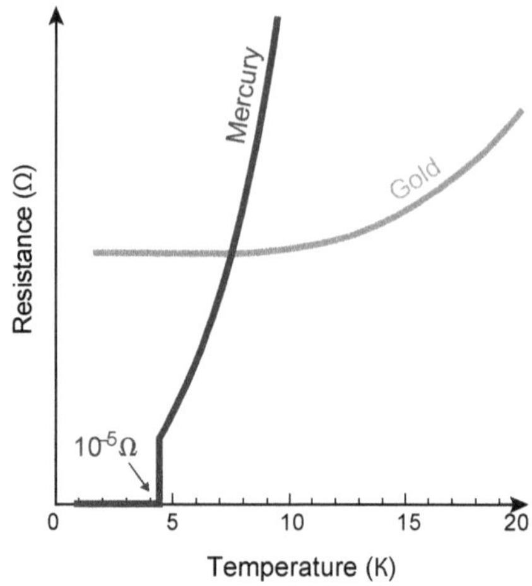

Figure 1. Electrical resistance R (in ohms) of a mercury sample versus temperature T (in kelvin). Below the critical temperature $T_c = 4.15$ K, the resistance disappears. This phenomenon — called superconductivity — was discovered on April 8, 1911. Note that not all metals, even at the lowest temperatures, transit into a superconducting state. For example, gold, which is an excellent conductor, does not have this property.

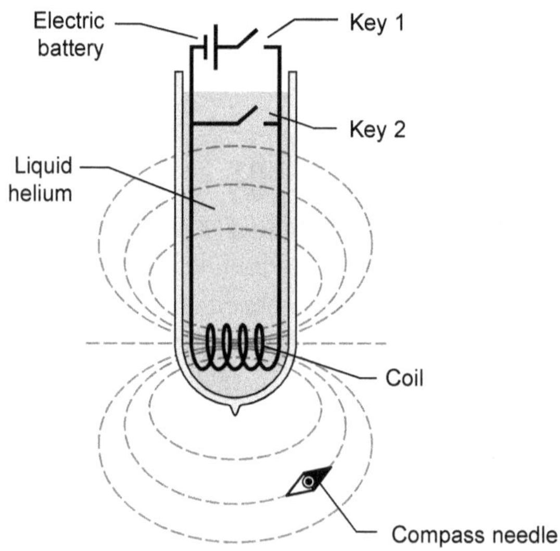

Figure 2. Experiment of Kamerlingh Onnes, proving the absence of current damping in a superconductor. The electric battery creates a current (constant) in the circuit, while the upper key remains closed. Then it is opened, thereby disconnecting the battery, and at the same time, the lower key is closed. The presence of current in the superconducting coil is manifested by its effect on the magnetic needle, which is oriented along the lines of the magnetic field.

While the theoretical explanation of superconductivity took a long time to come, experimental research went ahead. Besides mercury, superconductivity has also been found in other metals such as lead and tin. Superconductivity occurs in them also at very low temperatures: the highest critical temperature among pure metals, as it turned out, belongs to niobium (T_c = 9.2 K, that is, –263.95°C!). Scientists understood the tempting prospects for practical applications of this phenomenon, such as the transfer of energy without losses or the creation of super-powerful electromagnets (see Chapter 25). However, two major obstacles arose along the way. First, the need for extremely low temperatures required constant cooling of the device. The second hurdle Kamerlingh Onnes soon faced was the sudden disappearance of the superconducting state when the current flowing through the sample became too strong. The same destructive effect was produced by a magnetic field exceeding a certain threshold. The experimentally observed magnitude of this destructive field, called critical, was small. So, for mercury, the critical magnetic field is 0.03 T (compare this value with the field created by conventional bar magnets: from 0.1 to 1 T).

The nature of electrical resistance

What are the microscopic reasons for the existence of resistance to the flow of electric current in normal metals?

Recall that the electric current is due to the movement of free electrons under the action of a potential difference applied to the ends of the conductor: an electric field arises in the conductor, and electrons rush into the region with the highest potential. At an arbitrary point of the electrical circuit, the electrons on average have a velocity parallel to the axis of the conductor and equal in magnitude to v*; if the cross-section of the conductor is* S*, then the intensity of the electric current is* $I = nevS$*, where* e *is the electron charge and* n *is the number of electrons per unit volume.*

If the metal were an ideal crystal, then at zero potential difference, the electron would propagate at a constant speed, as in a vacuum. This follows from a theorem proved by the French mathematician Gaston Floquet (1847–1920), and applied to electrons by Felix Bloch (1905–1983). However, real metals almost always contain various defects (for example, impurity atoms embedded in the crystal lattice), which break the lattice symmetry and scatter electrons. After a series of such scattering, the electron deviates from its original direction, and its velocity, averaged over all

(Continued)

(*Continued*)

particles, becomes equal to zero. This is a semiclassical picture of the origin of electrical resistance of metals at low temperatures. When the temperature rises, another mechanism is added to the scattering of electrons by impurities and other lattice defects: this is the scattering of electrons by thermal vibrations of the lattice ions. In short, the electrical resistance R consists of two parts: one does not depend on temperature but does depend on the concentration of impurities and the degree with which they scatter electrons; the other depends on the temperature.

Thus the cause of the electrical resistance R in metals is the interaction of electrons with the crystal lattice vibrations and lattice defects. To create and maintain an electric current I in a circuit, it is necessary to maintain a potential difference V in it, that is, to spend energy. The corresponding power is released in the conductor in the form of heat and, according to the Joule law (see Chapter 16), is equal in magnitude to RI^2.

Meissner–Ochsenfeld Effect

In 1933, German physicists Walter Meissner and Robert Ochsenfeld, studying the effect of an external magnetic field on a superconductor, found that it did not penetrate into a superconductor placed in a magnetic field. This phenomenon, called the Meissner–Ochsenfeld effect, is associated with the appearance of non-dissipative currents on the surface of the superconductor, which, creating their own magnetic field in the bulk of the superconductor, compensate for the external field (Fig. 3). Everything happens as though the superconductor "expels" the magnetic field out of its volume.

A reader familiar with the phenomenon of electromagnetic induction (see Chapter 16) might assume that a similar effect occurs only when a superconductor is placed between the poles of a magnet. In this case, the magnetic flux penetrating the cross-section of the superconductor would change, and then it would generate an electromotive force. As a result, an infinitely large current would have to appear in the superconductor, which would destroy superconductivity.

Recall that in Physics there is a general Le Chatelier's principle, which states that if a system in stable equilibrium is influenced from the outside, changing any of the equilibrium conditions (in our case, an external magnetic field), then processes occur in the system directed towards resisting change.

Therefore, it is easier for a superconductor to generate the finite non-dissipative current over its surface, which simply eliminates the effect of the external field, reducing to zero changes in the magnetic flux. Of course, this current

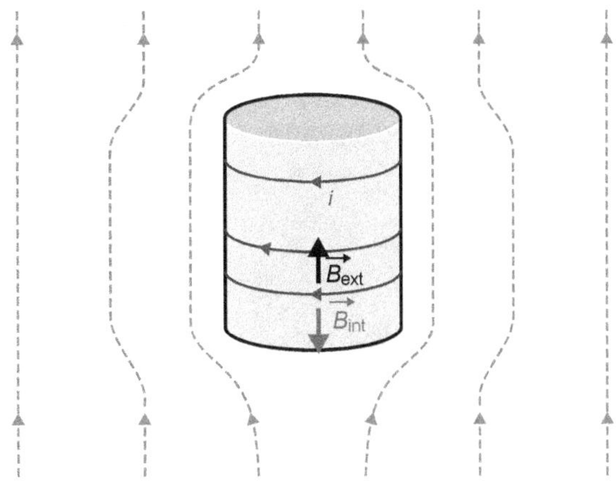

Figure 3. The Meissner–Ochsenfeld effect. An electric current is generated at the surface of a superconductor placed in an external field. This current creates a magnetic field inside the sample, which precisely compensates for the external field.

should not exceed the critical value that kills superconductivity. Therefore, the Meissner–Ochsenfeld effect takes place for fields that are not too strong.

Well, what if a bulk superconductor is placed in a magnetic field in its normal state at a temperature above the critical one, and only then cooled into a superconducting state? It would seem that in this case the non-zero magnetic flux that permeated the superconductor at a high temperature should remain in it. Therefore, with such an experiment, surface currents should not appear — this is exactly what Meissner expected. He set up a corresponding experiment, and, to his surprise, found that the magnetic field inside the superconductor became zero, which indicated the appearance of surface currents in it, and in the setting of the experiment.

What is the matter here? Why is the magnetic flux not frozen in the bulk of the superconductor? The reason for this behaviour is fluctuations — deviations of the system from equilibrium. Imagine that, in some small region of the superconductor, the existing magnetic field changes slightly. Then the magnetic flux will change slightly, and as a consequence, a small electromotive force will appear. But after all, the resistance is equal to zero: therefore, such a fluctuation would generate an infinite current that would kill superconductivity. To maintain its state, it is easier for a superconductor to generate a current of finite density over its surface, expelling all the magnetic flux from its volume. The emerging surface currents have another impressive manifestation: the magnetic field generated by them outside the superconductor is capable of repelling a magnet in such a way that the latter levitates over it (i.e., hovers) (Figs. 4 and 5).

Figure 4. Permanent magnet levitates over a superconductor.

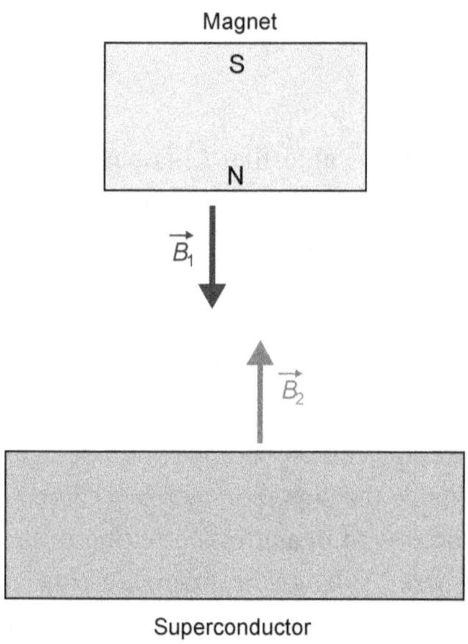

Figure 5. Explanation of the phenomenon of magnet levitation over a superconductor. In the presence of a field $\mathbf{B}_1$ generated by a magnet, a superconducting current appears on the upper surface of the superconductor, generating a magnetic field $\mathbf{B}_2$. This field is obviously opposite in direction to $\mathbf{B}_1$. The interaction of the magnet with the field creates forces acting on the magnet, which balances its weight. As a result, the magnet hovers over the superconductor.

Vortices of Abrikosov

The Meissner–Ochsenfeld effect usually occurs only if the external field is relatively weak. If the field is too strong, then the superconductor is not able to expel it and, as already mentioned above, it transits to its normal state. Thus, it seems that when placing a superconductor in a magnetic field, there are only two options: either the superconductor will go into its normal state by letting in the magnetic field, or the field in the superconductor will become zero.

Half a century after the discovery of the phenomenon of superconductivity, it was predicted that another scenario for the behaviour of a superconductor in a magnetic field can also occur. The above observation is valid only for the so-called type I superconductors, such as mercury, lead and aluminium. Soviet theoretical physicist Alexei Abrikosov, in his later famous 1957 work, showed that for some superconductors, which he called superconductors of the second kind (today, they are often called type II superconductors), there is a third possibility. Namely, if the external magnetic field is strong enough, then it can penetrate into such superconductors in the form of very thin tubes parallel to the field repetition of penetrate into the superconductor. Superconducting non-dissipative currents flow around these tubes, which form a kind of vortices. At a certain distance from the tubes, the magnetic field is equal to zero — superconducting currents screen it with their own field. In this case, the property of superconductivity does not disappear, although the tubes themselves are not superconducting: as a result, part of the sample volume remains superconducting, and a magnetic field penetrates into the other.

Experimental observation of Abrikosov's vortices turned out to be quite easy, although it was carried out 10 years after the publication of his theoretical work. The easiest way is to sprinkle the surface of the superconductor with iron filings or particles of another ferromagnetic material. Then these particles will begin to accumulate at the head of the tubes (Fig. 6(a)). One could use the neutron diffraction method: the vortex lattice is actually similar to the crystal lattice (see Chapter 9). In this way, one can confirm that the tubes penetrate inside the volume of the superconductor.

The behaviour of a type II superconductor in a magnetic field depends on the field intensity. When applying the magnetic field acting on the superconductor from zero, vortices are not initially observed. The magnetic field, due to the emerging surface currents, is completely expelled out of the sample, just as in the case of a type I superconductor. Thus, in sufficiently weak fields, the full Meissner effect takes place. At a certain critical field B_{c1}, the first vortices appear in the bulk of the superconductor, first in a small amount. By means of these vortices, a magnetic field begins to penetrate into the superconductor. With increasing field intensity, the number of vortices also increases. Ultimately, at a field B_{c2} (Fig. 6(b)), the vortices fill almost the entire volume of the sample, and superconductivity disappears.

Note that most type I superconductors (which do not have a mixed vortex phase) can be converted into type II superconductors by adding impurities. For

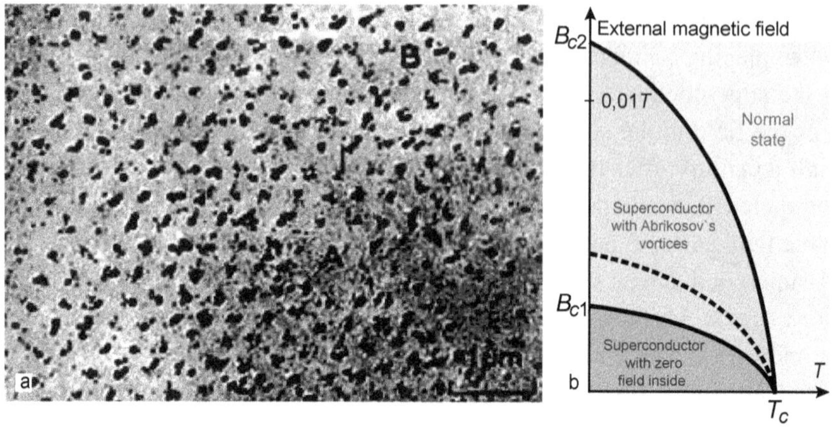

Figure 6. (a) The Abrikosov vortex lattice in a type II superconductor visualised by decorating its surface with cobalt particles, 1967. The particles are collected at the inputs of the magnetic field lines into the superconductor, i.e., at the outlets of vortices at the surface. The vortices repel each other and form a more or less regular lattice predicted by Abrikosov. (b) The phase diagram of the indium–bismuth alloy InBi (with 4% Bi) in the coordinates of temperature T and magnetic field B. The dotted curve defines the temperature dependence of the critical field of pure indium, which is a type I superconductor.

example, pure indium, a silvery metal, is a type I superconductor, but when 4% bismuth is added to it, it becomes a type II superconductor.

For his outstanding discovery, Alexei Abrikosov was awarded the 2003 Nobel Prize in Physics (he shared it with Vitaly Ginzburg and Anthony Leggett). As mentioned above, his theoretical prediction was 10 years ahead of the experimental confirmation of the existence of type II superconductors. But today, they are mainly used in medicine, transport, transmission of energy over distances, and the creation of super-powerful magnetic fields. For the first time in the history of superconductivity, theory outstripped experiment. Nevertheless, until 1957, the origin of this mysterious phenomenon could still not be explained.

Superfluidity: New Hopes

In 1938, Soviet physicist Pyotr Kapitsa (1894–1984) discovered that at temperatures below 2.18 K, the flow of liquid helium experiences no friction when passing through very narrow capillary tubes. This phenomenon, called "superfluidity", gave scientists hope for understanding the nature of superconductivity: after all, the similarity between an electric current flowing without resistance and a non-viscous hydrodynamic flow is obvious. Let us take a closer look at the latter.

As we have seen, helium at atmospheric pressure does not solidify even at the lowest temperatures (see Chapter 22): we explained this by the zero-point oscillations of its atoms having small mass, and the interaction between them is weak. Simply put, a superfluid state can be viewed as some kind of compromise between the "desire" of atoms to condense into a crystal and their quantum "neccessity" to move. As a result of the action of the forces of attraction between helium atoms, at low temperatures, the latter pass into a certain condensed state, however, unlike atoms of other elements, they do not form a crystal.

What characterises this condensed phase? The state of particles in it is of quantum nature, so they should be characterised according to the laws of the quantum world, namely by the wavefunction $\Psi(x,y,z,t)$ (see Chapter 22). It turns out that, at temperatures below 2.18 K, the macroscopic number of helium atoms accumulates in the same quantum state, and is described by the same wavefunction. These atoms form the so-called "superfluid condensate". When it flows with velocity $\mathbf{v}$, then its wavefunction corresponds to the wavefunction of some quantum particle moving with the same velocity $\mathbf{v}$. In a normal liquid, the particle slows down due to viscosity, i.e., interactions with the environment; in superfluid helium, on the contrary, all the atoms of the condensate are interconnected into a single whole and at not too high speeds do not interact with the environment and therefore cannot slow down! The flow of superfluid helium is a collective phenomenon: atoms move in it all together, like sheep in a herd. Even if a sheep wants to go back, it cannot do it!

The wavefunction $\Psi(x,y,z,t)$, which describes the superfluid condensate, is approximately determined by the solution of the Gross–Pitaevskii equation, which is similar to the already familiar Schrödinger equation that determines the motion of quantum particles in the microworld.

From Superfluidity to Superconductivity

A theory describing the properties of superconductors similar to condensation in superfluid helium was proposed by Soviet physicists Vitaly Ginzburg (1916–2009) and Lev Landau (1908–1968) in 1950. In contrast to the Gross–Pitaevskii equation, Ginzburg–Landau theory proposes two equations: one for the wavefunction of a superconducting condensate and the second one for a magnetic field. The latter, as we know, plays an extremely important role in the life of a superconductor but does not affect helium atoms in any way (since they have neither an electric charge nor a magnetic moment). The Ginzburg–Landau equations have proven to be an extremely effective tool for studying superconductivity. For example, Alexei Abrikosov predicted the existence of type II superconductivity, the existence of quantum vortices, etc. on the basis of the Ginzburg–Landau equations.

Despite the more powerful microscopic methods for describing superconductivity that appeared later, the Ginzburg–Landau equations remain very useful for researchers today, over 70 years after they were written. It was proved that near the transition from the normal to the superconducting state (precisely in the temperature range for which the equations were derived by Ginzburg and Landau), they exactly coincide with the results of the microscopic theory. Nevertheless, at the time of their discovery, the Ginzburg–Landau equations were exclusively "phenomenological" in nature, that is, they predicted and explained the available experimental facts without going into their microscopic nature.

Isotope Effect and the Role of the Crystal Lattice

It should be noted that the drawing of the analogy between the phenomena of superfluidity and superconductivity carries some difficulties. We have already said that all superfluid helium atoms in a condensate are in the same quantum state. However, this is only possible for certain types of particles called bosons. For example, photons are bosons, so the number of photons with a given energy and propagating in a certain direction is not limited. Helium atoms are also bosons, and they are electroneutral. Superconductivity, however, is obviously somehow related to the charge-carrying electrons, which are fermions. Unlike helium atoms, they obey the exclusion principle (Pauli's principle), according to which two electrons cannot be in the same quantum state. Therefore, it was not possible to simply rewrite the theory of superfluidity for an electron liquid in a metal.

However, it turned out that, under certain conditions, two fermions can be combined into a single "particle" which will no longer follow the Pauli principle. To do this, it is necessary to have some kind of attraction between them — "glue", that will connect them into a composite boson. It is due to this pairing of electrons in a metal that the phenomenon of superconductivity arises.

What kind of attraction is causing such a pooling of electrons? Its existence is not at all obvious: in fact, as you know, two particles of the same charge must repel each other! However, as German-English physicist Herbert Fröhlich discovered in 1950, if these electrons are not in a vacuum, but in a crystal, then such an attraction can take place. Indeed, in a crystal, ions are ordered into a crystal lattice, which can be deformed. The force of attraction between two electrons in the presence of this lattice is related to its elasticity. The presence of an electron in it causes local deformation, which contributes to the attraction of the second electron.

As a loose analogy, we can give the example of two balls lying on a rubber mat. If these balls are far from each other, then each of them deforms the rug, forming a hole around itself. If you put one ball first, and then another not too far from it, then their holes will merge into one, and the balls will roll together to the bottom of the common hole.

In metals, this attraction occurs due to the deformation of the crystal lattice. The discovery of the isotope effect in 1950 (Fig. 7) greatly influenced the identification of the role of elastic lattice vibrations (or phonons) which is so important for explaining the phenomenon of superconductivity. It turned out that two isotopes of the same metal have different critical temperatures, values that are inversely proportional to the square root of the isotope mass! This property is reminiscent of the fact that the vibration frequency of a ball fixed at the end of a spring depends on its mass (see Chapter 12).

The discovery of the influence of the properties of the crystal lattice of a substance on its superconducting properties played a decisive role in understanding the origin of this phenomenon and creating its microscopic theory. Fröhlich was already on the right track, but the electron–phonon attraction he found turned out

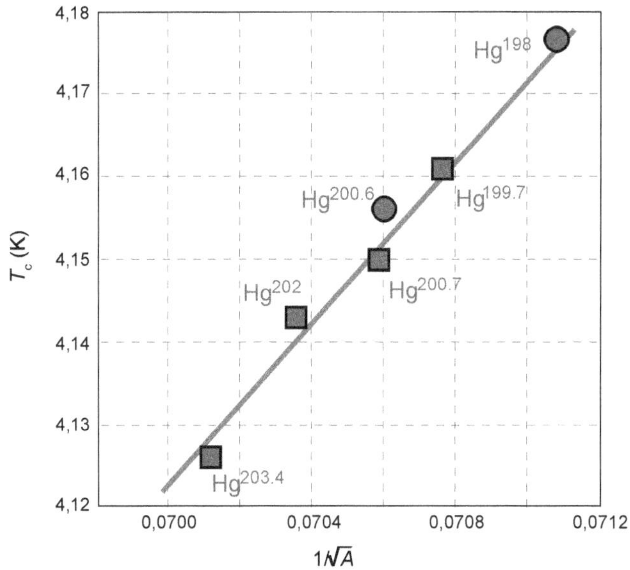

Figure 7. Critical temperature of various isotopes of mercury. The mass number A corresponds to the total number of nucleons (protons and neutrons). Recall that two isotopes have the same number of protons (mercury has 80) but a different number of neutrons and, therefore, have different masses. The molar mass of each isotope is indicated next to the designation, in g mol^{-1}. From C. A. Reynolds *et al.*, Superconductivity of isotopes of Mercury, *Physical Review* **78**, p. 487 (1950).

to be rather weak compared to the electrostatic repulsion between electrons and he could not explain how it could provide the formation of composite bosons.

Density of electronic states in a metal... which becomes a superconductor

In order to understand the meaning of the gap, it is necessary to introduce the concept of the density of electronic states. In an isolated atom, electrons can occupy only discrete energy levels (see Chapter 22). In a solid consisting of many atoms, these quantum states form a dense staircase that occupies an entire band on the energy scale. There can be several such bands, the last of them is called the conduction band. The number of electronic states corresponding to the energy interval between ε *and* $\varepsilon + \delta\varepsilon$, *for small* $\delta\varepsilon$, *turns out to be proportional to the width of this interval* $\delta\varepsilon$. *Thus it can be denoted as* $\rho(\varepsilon)\delta\varepsilon$, *where* $\rho(\varepsilon)$ *is some function of energy, called the density of states.*

In a normal metal, the density of states in such a band changes continuously. Each of these states can be occupied by an electron or remain free. At a temperature $T = 0$ *K, all low-energy states with energies* ε *less than a certain value* ε_F *(Fermi energy) are occupied by electrons, while states with energies* $\varepsilon > \varepsilon_F$ *are empty (Fig. a). In a superconductor at low temperatures, an energy gap appears in the dependence of the density of states near the Fermi energy (Fig. b). Such an occurrence is very unusual! Indeed, in solid-state physics, it is well known that if the Fermi energy of a system corresponds to the gap in the energy spectrum, and not to the conduction band, then this substance is a dielectric (see Chapter 28). In the case of a superconductor, the opposite is true: when a gap opens near the Fermi level, its resistance disappears!*

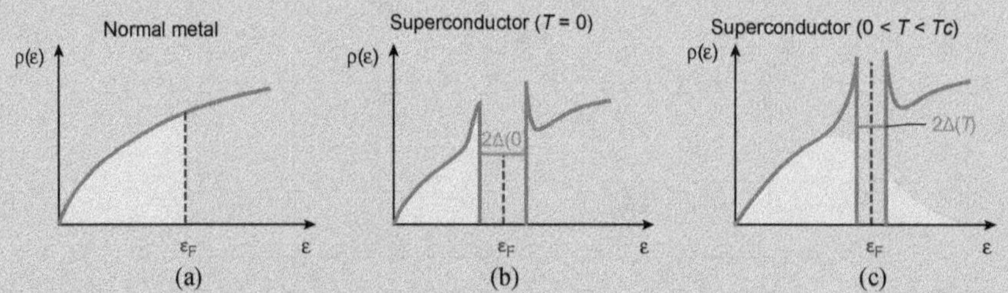

(a) Density of electronic states of a normal metal in the region close to the Fermi energy. The states of the shaded area are occupied at absolute zero.
(b) Density of states of a superconductor at absolute zero.
(c) The density of states of one and the same superconductor at a non-zero temperature below the critical one; gap is narrowed, and electrons appear in the conduction band. (The proportions are not respected; Δ is usually much less than the Fermi energy ε_F.)

BCS Theory

Seven years after the publication of the Ginzburg–Landau theory, American physicists John Bardeen, Leon Cooper and Robert Schrieffer constructed the theory of superconductivity (the so-called BCS theory), which gave a consistent microscopic explanation of this mysterious phenomenon and removed all contradictions existing at that time.

In a "normal" metal (that is, not a superconductor), at absolute zero, as already mentioned, the electrons occupy all energy states up to a certain value ε_F, called the Fermi energy (the latter depends on the concentration of electrons in the metal and the symmetry of its lattice). Each state is occupied by only one electron, in accordance with the Pauli exclusion principle. The Fermi energy in metals is usually on the order of several electronvolts.

As for the energy of the electron–phonon interaction, the corresponding energies are related by the so-called Debye frequency W_D and do not exceed 0.1 eV. For the Coulomb repulsion of electrons, one could assume that its characteristic value is $\sim e^2/(4\pi\varepsilon_0 a)$, where a is of the order of the interatomic distance. It is easy to estimate that this quantity also turns out to be of the order of an electronvolt. What happens in a superconductor? How, then, to arrange the attraction between electrons?

Bardeen, Cooper and Schrieffer noticed that the electrons in a metal are surrounded by lattice ions and other electrons, so the repulsion between them is greatly weakened and turns out to be of the same order as the attraction arising from the electron–phonon interaction. Therefore, the total electron–electron interaction for some metals turns out to be positive, while for others it is negative. The former (for example, gold and platinum) do not go into a superconducting state at any temperature since their electrons cannot form composite bosons. The latter, where the electron–phonon interaction wins in the competition of interactions, become superconductors at a certain temperature.

To calculate it, as well as to describe other properties of a superconductor, one can imagine electronic energy states in a metal filling an imaginary orange with a size corresponding to ε_F. The thickness of its skin corresponds to the energy associated with the Debye frequency, namely $\hbar W_D$. It is clear that electrons that occupy the energy levels deep in the orange, with energies much lower than ε_F, will remain at the same quantum states in spite of any electron–phonon interactions. But the electrons in the "orange peel", thanks to the Fröhlich interaction, can form composite bosons. In this case, it turns out that it is more advantageous for them to unite in pairs with opposite spins so that the total spin of such a pair, called the "Cooper pair", is equal to zero. The electron velocity vectors must also be opposite, otherwise the weak electron–phonon interaction will not hold them together.

What is the "size" (or a kind of average distance) of such a pair? Let's estimate it. The speeds of electrons in the "orange peel" are determined by the value of ε_F and turn out to be very large: $v_F \approx 10^6\,\mathrm{m\,s^{-1}}$. The characteristic of superconductivity energy scale is determined by the critical temperature: $k_B T_c \sim 10^{-22}$ J. Using dimensional analysis one can construct, from these values and Planck's constant (characteristic of the quantum world, as we have seen), the characteristic for superconductivity length $\xi \sim \hbar v_F / k_B T_c \sim 10^{-6}$ m. That is, the Fröhlich interaction has a wide range; it acts in the micrometre "size". Thus two electrons moving with the opposite velocities of the order v_F in magnitude form a pair of about a micrometre in size. The latter is called a "Cooper pair", by the name of its discoverer. Since this distance is large in comparison with the interatomic distances, then, as we suggested above, the Coulomb repulsion between two electrons is strongly reduced due to the screening effect of other electrons and lattice ions. As a consequence, Cooper pairing turns out to be energetically favourable for some metals at low temperatures.

According to the metaphor proposed by Schrieffer, the Cooper pair should not be imagined as a double star formed from electrons, but rather as a pair of dancers who came together to the disco, who sometimes come closer, sometimes move away from each other, but still dance together, regardless of whether they are separated for a moment by other dancers.

To "break" a Cooper pair into two electrons that make it up, it is necessary to expend some energy Δ, called the superconducting gap. As the superconductor heats up, the gap narrows, and more and more pairs break apart. The gap ultimately vanishes when the critical temperature is reached: for electrons, it is no longer energetically convenient to form pairs, and the superconducting state disappears.

Thus, the Bardeen–Cooper–Schrieffer (BCS) theory provided the long-awaited explanation of the microscopic mechanism of the phenomenon of superconductivity. In addition, it substantiated and allowed to calculate the values of the coefficients in the phenomenological Ginzburg–Landau equations, which to this day remain a very convenient and universal apparatus for describing the phenomenon of superconductivity.

Long Way to High Critical Fields…

After the theory of superconductivity was finally created by the middle of the 20th century, physicists armed with it began to search for new superconducting systems with high values of critical parameters (B_{c2} and T_c). Even at the dawn of research, having tested the elements of the periodic table, they expanded the search for new superconductors, moving on to the study of metal alloys (this is the name of macroscopically homogeneous metallic materials consisting of a mixture of two or more chemical elements with a predominance of metallic components). To produce high-quality alloys, researchers have developed an arsenal of methods, from rapid quench arc welding to spraying films onto a hot substrate. It should be noted that the first superconducting alloy was discovered long before the creation of any theories of superconductivity, in 1931. The most important impetus in the search for compounds with high critical fields was given by the work of Alexei Abrikosov already mentioned above. The idea expressed in it about the possibility of increasing the critical field B_{c2} by introducing impurities scattering electrons into the crystal lattice indicated the direction of the search for new superconducting systems suitable for creating super-powerful magnets. As a result of these searches, already in the 1960s and 70s, alloys Nb_3Se and Nb_3Al were created, the critical temperature of which is about 18 K, and the critical fields are higher than 20 T. With the creation of the $PbMo_6S_8$ alloy, the critical field reached a record 60 T at a critical temperature of 15 K. Among the discovered type II superconductors, some are able to withstand enormous electric current densities and remain superconducting in giant magnetic fields. The creation and practical use of superconducting cables based on them was a very difficult technological problem due to the fact that these materials are rather fragile, and their current-carrying properties are unstable. And yet, one of the obstacles to the widespread industrial use of superconductivity has been overcome: today, the critical fields of new superconductors reach values thousands of times higher than the first ones discovered during the Kamerlingh Onnes era.

... and High Critical Temperatures

In contrast to advances in the creation of superconducting systems with high critical fields, the problem of a noticeable increase in the critical temperature of superconductors remained unsolved even in the early 1980s. Since 1973, the record holder has been the Nb_3Ge alloy, in which the superconducting transition temperature reached 23.2 K. Unfortunately, it remained much lower than the boiling point of a cheap and widely used cryoagent — liquid nitrogen, which reaches 77 K at atmospheric pressure. In addition, T_c only slightly exceeded the boiling point of liquid hydrogen (20 K), which made it hard to use this inexpensive gas for cooling superconducting devices based on Nb_3Ge significantly below T_c. Thus, liquid helium remained necessary for the functioning of all superconducting devices existing at that time, even though it was very expensive to produce! The theoretical estimate of the maximum possible critical temperature offered by the BCS theory did not inspire optimism. Namely, the possibilities of the electron–phonon attraction mechanism were limited by the conditions of stability of the crystal lattice, and it turned out that the critical temperature cannot exceed 40–50 K. In search of alternatives, back in 1964, Vitaly Ginzburg and, independently of him, the American William Little, suggested that superconductivity can occur not only due to the electron–phonon interaction but also through some other mechanisms.

Thus Little suggested looking for high-temperature superconductivity in quasi-one-dimensional compounds, that is, in long polymer chains with easily polarizable side branches. However, researchers failed to succeed in synthesising such materials. In the decade between the 70s and 80s, superconducting materials other than metals or their alloys were discovered. These were the so-called oxides, which became superconductors at fairly low temperatures (about 10 K). However, no one expected this from them either: at ordinary temperatures, these compounds are poor conductors, with a low concentration of free electrons. The discovery of superconductivity in these new materials put two researchers from Zurich on the trail of a great future discovery.

Long-awaited discovery

The discovery by Müller and Bednorz of high-temperature superconductivity was completely uncharacteristic for modern physics. First, it was carried out by two researchers who worked alone and did not have much funding. The materials they used contained only readily available elements (not rare isotopes). Knowing what to do, these superconductors could be created in a day's work in any university laboratory. What a contrast, for example, to high-energy physics, in which discoveries require equipment worth billions of euros, and the list of authors of the article takes a whole page! Müller and Bednorz reminded us how important talent and personal initiative are, even if their discovery was largely due to their awareness of the available advances in the science of superconducting materials. Subsequently, it turned out that some high-temperature superconductors had already been synthesised before, but their creators did not choose (or did not have the opportunity) to measure the electrical resistance of the materials obtained at sufficiently low temperatures.

(*Continued*)

(Continued)

Müller and Bednorz chose the German Zeitschrift für Physik to publicise their discovery. A European discovery published in a European scientific journal, what is so special? But no: this event was unique in recent decades. European physicists are already accustomed to publishing their most important discoveries in the American press. In 2005, an American researcher proposed identifying the most cited papers in physics. The article describing the discovery of Müller and Bednorz, which had been waiting for three-quarters of a century, did not appear on his list. Amazing! In fact, he limited himself to articles published only by the American Physical Society and cited only in American journals. But if the article by Müller and Bednorz were on his list, it would take the second place in citations. This curiosity indicates that almost all modern physicists tend, and prefer, to publish articles in the USA — to such an extent that one can afford to do a statistical analysis ignoring the European scientific press. Moreover, this story shows that articles published in Europe, if they are very good, are read and cited no less than those that appear in the American press.

On the Shores of Lake Zurich

The long-awaited breakthrough was made by Karl Alexander Müller and Johannes Georg Bednorz when they worked at the IBM laboratory in Switzerland (where the first tunnel microscope was created, see Chapter 28). In the winter of 1985–86, scientists synthesised a compound that became a superconductor at 35 K! Its chemical formula can be written as $La_{2-x}Ba_xCuO_{4-\delta}$, where x and δ are specific non-integers. The article, published in 1986, was titled very carefully: "On the possibility of high-temperature superconductivity in the La–Ba–Cu–O system". The possibility soon became a reality, and for the discovery of superconductivity in this oxide, Müller and Bednorz were awarded the Nobel Prize in Physics already in 1988. They (see Panel on page 350) paved the way for mankind to reach superconductivity "at high temperatures" (it is called that, although these temperatures are well below 0°C). This baton was taken up by other researchers. So, literally a few months later, replacing barium with strontium, raised the critical temperature record to 45 K. The latter did not last long: a month later, it was found that at high pressure, the critical temperature of the compound found by Müller and Bednorz rises to 52 K. In 1987, the American Paul Chu realised that the effect of high pressure could be achieved by replacing lanthanum atoms with smaller atoms of its neighbour in the column in the periodic table — yttrium.

Having synthesised the compound $YBa_2Cu_3O_{7-\delta}$, he reached a critical temperature of 92 K — and this has already exceeded the nitrogen threshold (the boiling point of liquid nitrogen)! Subsequently synthesised new superconductors reached critical temperatures of 125 K and even (at very high pressures) of 165 K (−108°C).

High-Temperature Superconductivity: A New Mystery

Over the past third of a century, physicists have found a huge number of new superconducting substances, the critical temperature of which exceeds the record 23 K for 1973 (Fig. 8). They are divided into several groups: perovskites, pnictides, MgB_2, organic superconductors and hydrides. Copper oxides with yttrium and barium impurities, for example, $YBa_2Cu_3O_{7-\delta}$ (Fig. 9), were the first discovered and remain the most studied to date. All of them have a layered structure: copper atoms (Cu) and oxygen (O) form planes separated by other atoms, in this case, atoms Ba and Y. The movement of charge carriers is almost two-dimensional: they easily move in CuO_2 layers, but they rarely jump from one layer to another. The formed Cooper pairs are also mainly localised in planes. The mechanism of high-temperature superconductivity has not yet been fully understood. The key to understanding the phenomenon is probably the two-dimensional nature of the motion of electrons. All agree that here, as in the case of classical superconductivity described by the BCS theory, the phenomenon is due to the appearance of Cooper pairs. However, there is no consensus among scientists about the mechanism of interaction between charges leading to their Cooper pairing at such high temperatures. Currently, there are about 20 more or less conflicting theories.

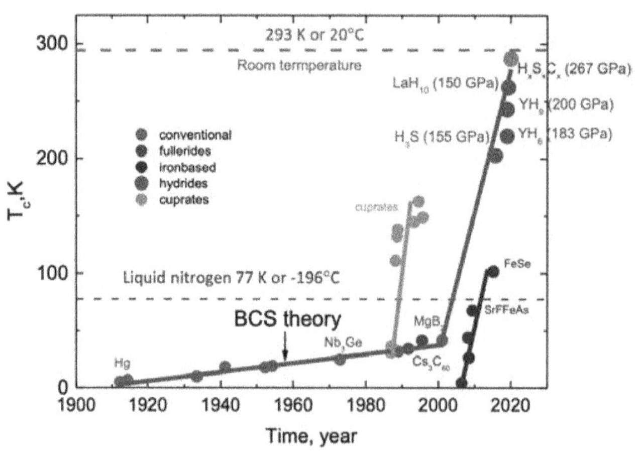

Figure 8. Superconducting transition temperature growth along time (courtesy of M. Eremets).

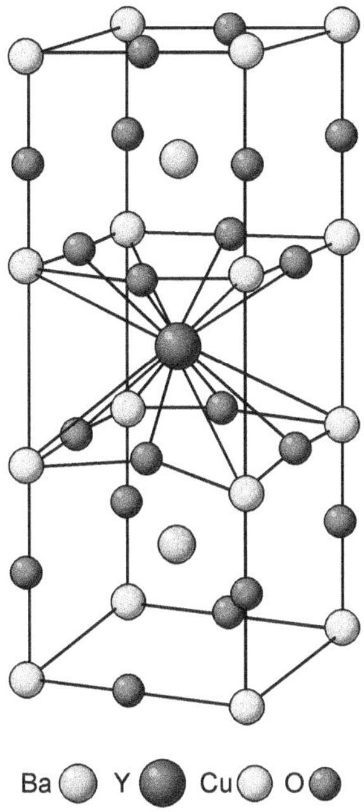

Figure 9. Crystal lattice of the superconductor $YBa_2Cu_3O_7$.

They are basically far from the BCS theory, which is based on the interaction between electrons through the electron–phonon interaction. Yet, Müller and Bednorz began to look for superconductivity in the $La_{2-x}Ba_xCuO_{4-\delta}$ compound precisely because, thanks to their intuition and some vague considerations, they expected the critical temperature to be especially high here! Many believe that Müller and Bednorz were just lucky: they discovered superconductivity exactly where they were looking for it, but mistaken motivation led them to this remarkable discovery. However, some recent experiments show that they may not have been wrong. For theoretical physics, the discovery of high-temperature superconductors is a mystery comparable to what the discovery of superconductivity in mercury was.

Dream of Perpetual Motion

In conclusion, let us return to the remarkable consequence of superconductivity. As we have already seen, the current established in a superconductor does not decay. This fact brings us back to the concept of "perpetuum mobile" — perpetual motion, a kind of holy grail, which alchemists, inventors and scientists have been seeking for many centuries, and which, nevertheless, is impossible according to the laws of classical physics! Without energy from the outside, the slightest friction will eventually stop any movement. Is superconductivity a manifestation of the perpetual motion of electrons? To answer this question, consider two different cases. The state of a superconductor through which a direct current flows in the absence of an external magnetic field, even under the most favourable conditions, at the lowest temperatures, is formally metastable — after all, its energy is higher than in a state without current. This means that such a current should eventually decay, just as a diamond crystal should eventually turn into graphite (see Chapter 23). How can this happen? Such a giant as Bohr took part in the discussion of this problem. The problem turned out to be that, due to the quantum nature of superconductivity, such damping cannot occur little by little; the superconducting current must decrease by macroscopic jumps, and it will take a long time to wait for such an event, perhaps even a time exceeding the lifetime of our Universe. In another example, consider a superconductor in the Meissner state, i.e., placed in a magnetic field below the first critical. Here, the direct current excited in a superconductor is indeed eternal. Is this really the perpetual motion that scientists of past centuries dreamed of? In a sense, yes, because the movement of charged particles is associated with current in classical physics. However, in quantum mechanics, this phenomenon has a significantly different meaning from its classical interpretation. In classical mechanics, observables are well defined, such as the position of particles, which depends on time. Therefore, in the presence of current, here, we can confidently talk about the movement of charges in space. The superconducting current under the conditions of an applied magnetic field is the reaction of the entire condensate to an external influence, which saves the superconducting state from destruction. A superconductor in such a current state is in thermodynamic equilibrium and can remain in it for an infinitely long time. Thus, superconductivity is a rare case of the manifestation of the laws of quantum mechanics in the macroscopic world around us.

Chapter 25

Applications of Superconductors

In the previous chapter, we outlined the phenomenon of superconductivity and narrated the long journey of scientists towards understanding its quantum nature. In this chapter, we will continue our story about the unusual properties of superconductors and focus on their applications that are little known to the general public.

Magnetic Flux Quantization in a Superconducting Ring

As we have seen, Niels Bohr was able to explain many of the properties of the atom by assuming that the velocity v of the electron in a circular orbit of radius R satisfies the relation $mvR = n\hbar$, where n is an integer, m is the electron mass, and $\hbar$ is Planck's constant (see Chapter 22). Remarkably, this rule can be generalised to describe the motion of any particle that makes a circular motion in a certain quantum state. Note that Bohr's postulate is not easily applied to the motion of an electron in an atom since the latter cannot be described by a well-defined orbit (otherwise, the uncertainty relations would be violated (see Chapter 22)). On the other hand, as Fritz London (1900–1954) showed in 1948, it can be successfully used for the superconducting current flowing in a circular ring of radius R (Fig. 1).

Bohr's quantization rule, applied to the motion of Cooper pairs in a superconducting ring (see Chapter 24), has an unexpected consequence: the magnetic flux crossing the superconducting ring is quantized — just like the radius of the orbit in an atom! More precisely, the magnetic flux Φ turns out to be an integer multiple of the "magnetic flux quantum"

$$\Phi_0 = \pi\hbar/e = h/(2e) = 2\cdot 10^{-15}\,\mathrm{T}\cdot\mathrm{m}^2,$$

where $-e$ is the electron charge (see Panel on page 357).

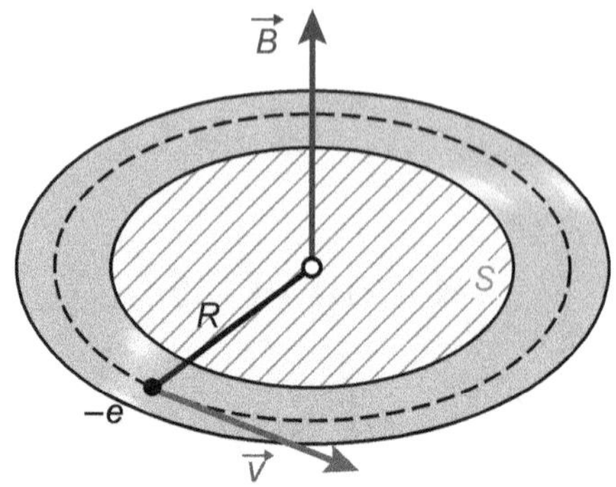

Figure 1. Current in a ring of radius R. The flux of magnetic induction through the ring $\Phi = BS$, where S is the area covered by the ring, and B is the magnetic field. Applying Bohr's quantization rule to a moving charge reveals that this flux is quantized.

The flux quantum Φ_0 is an extremely small quantity, so small that the first experimental test of the London hypothesis was carried out by American physicists Bascom Deaver and William Fairbank only 13 years later, in 1961.

It is noteworthy that while the quantized physical quantities that we have mentioned so far belong to the microscopic world, scientists can measure the magnetic flux quantum in relatively large, almost macroscopic samples (that is, visible to the naked eye). An example of such "mesoscopic", i.e., intermediate between the micro- and macroworlds of objects, is the vortices of Abrikosov, which can be located from each other at distances of micrometres. We recall that these vortices arise in a type II superconductor placed in an external magnetic field (see Chapter 24). Each Abrikosov vortex is a carrier of a magnetic flux quantum Φ_0. Another similar example is a superconducting ring, which makes it possible to observe discrete changes in the magnetic flux penetrating it literally in one quantum Φ_0. (see the following). Such observations are analogous to the experiment of Deaver and Fairbank in 1961. Instead of a ring, they used a superconducting cylinder.

What happens if one places the ring in a alternating magnetic field B? For a "normal" conductor, the magnetic flux Φ penetrating the ring changes with the field. As a result, an electromotive force of induction $d\Phi/dt$ arises, which generates a current I_N in the ring, such that $RI_N = d\Phi/dt$, where R is the resistance of the ring. For a superconductor, as we already know, the flow must remain unchanged, otherwise its change would cause an infinite current that destroys the superconducting state. Therefore, a current I_S will also flow along the superconducting ring with a change in the magnetic field, such that the total flux Φ crossing the ring (which is the sum of the external flux $\Phi_{ext} = BS$ and the intrinsic induction flux $\Phi I = LI_S$, where L is the inductance of the ring) does not change.

Flux quantization in a superconducting loop

By resorting to a simplified approach, let us find the formula describing flux quantization in a superconductor. Let us refer to the case of a circular ring with zero resistance and containing free charged particles. By increasing the magnetic field, the magnetic flux Φ also increases, which leads to the rise of an electromotive force of induction (see Chapter 16):

$$V = -\frac{\Delta\Phi}{\Delta t},$$

where $\Delta\Phi$ is the flux change that occurs during the time interval Δt. In this case, the electric field induced in the ring is given by

$$E = \frac{V}{2\pi R}.$$

Now let us consider a charge q of mass m moving along a ring with a speed v (Fig. 1). The force acting on it is equal to qE, and, according to the basic principle of dynamics, the acceleration experienced by the charge is equal to qE/m. This acceleration should be equal to the ratio of the increase in speed Δv to time Δt, therefore

$$\frac{\Delta v}{\Delta t} = \frac{qE}{m} = -\frac{q}{m}\frac{\Delta\Phi}{2\pi R\Delta t}.$$

Thus, one finds $\Delta\Phi = \frac{2\pi}{q}(mR\Delta v)$. Under the assumption that Bohr's quantization rule $mRv = n\hbar$, mentioned at the beginning of this chapter, is applicable to the motion of the charge under consideration in a superconducting ring, we find that the smallest non-zero value of $mR\Delta v$ is equal to $\hbar$. Hence it follows that the smallest non-zero value of $\Delta\Phi$ is equal to $2\pi\hbar/q$. The charge of the Cooper pair is $2e$; thus we arrive at the formula for flux quantization in a superconducting ring given in the text.

Tunnel and Josephson Effects

The situation becomes even more interesting if the superconducting ring is interrupted by a thin layer of dielectric material. Let us assume that we managed to create such a thin layer in a ring made of normal metal (Fig. 2). Intuitively, it

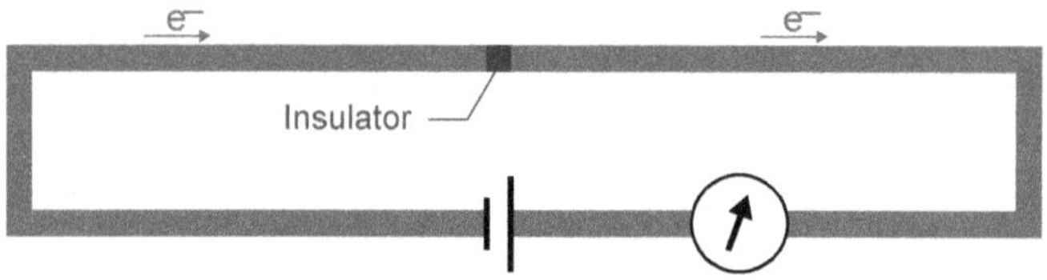

Figure 2. Tunnel effect in a conducting ring containing a dielectric barrier. If the dielectric layer is not too thick, then the electrons, due to the tunnel effect, pass through it with some probability. The current flowing in the ring is measured with an ammeter.

seems that the current will not be able to flow in it. Nevertheless, if the thickness of the dielectric layer is not too large (for example, about a micrometre), the current in such a structure can flow! Some electrons "magically" manage to pass through the dielectric. This purely quantum phenomenon is called *tunnel effect*. For example, when studying objects with a tunnelling microscope, the current passes between the tip of the device and the surface of the sample through a vacuum gap of about a nanometre in size (see Chapter 28). Not only electrons but also other more massive particles, even such as atomic nuclei, can tunnel with some probability through "walls", in the quantum world. True, with an increasing mass, the more difficult it becomes for them to perform these miracles.

What happens in the case of a superconducting ring? It turns out that a peculiar tunnelling effect takes place here too. Cooper pairs manage to overcome a dielectric layer several nanometres thick or a normal metal layer a dozen nanometres thick. It would seem that for such composite bosons, the tunnelling mechanism should consist of sequential tunnelling through the wall first of one and then for the second electron. The superconducting current correspondingly must be proportional to the square of the one-electron tunnelling probability through the barrier, hence it should be small. However, another curiosity about superconductivity is the fact that, as we already know, the pairs are rather smeared objects in space. Thus one Cooper pair tunnels through the barrier with approximately the same probability as that one for the single electron that belongs to it. We can say that both electrons somehow tunnel *coherently*, simultaneously. This phenomenon was predicted in 1962 by the Englishman Brian Josephson (then only a 22-year-old graduate student at Cambridge University), and the "sandwich" he invented — a dielectric layer between two superconductors — is called the *Josephson junction*.

For this discovery, in 1973, Brian Josephson was awarded the Nobel Prize in Physics. Why was such a prestigious award given for the "simple" demonstration that a superconducting current has the same property as a conventional one? First, it was completely unexpected that Cooper pairs could tunnel through the barrier without breaking. Second, Josephson was able to predict the remarkable properties of the device he invented. Their description is beyond the scope of this book, but we will simply address one of its applications.

The Josephson junction is a quantum magnetometer capable of measuring ultraweak magnetic fields, for example, the one associated with the blood flowing through the heart. This allows early diagnosis of cardiovascular diseases.

Measurement of Very Weak Magnetic Fields

The simplest quantum magnetometer consists of a superconducting ring with the thinnest dielectric bridge (Fig. 3(a)). Let us imagine that this ring is placed in an external magnetic field, which, at first, like the current in the circuit, is equal to zero. Then the flow inside the ring is also zero. Let us start by increasing the external field. As long as the field is not too strong, the total magnetic flux Φ passing through the ring should remain unchanged and equal to zero. For this, it is necessary that the magnetic flux ΦI generated by the current I flowing through the ring at any given moment compensates for the change in the external flux Φ_{ext}. As the magnetic field grows, this current also increases until the critical value of I_c is reached (due to the choice of the resistance of the dielectric layer, it is possible to reach this at $\Phi_{ext} = \Phi_0/2$).

As soon as the current becomes equal to I_c, superconductivity in the vicinity of dielectric layer is destroyed, and the flux quantum Φ_0 enters into the circuit.

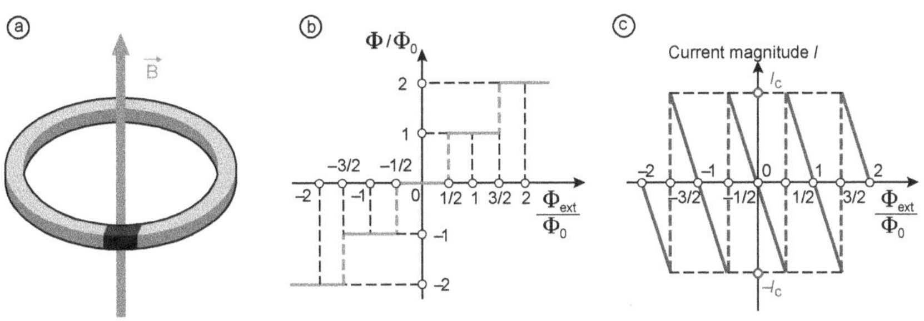

Figure 3. (a) Superconducting ring including a Josephson junction is placed in an external magnetic field. (b) With a monotonic increase of the external flux $\Phi_{ext} = BS$, the total magnetic flux Φ changes by jumps. (c) The magnitude I of the superconducting current cannot exceed the critical value I_c, determined by the properties of the contact, and therefore the changes occur in a sawtooth manner. The current changes sign when the superconductivity in the contact is destroyed (here, we consider the case when the critical value of the current I_c is reached with an increase in the external flux by $\Phi_0/2$).

The total magnetic flux will increase by one quantum. Such a change is possible only due to the destruction of superconductivity in the region of the bridge, which is what makes the device so extraordinary! What will happen to the current? Its magnitude will remain the same, but the direction will change in the opposite sense. If before the entry of the flux quantum Φ_0, the current I_c completely screened the external flux, and then after its entry, it should amplify the external flux $\Phi_0/2$ to the value Φ_0. Therefore, at the moment of entering the flux quantum, the direction of the current abruptly changes to the opposite sense again. With a further increase in the external field, the current in the ring will begin to decrease, superconductivity in the ring will be restored, and the flux inside the ring will remain equal to Φ_0. The current in the loop will turn to zero when the external flow also becomes equal to $\Phi_0\Phi_0$, and then it begins to flow in the opposite direction. Finally, when the value of the external flux is $3\Phi_0/2$, the current will again become equal to I_c, superconductivity will collapse, the next flux quantum will enter, and so on (Fig. 3(c)).

The stepwise character of the current as a function of the flux makes it possible to measure the value of the external field with extraordinary accuracy. However, the problem of measuring the current in a ring with a tunnel contact remains.

SQUID Magnetometer

Often in a superconducting ring, instead of one, two Josephson junctions are created at once. Thus a "superconducting quantum interferometer" or SQUID (from the Superconducting Quantum Interference Device) is obtained (Fig. 4). Its

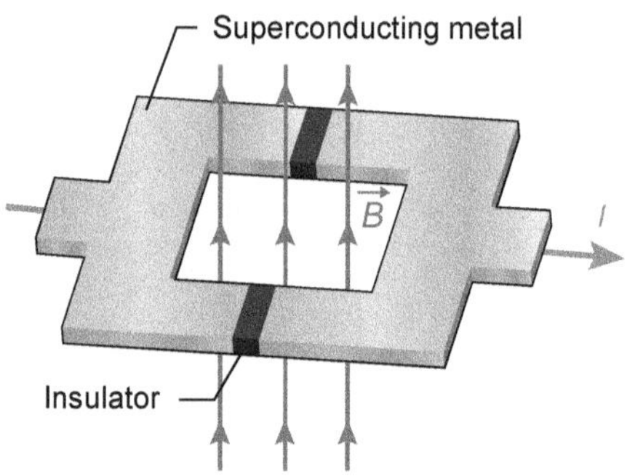

Figure 4. Principle of SQUID operation. The magnetometer consists of a superconducting ring with two Josephson junctions. The current *I* flowing in the SQUID is split into two branches. If the device is placed in an external magnetic field *B*, then these two currents interfere, resulting in a potential difference between the tunnel contacts, the measurement of which allows you to estimate the value of the field.

principle of operation is based on the interference of the wavefunctions of two superconducting condensates separated by Josephson junctions, which can be compared with the interference occurring in two adjacent Young slits in optics (see Chapter 3). With the help of sophisticated devices (generators and amplifiers), the SQUID can measure flux fluctuations much smaller than the quantum Φ_0. It is so sensitive that it detects magnetic fields from the heart or brain activity! These fields are 100,000 times weaker than the Earth's magnetic field (which is about $5 \cdot 10^{-5}$ T on its surface). The first attempts to use SQUID in medicine, such as magnetocardiography and magnetoencephalography (Fig. 5), date back to the 1970s. To minimise the influence of the Earth's magnetic field on the measurements, they were performed in a special room: the walls consisted of three layers of metal with high magnetic permeability, forming powerful magnetic shields, separated by two layers of aluminium, preventing the penetration of the electric field. Thus the Earth's magnetic field, inside the volume, decreased by 10,000 times. However, the creation of such premises was very expensive. Today, thanks to the current advances in technology in the field of superconductors, magnetometers no longer require a magnetic shield and are capable of measuring magnetic fields with an accuracy of 10^{-15} T! The only thing the patient has to do is remove all metal objects such as keys from their pocket.

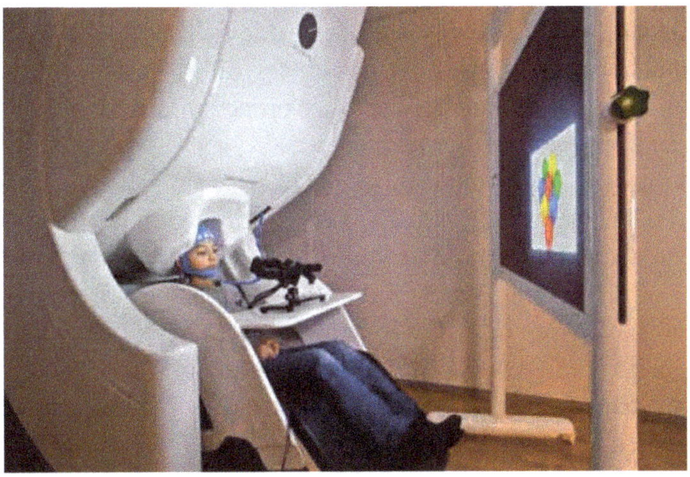

Figure 5. Magnetoencephalographic scanner, consisting of 306 SQUIDs that register even extremely weak magnetic fields generated by neural activity. The top of the scanner contains the liquid helium needed to cool the device. The child's neural activity is detected through visual and auditory stimulation (a black oculometer complements the apparatus).

Thorny Path to Records

As Kamerlingh Onnes assumed from the very beginning, the creation of strong magnetic fields is an obvious application of the remarkable properties of superconductors. Magnetic fields for industrial use are usually produced using electromagnets, that is, coils through which an electric current flows. The field strength depends on the current flowing in the coil and on the number of turns of the wire in it (see Chapter 16). However, a coil of ordinary conductive material has a resistance, and heat is generated in it when current flows, due to the Joule-Lenz effect. This dissipates a lot of energy, and to prevent the wires from melting, they must be cooled intensively! For example, in 1937, a field of 10 T was produced for the first time. The power required was so high that it was possible to carry on experiments only during the night when the need by other users was quite low: the coil's cooling system required a water flow of 5 l s^{-1}.

For a superconductor, these restrictions do not exist! At first glance, it is enough to make a coil from a superconducting wire and create a sufficiently strong current in it: since the superconductor's resistance is zero, it will not generate heat. And when the current is established, it will not be necessary to supply power to the circuit! It would seem that the game is worth the candle, despite the fact that the coil must be kept at the temperature of liquid helium. But, unfortunately, generally type I superconductors do not withstand magnetic fields to an extent for practical applications (see Chapter 24). The solution to the problem was the discovery of type II superconductors, which, as we already mentioned, can remain in a superconducting state up to very high magnetic fields. The magnetic field penetrates into their volume in the form of vortices with a normal core. However, a superconducting phase remains between the vortices, through which the superconducting current can flow without resistance.

Unfortunately, not everything turned out to be so simple. The fact is that when current flows Ampère's force acts on vortices (through which a magnetic field penetrates into the superconductor) in the direction perpendicular to both the magnetic field and the current. As a result, the entire lattice of Abrikosov vortices begins to move. The product of the Ampère's force vector with the vortex displacement vector implies work. Thus, the motion of the vortex lattice occurs with energy dissipation, and again the electrical resistance of the superconducting coil becomes non-zero!

Fortunately, vortex movement can be prevented. For this, it is sufficient that the superconductor contains microscopic defects. As a rule, they arise spontaneously as a result of heat treatment during the manufacture of superconducting alloys (Fig. 6). Single vortices "catch" on these defects, and then the entire lattice

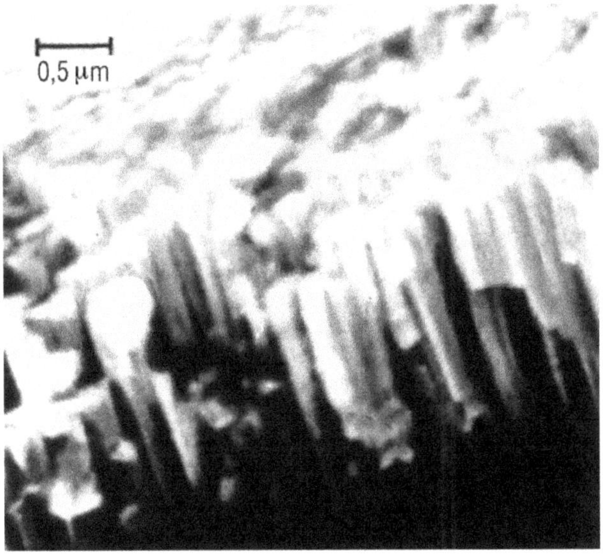

Figure 6. Electron microscopic image of a superconducting film of niobium nitride (NbN), obtained by sputtering a metal onto a glass plate. The columnar structure of the material is clearly visible. It is rather difficult for Abrikosov vortices to jump over the border of such grains.

of Abrikosov vortices no longer moves. It is clear that such a mechanism cannot withstand arbitrary an Ampère's force of any magnitude. However, as long as the current does not exceed a certain critical value, the electrical resistance of the superconducting wire remains zero. This phenomenon is called *pinning*: the vortices seem to be "pinned" on the defects. Usually, defects are harmful, but in this case they help! Nevertheless, the presence of defects, as will be shown below, plays also a negative role.

Due to the pinning phenomenon, many type II superconductors are used to generate strong magnetic fields. This, for example, is a tin–niobium alloy, in which current densities up to 10^5 A cm^{-2} can be achieved (compare this value with several hundred amperes per square centimetre for copper). In this case, the upper critical field B_{c2} for this alloy at low temperatures is 25 T.

Superconducting Cable Technology

To obtain strong magnetic fields, the creation of an alloy with suitable critical parameters is necessary, but not sufficient. You still need to make a cable from it! Tin–niobium alloy is fragile, and a cable made from it breaks at the slightest

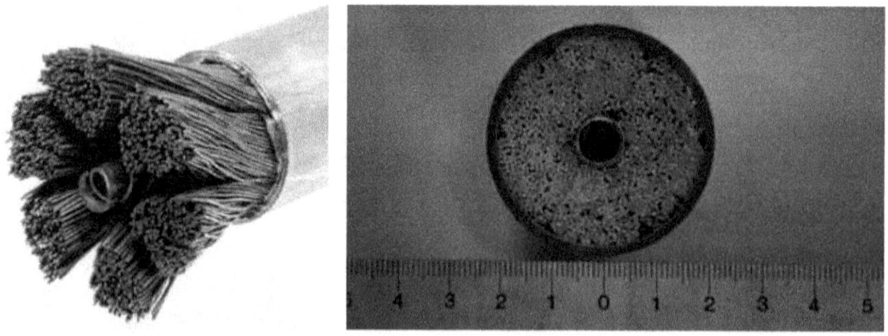

Figure 7. Tin–niobium cable (Nb_3Sn), composed of a plurality of superconducting filaments embedded in a copper matrix. Its diameter is about 4 cm. Cooling liquid helium passes through the hollow space in the centre of the cable. The cable carries a current of 68 kA.

twisting. This problem was solved by filling the copper tube with a powdered mixture of niobium and tin. Then this tube is stretched (drawn) in such a way as to obtain a wire, which is then heated. The powder melts to give the desired alloy of tin and niobium. The described process underlies the creation of the so-called composite superconductors. They are obtained by drilling parallel channels in a copper matrix and inserting superconducting fibres into them. The die is subjected to a drawing procedure, and the resulting wire, in turn, is reinserted into the holes of the next die, etc. By repeating this procedure several times, a cable containing millions of superconducting fibres is obtained (Fig. 7). For example, in the coil used for the *international thermonuclear experimental reactor* (**ITER**) (which will be discussed below), each cable consists of 900 superconducting fibres made of tin–niobium alloy Nb_3Sn and 522 copper strands with a diameter of 0.8 mm, which are divided into six “petals”. Each of the superconducting fibres consists of approximately 9,000 Nb_3Sn strands a few micrometres in diameter embedded in the copper matrix. The total number of strands in the cable exceeds 8 million. Of course, the same can be done with any other alloy, for example, the niobium–titanium alloy NbTi, which is more common and less expensive than the Nb_3Sn alloy.

Why are copper and superconducting threads combined? The point is that using a pure superconductor cable is risky. Superconductivity can suddenly disappear in some place, for example, due to defects added for vortex pinning. In this case, the corresponding section of the cable, under the influence of the strongest current flowing through it, heats up quickly, and if the released heat is not discharged in time, then the entire cable can completely transit into a normal state. This will lead to catastrophic consequences, from severe damage to the cable to the destruction of nearby objects. The presence of copper, a good conductor of heat, prevents such a disaster.

What About High-Temperature Superconductors?

After the discovery of superconductors with high critical temperature by Müller and Bednorz, scientists continuing the research in this new field hoped soon to work out miracles, because cheap liquid nitrogen could be used for cooling, and the critical fields promised to exceed 100 T. But, in practise, the implementation of their plans was far from easy. The difficulties in creating superconducting cables based on new materials turned out to be similar in many respects to those that arose when using traditional superconductors, for example, the Nb_3Sn alloy, such as the high fragility of materials, and problems associated with the pinning of the Abrikosov vortex lattice. The problem was further complicated by the looseness of the vortices along their axis, due to the weak coupling between the layers in the quasi-two-dimensional high-temperature superconductors. Nevertheless, good results have been achieved by creating composite materials based on superconducting oxides and silver, and some superconducting cables based on YBaCuO are already reaching the preindustrial stage.

Where Do Superconductors Work?

Today, the magnetic fields generated by superconducting magnets reach values of several tens of tesla. Often these magnets have hybrid structure: the outer superconducting coil creates its own magnetic field, and the inner one, with a copper winding, further enhances it in its volume. Such coils are used, for example, at the French National Laboratory for High Magnetic Fields in Grenoble, where they create continuous magnetic fields reaching almost 40 T (this is where the quantum Hall effect was discovered, see Chapter 28). More recently, researchers at the USA National High Magnetic Fields Laboratory have developed the world's most powerful superconducting magnet, capable of creating a magnetic field with a record of 45.5 T. In another branch of the French National Laboratory in Toulouse, even higher-pulsed magnetic fields, reaching 100 T, are produced. However, this is obtained by other methods, without resorting to superconductivity.

The use of superconducting coils is not limited to Physics laboratories. They are used daily in hospitals for MRI examinations (see Chapter 27), which require intense and uniform fields. Let us mention two more important directions of using superconducting magnets: in the particle accelerators in the study of elementary particle physics and as an important element of prototypes of thermonuclear reactors.

LHC at CERN

In the spring of 2012, a new elementary particle was discovered — the Higgs boson — or rather, according to CERN researchers, "a particle compatible with the Higgs boson" (physicists are careful people!). The existence of this particle was predicted theoretically a long time ago, and also it fully agreed with existing experimental observations. Proof of the existence of this mysterious boson would explain why elementary particles have mass. To verify the existence of the Higgs boson and to be able to carry on other basic research, an underground ring 26.66 km long was built near Geneva — the Large Hadron Collider (LHC). Protons in it are accelerated to speeds very close to the speed of light, using the strongest magnetic fields, directing them along a circular path. This field is generated by several thousand superconducting magnets installed along the ring. The use of conventional magnets would require expensive cooling devices, which, due to their bulkiness, would be impossible to place in the tunnel. The niobium–titanium superconducting wires from which the magnet coils are made are capable of carrying currents of up to 12,000 A. It takes weeks for this entire cyclopean device to cool down to temperatures below 2 K.

ITER: Energy of the 22nd Century?

Another machine that uses superconductors is the International Thermonuclear Experimental Reactor (ITER), which is currently under construction in Cadarache, near the gorge of Verdon in France. ITER is designed to generate energy by nuclear fusion. Recall that the nuclear fusion reaction consists of the fusion of two light nuclei (for example, deuterium (^2H) and tritium (^3H)), a heavier nucleus is formed. In the process of this reaction, as in the fission of heavy nuclei (see Chapter 13), energy is released. Nuclear fusion requires a very high temperature (100 million degrees!). It is through the fusion of nuclei that the energy of the Sun is generated. Ionised particles form "plasma", a hot gas that must be kept in the chamber without allowing it to touch the walls. In the case of ITER, such confinement is provided by a magnetic field acting on charged particles moving in the toroidal chamber (Fig. 8). In short, the mechanism is the same as in particle accelerators such as the LHC. However, in the latter, the proton beam is very narrow, and the radius of their trajectories is gigantic (about 10 km). The radius of the fusion reactor is much smaller, and yet the volume in which it is necessary to keep the plasma at several hundred million degrees is still 840 m^3.

The ITER design is a "tokamak" (toroidal chamber with magnetic coils), a type of device invented in the 1950s by two Russian physicists: Andrei Sakharov

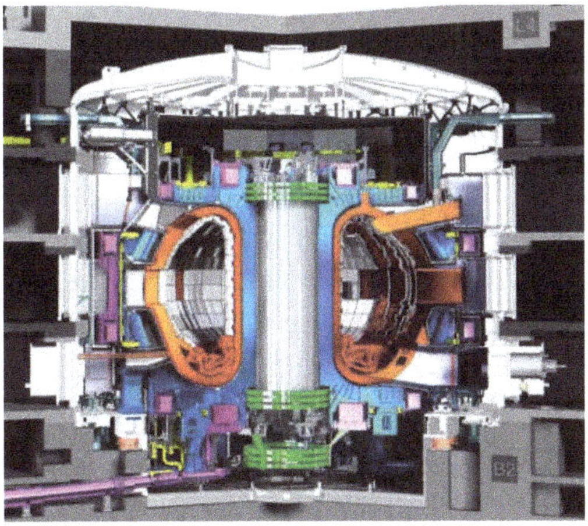

Figure 8. Model of the ITER reactor, sectional view. It has a height of a five-storey building, with a diameter of about 30 m. In the centre, there is a solenoid that accelerates charged particles. There are 18 coils of the toroidal field, which hold the plasma in the chamber. Six additional coils of the poloidal field (that is, directed along the lines passing through the poles of the spherical coordinate system) prevent the incandescent plasma from touching the walls and ensure its stability. For the manufacture of various coils in such an installation, more than 500 t of superconducting alloy Nb_3Sn were spent.

(1921–1989) and Igor Tamm (1895–1971). The first tokamaks used conventional electromagnets, which consumed enormous amounts of energy. The magnetic fields required to confine the plasma are of the order of 10 T, i.e., they are quite moderate and allow one to use superconducting magnets to create them, thus providing significant energy savings. The purpose of the creation and commissioning of the ITER is "to demonstrate the possibility of using thermonuclear fusion as a potential source of energy". Specific results, that is, cost-effective and safe production of electricity by nuclear fusion, are expected, according to the most optimistic forecasts, by 2040.

And Future Applications...

Let us talk about some interesting ideas for using superconductors in addition to creating high magnetic fields. For example, the levitation effect arising from the Meissner–Ochsenfeld effect (see the previous chapter) is used to create high-speed MAGLEV trains moving due to magnetic levitation. Such trains float above the rails, thanks to superconducting magnets installed in the cars, interacting with magnets placed along the rails on the ground. The speed record belongs to a

Japanese MAGLEV, tested in 2015 on an experimental section of the route between Tokyo and Nagoya. During the tests, the train accelerated to 603 km h^{-1}.

Another application of superconductors is energy storage, which is an important task for the use of solar, wind and other power plants that generate energy at a variable rate (Chapter 13). Indeed, the excess energy accumulated during production peaks should be stored in some way and then released as needed. One solution is to generate current in a superconducting coil. The accumulated electromagnetic energy in this case is $LI^2/2$, where I is the strength of the current flowing in the coil and L is its inductance. At the moment, practical applications of this energy storage method are restricted due to the energy costs required for cooling. Experimental lossless power transmission is already being practised, for example, on Long Island in the United States and Essen in Germany: superconducting cables several hundred metres long are replacing high-voltage power cables for entire neighbourhoods.

Scientists are working on creating an element base for quantum computers (see Chapter 28), based on quantum superconducting processors resorting to Josephson junctions. So, this area of science is in the active phase of its development. Undoubtedly, in the decades to come, we could expect the emergence of many new areas of application of superconductivity.

Chapter 26

Snowballs and Bubbles in Liquid Helium

Helium is the second element in the Mendeleev periodic table, and because of its eccentric properties, possibly it is the element that troubles scientists most. It has cause many headaches and sleepless nights, but the beautiful mechanisms of its particular properties has provided many satisfying and pleasant moments.

The Achievement of Liquefying Helium

From previous pages, the reader already knows that helium can transition to the liquid state only at very low temperatures, and furthermore it cannot become solid at any temperature under standard atmospheric pressure. At variance, helium can become superfluid, namely completely losing its viscosity, under particular conditions. No other element has these properties.

Helium was first liquefied in the year 1908 by Heike Kamerlingh Onnes in a laboratory at the University of Leyden, more precisely in July (Fig. 1). For a rather long time, Kamerlingh Onnes was competing with several other researchers who were trying to liquefy the gas. Helium was the only element refusing to become liquid! In March of 1907, Kamerlingh Onnes thought that he was able, not only to liquefy helium, but also to make it solid. In fact, after a sudden variation of the pressure, he had observed the formation of a white cloud inside the gaseous state, erroneously thinking that it was solid helium. Full of enthusiasm he sent a telegraphic message to his English colleague Sir James Dewar (the first scientist to succeed in liquefying hydrogen) declaring "converted helium into solid".

The international press celebrated the event. Unfortunately, the believed solid helium, that white cloud, turned out to be droplets of solid hydrogen that had

Figure 1. Kamerlingh Onnes (sitting on the right side of the picture) and his co-workers in the laboratory in Leyden where helium was liquefied for the first time. Onnes was rewarded with the Nobel Prize in Physics in the year 1913 for his studies on the properties of matter at low temperature.

entered the complex system of cans and capillaries of the apparatus! Deeply frustrated Kamerling Onnes was derided by most people, claiming that instead of solid helium, he had discovered the *halfium* ("half" in Dutch means "one half", while the meaning of "heel" — a play on "hel" in "helium" — is the "whole"). Moral comments to take away from this event: (1) major honours can fall down; (2) it is better not to imitate even the greatest scientists in announcing some discovery too early! Anyway, on 10th July, helium was indeed liquefied.

Thanks to this major technological success a quite novel field of studies was opened to researchers. Resorting to liquid helium, it became possible to carry on a variety of experiments in the temperature range close to absolute zero. In particular, by using liquid helium, Kamerlingh Onnes could discover the phenomenon of superconductivity in mercury at a temperature of around 4 K (see Chapter 24).

At this point, we shall describe a less well-known history: the transport of electrical charges in liquid helium.

Electric Charges in Liquid Helium

In general, in common liquids, some number of electrically charged particles are always present. For instance, in water, at ordinary temperature, a non-negligible fraction of the molecules H_2O dissociate into H^+ and OH^- ions. In practise, the positive ion is bonded to a water molecule to form H_3O^+. In liquid helium, this type of dissociation is completely absent, and no "free" electric charges are present. To a very large extent, most helium atoms are in their lowest energy quantum state. In order to promote one atom to an excited energy state, one has to provide an energy amount of around 20 eV (1 eV = 1.6×10^{-19} J). According to the Gibbs–Boltzmann equation, the probability that, at temperature T, one atom can acquire an energy E is given by $\exp[-E/(k_B T)]$, where k_B is the Boltzmann constant (see Chapter 7). At standard atmospheric pressure, helium can be found in the liquid state only below 4 K. At this temperature, $\frac{E}{k_B T} = 58.000$, and therefore the probability $\exp(-58.000)$ is so small that it can be safely considered zero. Even at room temperature, the probability of finding a helium atom in an excited state is negligible. *A fortiori*, the probability to find the latter in the ionic state He^+ is insignificant. However, some kinds of charged particles can artificially be introduced into liquid helium, and consequently weak currents can be detected in it when an electric field is applied.

For instance, one can introduce nuclei of He^+ by irradiating the liquid helium with α particles. By resorting to β radiation, one can also introduce electrons.

Why would one wish to violate the neutrality of the poor helium? This is because this violation implies the occurrence of unexpected phenomena that required serious efforts by physicists to be explained.

We will first address the case in which the charge carriers are positive ions. This was the first case where the mysterious events were registered. Later, in helium, even more miraculous behaviour of the negative charge carriers was discovered.

Structure and Transport of Positive Ions

Slightly before the 1960s, physicists got involved in this subject and asked themselves some questions. They were studying the mass of the charge carriers in liquid helium by detecting their trajectories in uniform magnetic fields. A charged particle having a given initial velocity must move along a helical trajectory, and, from the radius of the latter, the particle mass can be derived. The results were found to be totally unexpected. The mass of the charge carriers, both positive and negative, turned out to be 10,000 times greater than the mass of a free electron!

Another surprising property directly involves the mobility of the He^+ ions in the liquid, namely the relationship between their velocity and the force causing the motion. The mobility of the isotope ^3He in the more abundant ^4He was already known. The mobility of the ion He^+ was expected to be of the same order of magnitude. However, it was found that, for He^+ ions, this value is about 100 times less. How do you explain this new helium quirk?

The solution was found by U.S. physicist Kenneth Robert Atkins and described in his article in 1959. According to his theory, the presence of the He^+ ion creates a perturbation in the surrounding helium atoms. This positive ion attracts its electrons to itself and at the same time repels its nuclei (this phenomenon is called the polarisation of atoms). Due to the small difference in distances, attraction prevails over repulsion, so atoms approach the He^+ ion: their concentration increases as they approach the He^+ ion, and the pressure around it increases. As already mentioned, at low temperatures and pressures of 25 atm,[1] helium solidifies. The calculation shows that this pressure is reached at a distance $r_0 = 0.7$ nm from the He^+ ion (Fig. 2) (to have an idea of the scale: the radius of the helium atom is 0.13 nm). Thus a kind of snowball grows around the ion: a ball of solid helium with an ion in the centre! When a potential difference is created in the liquid, this snowball, having a charge in the centre, begins to move in the direction of the electric field. In its movement, the snowball is not alone: it carries along a kind of "retinue" of polarized atoms of liquid helium.

This model allowed physicists to explain the available experimental results, including why the mass of the carrier of a positive charge exceeded the mass of an He^+ ion by more than 10 times. According to Atkins, this mass, in addition to the mass of the solid helium snowball itself, includes two additional terms. First,

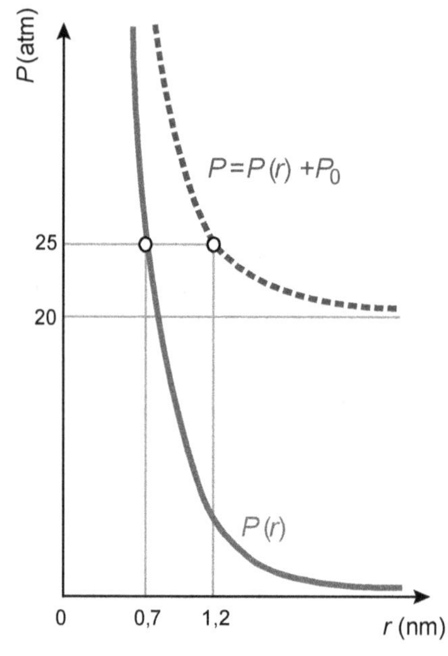

Figure 2. Local pressure as a function of the distance r to the He^+ ion at standard atmospheric pressure (solid curve) and at an external pressure P_0 equal to 20 atm (dashed line). The dotted curve is obtained by vertically displacing the solid curve.

[1] It is reminded that 1 atm, a unit of pressure, corresponds to 100 kPa.

to the mass of the snowball, one should also add the mass of the "retinue", the mass of the polarised atoms involved in the motion. Calculation shows that the latter is $28m_0$, where $m_0 = 6.7 \cdot 10^{-27}$ kg is the mass of the helium atom ^{4}He. Secondly, when a body moves in a liquid, it pushes the layers of liquid around itself, which requires energy. Therefore, to accelerate a body when it moves in a liquid, some additional force is required in comparison with the one which would be necessary when it is accelerated in a vacuum. Thus an object in a liquid behaves as if it had a mass $m+\delta m$ greater than its actual mass m. The excess δm is the "added mass" we talked about back in Chapter 15 when discussing the motion of bubbles in water. For our snowball moving in helium, the corresponding correction turns out to be $15m_0$. Finally, the mass of the snowball itself is the product of its volume and the density of solid helium,[2] which gives $32m_0$.

Thus, summing up all three terms, we find that the mass of a positive ion moving in liquid helium is $75m_0$ — a value approximately equal to the value found from the analysis of experiments. In the above discussion, we used the concepts of classical physics, which easily describe the motion of positive charges. However, for negative charges, everything turns out to be much more complicated.

And How Is the Carrier of a Negative Charge Arranged?

We have already pointed out that liquid helium in the equilibrium state does not contain free charges. If an electron is forced into it, then the latter will cause local shocks. To talk about this, let us digress and discuss the electronic structure of atoms. There is an important law in the quantum world: this is the Pauli exclusion principle, which does not allow two electrons to be in the same quantum state at once (see Chapter 22). For example, the helium atom has two different states with the same minimum energy, which are occupied by two electrons. There are other energy states for electrons, but they correspond to much higher energies (minimum 20 eV), and they remain unfilled. Thus it is impossible to create a He^- ion by adding a third electron to the neutral helium atom. And yet, being accelerated to relatively modest energies of 0.5 eV, electrons penetrate through the surface into the bulk of liquid helium!

Three Italian physicists, G. Careri, U. Fasoli and F. S. Gaeta, suggested that when an electron penetrates into the volume of liquid helium, the latter does not

[2] This helium density is about 1,800 kg m^{-3}, under a pressure of 7 million pascals (70 atm). This value is about 14 times more than at the standard atmospheric pressure (1 atm) when the helium density is 125 kg m^{-3}.

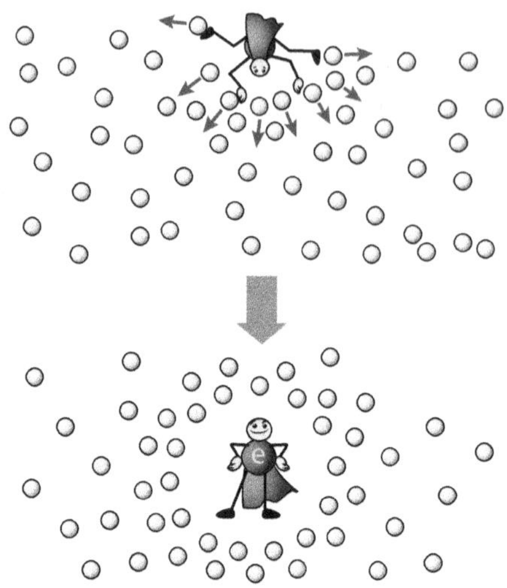

Figure 3. Due to the laws of the quantum world, an electron cannot get too close to helium atoms. It pushes away the atoms and generates a kind of cavity around itself.

at all try to "adjust" to a free energy level in one of the atoms, "paying" 20 eV for this. No, it simply remains itself, and pushes the surrounding helium atoms apart, creating a cavity for itself and spending only 0.5 eV for this (Fig. 3). The resulting "bubble" is the carrier of the negative charge.

What is the radius R of this bubble? Its size is due to the balance between the forces of surface tension and the pressure of an electron on the surface. On the one hand, the formation of a bubble requires the expenditure of energy E_1, which is greater the larger the bubble volume (surface energy, see Chapter 6). On the other hand, an electron in a bubble is continuously moving and has a kinetic energy E_2, which, due to the uncertainty principle, is greater the smaller the bubble is. The bubble radius R will be such that it minimises the total energy $E_1 + E_2$. It is easy to estimate energies E_1 and E_2. The first value is $E_1 = 4\pi\sigma R^2$, where σ is the surface tension of liquid helium. The energy E_2 can be found from the uncertainty principle (see Chapter 22): according to it, the electron momentum $p = m_e\nu$ is approximately $\hbar/R$, so the kinetic energy $E_2 = m_e\nu^2/2$ turns out to be of the order of $\hbar^2/(2m_eR^2)$, where $\hbar$ is Planck's constant, m_e is the mass of an electron and ν is its speed. By minimising the total energy $E_1 + E_2$, one can find that in a state of equilibrium $R^4 \approx \hbar^2/(m_e\sigma)$. An accurate calculation gives the value of $R \approx 2$ nm for the bubble radius. It practically does not have its own mass, because the electron mass is negligible compared to the added mass $\delta m = 2\pi\rho R^3/3$ (see Chapter 15), where ρ is the density of liquid helium at ordinary pressure. Here, it should be noted that the electron, like the He^+ ion, also polarises the helium atoms around the bubble, so the mass of the "retinue" accompanying the bubble when it moves in an electric field should be added to δm. However, due to its large radius compared to the snowball, the effect of polarisation of the surrounding helium is weak, and the corresponding mass turns out to be negligible compared to the added one, $\delta m = 245m_0$, which determines the effective mass of the negative charge carrier in liquid helium.

The Effect of Pressure

What happens if liquid helium is subjected to external pressure? First of all, we are interested in carriers of positive charges, our famous "snowballs". The higher the external pressure P_0, the faster the pressure of 25 atm is reached near the He^+ ion (Fig. 2). As a result, the size of the "snowball" becomes larger and larger with increasing external pressure (curve $r(P_0)$ in Fig. 4).

What happens this time, with an increase in external pressure, for a bubble that is a carrier of a negative charge? Like any other bubble, it contracts with increasing external pressure (Fig. 4, upper curve). When P_0 reaches about 20 atm, the bubble radius $R(P_0 = 20$ atm) becomes equal to the radius of the snowball $r(P_0 = 20$ atm) = 1.2 nm. One might think that with a further increase in pressure, the bubble will continue to shrink, and R will decrease. But not at all! The point is that the total pressure on the bubble surface actually turns out to be greater than the external P_0 since it is necessary to add the induced pressure to it due to the attraction by the electron of the liquid helium atoms polarised by it from its "retinue". It turns out

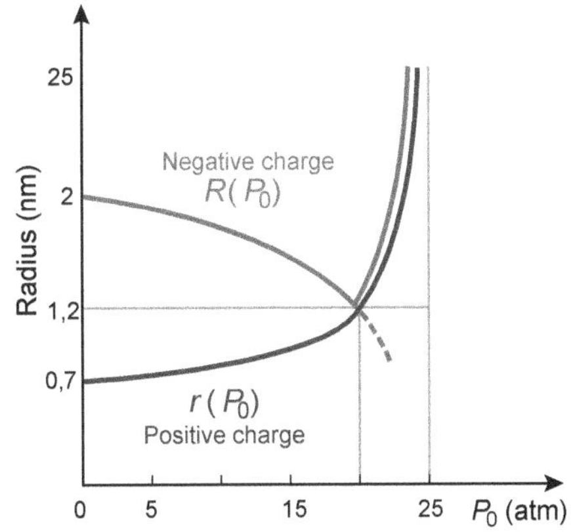

Figure 4. Evolution of the radius r of the positively charged "snowball" and of the radius R of the negative bubble in liquid helium as a function of the external pressure P_0.

that at an external pressure of 20 atm, the pressure on the bubble surface reaches those 25 atm that are necessary for the solidification of helium. Thus, the bubble surrounds itself with a shell of solid helium and becomes a kind of ice "nut", inside which an electron moves randomly! A further increase in external pressure leads to a thickening of the "shell" outside, up to the complete solidification of liquid helium. The inner radius of the "nut" practically does not change when the pressure rises above 20 atm. Thus, charged bubbles in liquid helium are the centres of its freezing as the external pressure approaches the critical 25 atm. Remember how the steam bubbles in the kettle serve as the nucleus for boiling.

Let us address a few more words to what happens at pressures above 25 atm with charge carriers in solid helium. They remain the same: bubbles with a negative charge, inside which an electron moves, and He^+ ions, whose "snowballs" are now infinitely large. It is clear that the mobility of charge carriers in solid helium is much lower than in its liquid phase. Could you imagine that helium had such amazing properties? As Lev Landau noted, the quirks of helium open a window to the quantum world for us.

Chapter 27

MRI Looks Inside Us

"I hold a mirror up to you, where you will see what's deep inside you..." speaks Shakespeare through the lips of Hamlet. Modern medicine has numerous resources to monitor what is happening inside the human body. For example, for more than a century, X-rays have been used to scan it, and organs and tissues are currently examined using ultrasonic waves. More recently, another discovery has revolutionised medical diagnostics — magnetic resonance imaging (MRI).

The Invention of MRI

MRI is based on the phenomenon of nuclear magnetic resonance (NMR), that is, the ability of some nuclei, when placed in a magnetic field, to absorb radiation of a certain frequency.

The first signals of NMR were recorded in 1946 independently by two groups of American physicists led by Felix Bloch (1905–1983) and Edward Purcell (1916–1997).

At that time, researchers faced enormous technical difficulties, and they had to create on their own all the necessary equipment for their laboratory experiments. For example, the magnet used in Purcell's experiments was taken from a recycled appliance at the Boston tram company! In addition, it was incorrectly calibrated so that the actual magnetic field was stronger than the one required to resonate at 30 MHz generated by the RF transmitter. Therefore, Purcell and his young collaborators failed to get the desired signal. After several days of unsuccessful experiments, an extremely disappointed Purcell resigned himself to defeat and turned off the current supplying the electromagnet. While the magnetic field was decreasing, the researchers looked sadly at the oscilloscope screen, on which they had hoped to see the signal for days. Since they did not turn off the radio frequency generator, when the value of the decreasing magnetic field nevertheless reached a value corresponding to the resonance, the expected signal was briefly displayed on the screen. For the discovery of the phenomenon of NMR, Purcell and Bloch in 1952 shared the Nobel Prize in Physics.

Magnetic Moment and NMR

Not all nuclei are suitable to realise the occurrence of NMR. In this chapter, special attention will be paid to hydrogen nuclei (protons), which yield a significant part of the mass of the human body. For us, it will be important that these nuclei have a magnetic moment (see Chapter 22).

What is the mechanism underlying atomic magnetism? The clearest example is an electron orbiting around a nucleus (Fig. 1). It is equivalent to an electric current in a metallic loop, and therefore it generates a magnetic field. In addition, when exposed to an external magnetic field, it will react in a certain way — how? We will see later.

The magnetic moment of a particle is not necessarily related to its rotation around an external point. The electron, proton and neutron all have their own magnetic moments, called *spin*. The word "spin" is synonymous with the verb "to rotate". This is because Louis de Broglie once believed that the spin of particles is associated with "a kind of internal rotation". However, such "internal rotation" has not been found in nature. Today physicists believe that an electron, despite having a spin, is point-like and has no internal structure. Therefore, the spin of an elementary particle should be better perceived as its innate property, such as mass or charge.

The vector of the magnetic moment in the microworld obeys the rules of quantum mechanics. According to them, when a particle is placed in an external magnetic field, the component of the magnetic moment parallel to the field can take

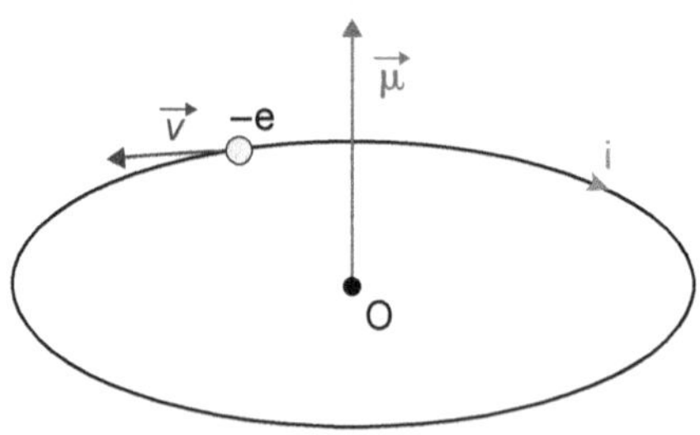

Figure 1. An electron rotating around the point O with speed v creates a magnetic moment parallel to the axis of rotation.

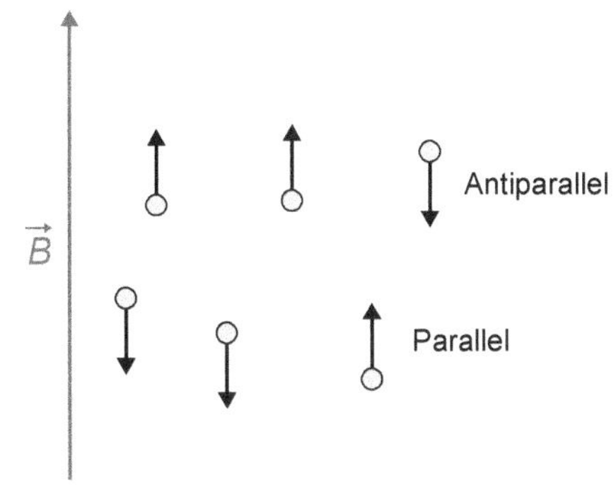

Figure 2. Possible states of the magnetic moment of a proton in a magnetic field. In the state of equilibrium, it is either parallel (with a certain probability) or opposite (with a lower, temperature-dependent probability).

on only a finite number of values. In particular, for the spin of the proton (as well as for the electron and neutron), only two of its projections are possible (see Chapter 22): the magnetic moment $\boldsymbol{\mu}$ can be oriented relative to **B** only parallel to the field, which is energetically more favourable or antiparallel (Fig. 2). The difference in energies corresponding to these two directions is

$$\Delta E = 2\mathrm{B}\mu. \tag{1}$$

It is this value of ΔE that is measured by irradiating the sample with an electromagnetic field of the appropriate frequency.

The Principle of Proton NMR Spectroscopy

Let us consider a proton initially in the state with its magnetic moment projected along an applied constant magnetic field. When irradiated with an electromagnetic wave of the corresponding frequency, it can absorb a quantum of energy equal to ΔE, while passing into the state with the opposite projection of the moment.

Thus, the condition for the absorption of such a quantum of radiofrequency implies

$$hv = 2B\mu, \tag{2}$$

where v is the correspondent frequency and $h = 6.63 \cdot 10^{-34}$ J · s is Planck's constant. The requirement of that frequency is not strict: absorption remains noticeable even if the values of the right and left sides of the equation differ slightly. But

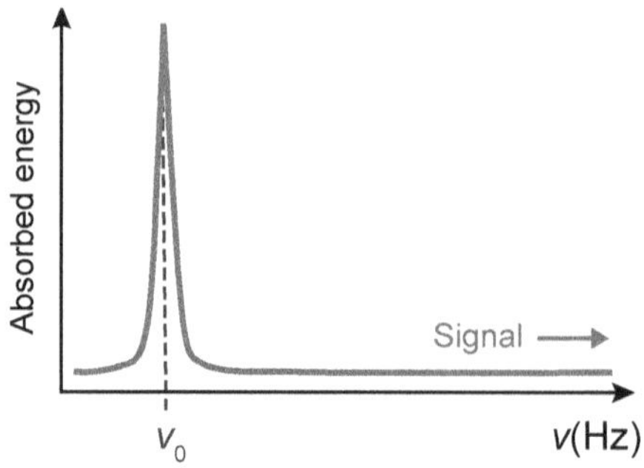

Figure 3. Energy absorbed by the magnetic moment of the nucleus, depending on the frequency of the radiation incident on it.

in a precise magnetic field *B*, when the radiation frequency υ is swept, there is a sharp absorption maximum ("peak", usually called signal) around a certain frequency: the **resonance frequency** in the physical sense (see Chapter 11) (Fig. 3).

This is the principle of operation of NMR spectrometry, which basically studies the absorption spectrum of a sample. Instead of changing the frequency, the field *B* can be swept while the frequency is kept constant at a given value. At the resonance condition, sharp maxima are observed. For a real sample where many protons are present, a slight distribution of the local magnetic fields is related to the interactions of the protons between themselves. This explains why the signal is not infinitely sharp (see figure) but is rather a distribution of many "rays" very close to each other. Other possible sources of the broadening of the resonance line or a given "structure" of the signal are the non-homogeneity of the magnetic field over the whole sample or the diamagnetic effects related to the electronic currents. These aspects will be better clarified in the subsequent sections.

The magnetic moment of the proton is $\mu = 1.41 \cdot 10^{-26}$ J T^{-1}. Thus, with an external magnetic field of 1 T, absorption will occur at a frequency close to 42 MHz. The corresponding wavelength $\lambda = c/\upsilon$ is 7 m, which is about 50 times the radiation length in a microwave oven (see Chapter 16) and belongs to the radiofrequency range.

The early NMR spectrometers, say in the 1960s, used relatively weak magnetic fields. Today, many spectrometers operate with fields so high that proton magnetic resonance occurs at frequencies up to 900 MHz. Among other things, strong magnetic fields have greatly improved the resolution of the related spectra. Obviously, the first electronic devices used by Bloch and Purcell to detect NMR were significantly different from the modern ones, which operate by resorting to

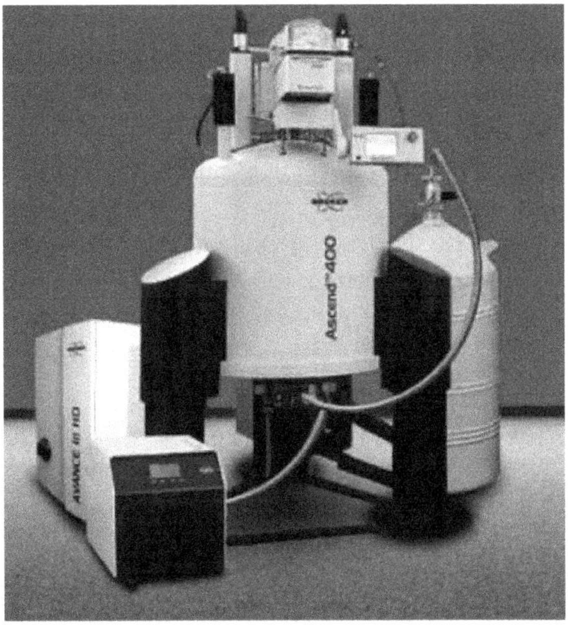

Figure 4. Medium-size NMR spectrometer for laboratory measurements. The sample is placed in a tube that is often settled from the top of the device. The liquid nitrogen reservoir on the right cools the superconducting magnet.

powerful superconducting magnets (Fig. 4). However, the measurement protocol remains basically similar: the coil driven by a powerful generator creates a radiofrequency electromagnetic field of the order of one-thousandth of a tesla. The sample to be studied is positioned on the axis of the coil perpendicular to the static magnetic field. As already mentioned, the latter is generated by a "classical" or superconducting electromagnet and is in the range of 1 to 23 T (see Chapter 24).

How NMR Spectroscopy Works

How to interpret a real NMR spectrum? In Eq. (1), the nuclear magnetic moment μ and the applied magnetic field B are present. Having found the resonant frequency, it is possible to calculate ΔE... which, it would seem, is of little interest! However, the magnetic field in the substance acting on a given hydrogen nucleus is not exactly the magnetic field that is applied to the sample. It is necessary to take into account the corrections due to the presence of other nuclei and electrons in the substance: depending on the circumstances, they can represent a kind of "screen" for the external field or, on the contrary, amplify it. Depending on the motion of electrons in the immediate vicinity of the nucleus under consideration, the resonance can shift in frequency. Usually, to get rid of the magnetic field value, the positions of the different resonance lines are given by a dimensionless number: the **chemical shift**. Thus, chemical shift becomes a characteristic of the environment around the core.

For example, the NMR spectrum of ethanol CH_3–CH_2–OH (Fig. 5) has three groups of signals. These groups correspond to the hydrogen nuclei of the molecular groups CH_3, CH_2 and OH, respectively. The complex structure of the signals corresponding to the CH_3 and CH_2 groups is due to the interaction between the magnetic moments of hydrogen atoms belonging to the same group of a given carbon atom.

And that is not all! The relative intensity of the resonant signals also provides information about how many nuclei are involved in resonance. In addition, in a given molecule, the hydrogen nucleus can be replaced by one of its isotopes, which has a different magnetic moment, and thus information about each nucleus of the molecule can be obtained. This information involves the temperature: a study of the spectra obtained at different temperatures shows how the environment surrounding the core changes.

Thus, proton NMR spectrometry provides valuable information about the local environment around hydrogen nuclei, that is, it can serve as a method for

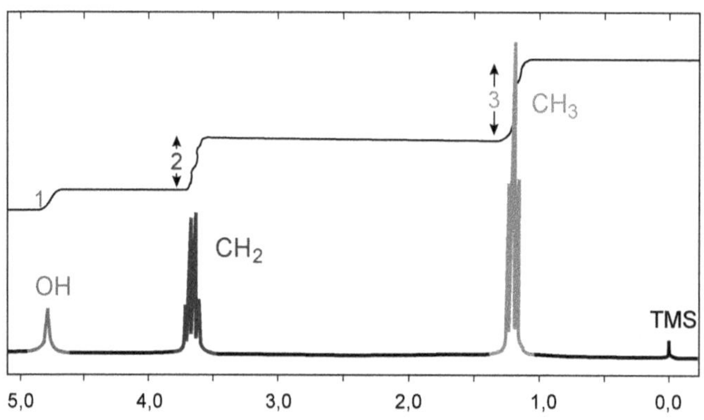

Figure 5. NMR spectrum of protons in ethanol (CH_3CH_2OH). The three groups of lines correspond to different protons, and therefore they have different chemical shifts. The integration curve (upper curve) reports the relative intensities of each group of lines. The zero origin for the chemical shifts corresponds to the resonance of the protons in tetramethylsilane $Si(CH_3)_4$, or TMS, that is frequently used as reference. The chemical shift of 3 ppm means that the distance in hertz of the signal from the reference (that of TMS) falls at $3 \cdot 10^{-6}$ from that reference line.

NMR at work in oenology: SNIF technology

NMR serves not only chemists but also connoisseurs of good wine! It can be used to determine the origin of the wine — this method is called SNIF. It was assumed that this abbreviation stands for Specific Natural Isotope Fraction (a specific fraction of natural isotopes), but it sounds close to the English word "sniff". NMR allows you to "smell" the wine in some way.

The SNIF method was invented by chemists Gérard and Maryvonne Martin in Nantes in the 1980s, initially to determine whether sugar was added to wine (see Chapter 14). In addition, this method provides information on the geographical origin of ethanol! Indeed, in different regions, the processes of photosynthesis and metabolism for hydrogen (^1H) and its isotope deuterium (^2H) proceed differently, which is how the specific NMR spectrum allows one to judge the origin of grapes. The magnetic moment of deuterium is less than that of hydrogen, thus the chemical shifts of their NMR signals are different. By measuring the intensity of the signals, the amount of deuterium is calculated, compared with the available map of the distribution of deuterium in wine-growing regions, and thus an idea of the origin of the wine is obtained. In nature, the proportion of deuterium is very small and is measured in parts per million (ppm). It is 90 ppm at the South Pole and an average of 160 ppm at the equator. Note that on Venus it is 16,000 ppm, that is, hydrogen there contains 1.6% of deuterium.

The origin of wine is indicated by another factor — the localisation of deuterium in various groups. Ethanol molecule CH_3–CH_2–OH can turn into CH_2D–CH_2–OH or CH_3–CHD–OH or CH_3–CH_2–OD (here D stands for deuterium).

By revealing the composition and origin of wines, the SNIF method makes life difficult for fraudsters: it becomes hard to add sugar, dilute wine or change the label!

elucidating the structure of molecules and their identification (see Panel above on this page). Recall that many other nuclei also have a magnetic moment and are studied in laboratories: this is the NMR spectrometry of carbon-13, phosphorus-31, etc.

Finally, it should be noted that NMR experiments can also be carried out with a substance being in a solid state. For example, valuable information can be obtained about microscopic magnetic and electrical interactions in crystals. However, these aspects are beyond the scope of this book.

Special Nuclear Resonance Method: The FID

Modern methods of NMR have become more effective: now there is no need to change the radiation frequency or the value of the external magnetic field.

The action of the RF field on the sample (leading to a resonant transition of the nucleus between the two states) is limited to a finite period of time: this is called an RF pulse. After the action of the pulse, the field remains constant and equal to B. Nuclear magnetic moments, which, before the impulse, were either parallel to the field or directed oppositely, are unbalanced by the impulse, and begin to rotate around with a uniform speed (Fig. 6). This rotation around the direction of the field, called "Larmor precession" (see Panel on page 387), generates an electrical signal that can be detected. Typically, the RF generator using the coil to excite the magnetic moments is abruptly turned off to stop the pulse. Then the radiofrequency receiver is instantly activated, which uses the same coil, but this time in order to record the signal associated with the precession of the nuclei. This signal, due to the electromotive force induced by the "free" precession of nuclei, is called Free Induction Decay (FID) (Fig. 7). Then the signal is processed by a computer, and the NMR spectrum is reconstructed on its basis. Thus, the same information as when analysing the absorption of radiation as a function of the frequency or intensity of the magnetic field can be obtained much more quickly.

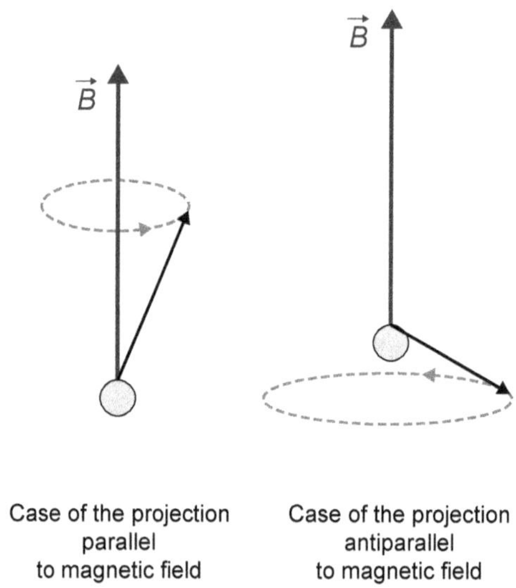

Figure 6. Precession of the magnetic moment. In the presence of the strong magnetic field **B** and at sufficiently low temperature, a nuclear magnetic moment can stay parallel to **B**. After the application of the radiofrequency pulse perpendicular to **B** and having frequency equal to the resonance frequency, the magnetic moment rotates around **B** with an angular speed which depends on **B**, and it can take any angle dependent on the pulse duration. In the figure, two possible angles of the magnetic moment with respect to **B** are shown.

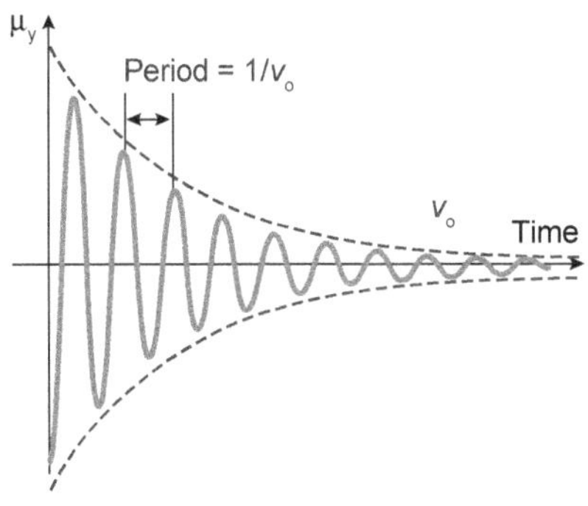

Figure 7. Precession signal of nuclei as a function of time. The μ_y component of the nuclear magnetic moment rotates at frequency υ_0 (given by the formula $h\upsilon_0 = 2B\mu$): its value is gradually damped due to relaxation phenomenon. The FID technique measures generally the envelope of this signal (in dashes).

From NMR to MRI

After the discovery of Bloch and Purcell, NMR spectrometry began to develop rapidly. Large NMR research groups have emerged in France and Italy. French physicists Anatol Abraham (1914–2011) and Ionel Solomon (1929–2015) and Italian physicist Luigi Giulotto (1911–1986) founded world-famous scientific schools in Paris and Pavia. Groups like these were the driving force behind the launch of the Ampère group, which contributed to scientific progress in this field. NMR is widely used in solid-state physics, chemistry, biology and metrology. And, of course, physicists did not delay to begin the use of NMR in medicine. The first two-dimensional image of two samples of water was obtained in 1973 by the American chemist Paul Lauterbur (1929–2007). In 1976, American scientist Raymond Damadian (1936–) presented the first NMR image of an animal tumour. Today, many hospitals are equipped with MRI machines for medical diagnostics (Fig. 8).

Principle of NMR Imaging

Let us describe the principle of image formation using MRI. This method, using a magnetic field inhomogeneous in space, was proposed by Paul Lauterbur (Fig. 9) in 1973. As a result, the resonance frequency of the nucleus, which depends on the value of the field **B**, turns out to be dependent on the position of the nucleus in space.

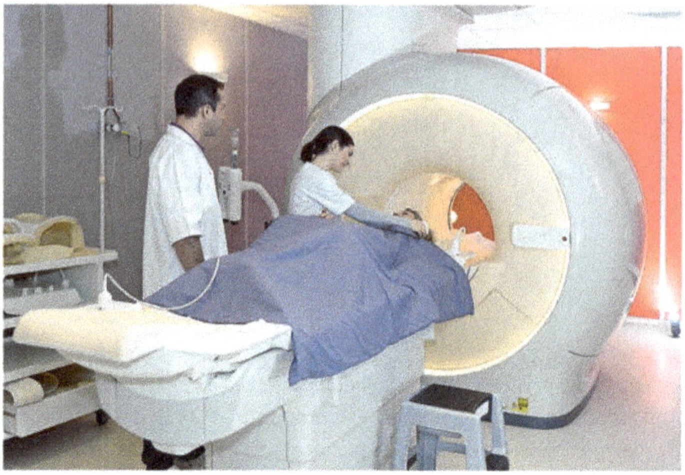

Figure 8. Commercial MRI machine. The patient is placed inside the working cylinder.

Figure 9. Paul Lauterbur (1927–2007). In 2003, he was awarded the Nobel Prize in Medicine for his contribution to the development of MRI.

Precession at the heart of NMR

The term "precession" in mechanics means a change in the direction of the angular momentum vector or, more simply, the direction of the axis of rotation of a rotating object. For example, the Earth rotates on an axis that shifts over time, resulting in a shift in the equinox dates. The top, shortly before falling, also demonstrates the phenomenon of precession: its axis of rotation deviates from the vertical (see Chapter 17).

Let us return to the subject of this chapter, the nuclear magnetic moment. A suitable representation of a certain atom is given by a punctual electric charge rotating around a fixed charge of opposite sign because of the effect of the electrostatic attraction (Fig. a)).

*Let this charge also be affected by a magnetic field **B** (Fig. b)). Now, in addition to the electrostatic force of attraction to the stationary centre, the moving charge is also affected by the Lorentz force, directed perpendicular to the field and the velocity vector and equal in magnitude to* $B \cdot v \cdot \sin\alpha$*, where* α *is the angle between **B** and v. In terms of the vector product (see Chapter 4), the Lorentz force can be written as* $\boldsymbol{F} = q\boldsymbol{v} \times \boldsymbol{B}$*. This expression is reminiscent of the Coriolis force that appears when writing the equations of motion for a body in a reference frame that itself rotates around an axis with an angular velocity* Ω*. As we already know, in this case, it is necessary to add a fictitious force to the balance of forces acting on the body, equal to* $mv \times \Omega$*, where* Ω *is the angular velocity vector parallel to the rotation axis (see Chapter 4). The expressions for the two forces, Coriolis and Lorentz, are very similar, specially if the vectors are parallel. In this case, you can even do a trick so that*

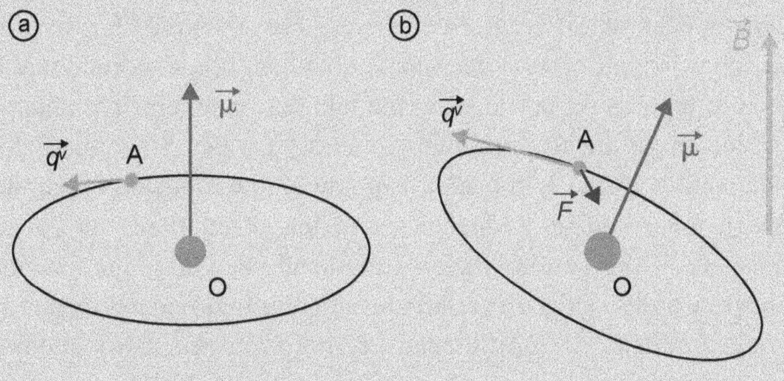

Larmor precession. (a) The charge q rotating around the fixed point induces a magnetic moment. (b) If an external magnetic field **B** is applied, the Lorentz force can be compensated by the Coriolis force if one refers to a rotating frame of reference. In this frame, the magnetic moment is fixed. At variance, it rotates for a motionless observer.

(*Continued*)

(Continued)

both forces compensate for each other! For this, it is enough to choose $\boldsymbol{\Omega} = -q\boldsymbol{B}/m$. *In other words, it is necessary to switch to a coordinate system that rotates around a vector with an angular velocity of* $-q\boldsymbol{B}/m$. *In this frame of reference, the magnetic field and the Lorentz force cancel out, and everything happens as if the charge A experiences only an electrostatic attraction. That is, its orbit in a rotating frame of reference turns out to be stationary, while in a fixed coordinate system it rotates with an angular velocity* qB/m. *And the magnetic moment rotates with it. So much for the precession! This result remains unchanged in quantum mechanics: the magnetic moment of a spin in a magnetic field is also subject to precession.*

Although there is precession in the system we just studied, it does not affect a magnetised rod (like a compass needle)! While in a magnetic field, being deflected from its equilibrium position and released without any initial velocity, the magnetic rod will oscillate without rotating around the field. In the end, its vibrations will damp out due to friction, and the rod takes the South–North direction.

To deal with the new problem statement, consider a simple one-dimensional case with groups of small spheres filled with water, located along the x-axis (Fig. 10). With a uniform magnetic field, they all give a signal at the same frequency. Now suppose that the additional coils create an x-dependent magnetic field, that is, the magnetic field has a gradient. Then the NMR signal for different groups will appear at different frequencies. For example, for five groups of spheres, a set of five absorption maxima is obtained. It is important that the intensity of each of them is proportional to the number of spheres participating in the resonance, that is, to the corresponding amount of water. Since the field gradient (i.e., the derivative dB/dx) is known, it is possible to establish an accurate correlation between the resonant frequencies and the position of the corresponding spheres in space. Thus, various signals can already be tied to the location of their sources in space and to judge the relative content of hydrogen along the x-axis. By creating field gradients along different axes, more complex distributions of hydrogen atoms can be analysed.

MRI applies the same principle, only in three dimensions! However, in space, everything turns out to be much more complicated than our one-dimensional model. To visualise the density distribution of hydrogen requires powerful computers to manipulate radiofrequency fields. It took years of research to improve the design of the magnetic field profile and develop methods for processing the received NMR signals. In a very simplified way, we can say that computer

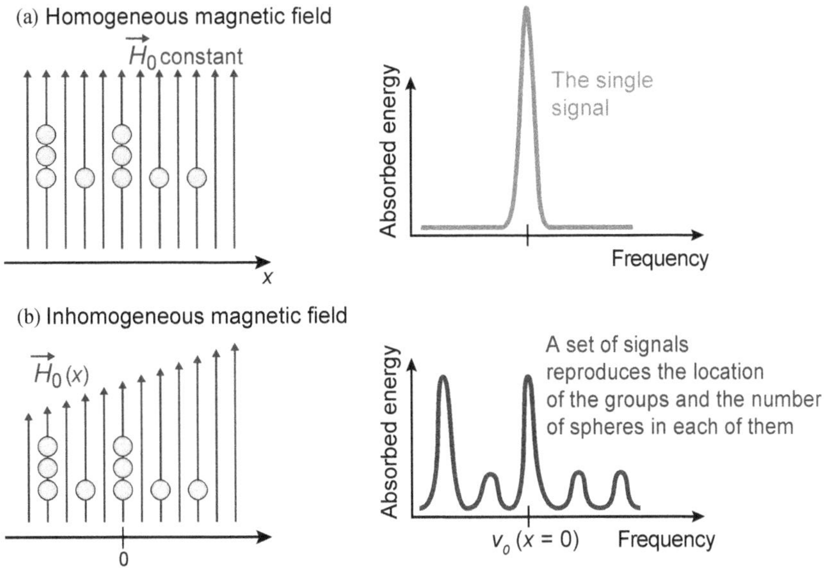

Figure 10. An example of NMR in one-dimensional space. Spheres filled with water in different quantities are located at different points in space. (a) By applying a uniform magnetic field, we obtain a single resonant NMR signal at the frequency determined by formula (2), that is, $\hbar w_0 = \mu B_0$. (b) In the presence of a field gradient, resonance signals from different points of the sample occur at different frequencies, and their intensity depends on the number of excited protons. Due to this, the NMR spectrum (that is, the collection of absorption signals) reproduces the location of the spheres filled with water in space.

processing makes it possible to display the distribution of hydrogen in space: the intensity of the signal emitted by a certain area of space is proportional to the number of hydrogen atoms in this area, which makes it possible to obtain information about the local density of tissue. Through the use of tomography methods, the patient's body is "sliced" along "sections" in such a way as to obtain a three-dimensional picture of one or another internal organ (Fig. 11).

Spin Echo

The materials under study usually contain inhomogeneities. It follows from this that the precession frequency for different nuclei is different, therefore, the FID signal, after the pulse is applied, becomes more and more distorted over time. This distortion can be corrected using a special technique called "spin echo". Its essence is as follows. The radio frequency pulse of duration t_1 created at the initial moment of time forces the magnetic moments of the spin to line up perpendicular

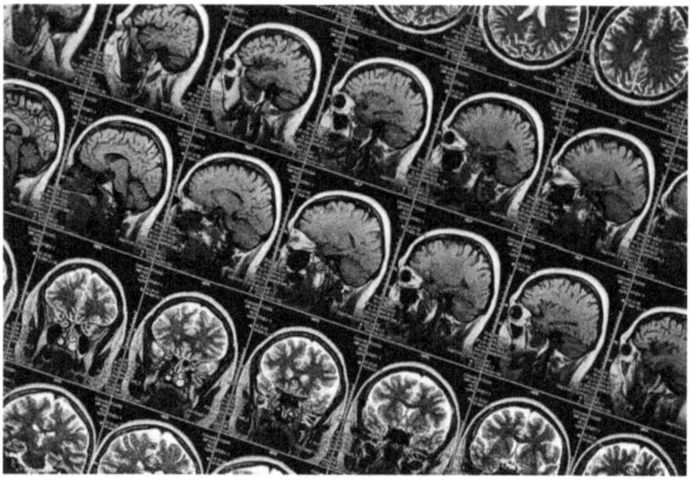

Figure 11. Tomography of the brain. Tomography is a three-dimensional generalisation of the example shown in Fig. 10. It allows you to get the same result as if the object (here, the brain) were dissected in layers, but bloodless, painless and without tissue damage!

to the external magnetic field. After the time t_D has elapsed (during which a certain change in FID has occurred), a second RF pulse is applied, twice the duration, namely $2t_1$. This pulse reverses the magnetic moments in the opposite direction (since two 90° turns are equivalent to inverting the direction). Magnetic moments, which rotated more rapidly during the time interval t_D and therefore were "ahead", now lag behind. However, as they continue to spin more quickly, they soon catch up. Therefore, during the next time interval t_D, all magnetic moments are aligned. Thus, a signal that has been altered, like an echo, is restored to its original form.

Striking Pictures

To appreciate MRI, it should be understood that it allows you to get real images of human organs, and not their "shadows", as in images obtained using X-rays (indeed, the receiver collects X-rays after passing through the body, where they are more or less absorbed by bones and tissues).

The human eye is sensitive to electromagnetic waves in the visible region (see Chapter 3). Unfortunately (or fortunately), the eyes are not able to perceive the radiation of the internal organs of our bodies: we see only the outer shell. Under NMR conditions, nuclei emit electromagnetic waves in the radiofrequency range (at frequencies much lower than visible light). Therefore, such waves, passing through the body, reach the measuring device, which, in combination with a high-performance computer, converts the received signals into a visible image already available to our eyes.

Physicists and mathematicians have largely contributed to this amazing achievement in medicine by understanding the quantum mechanical properties of nuclear magnetic moments, the theory of the interaction of matter and radiation, as well as the creation of digital electronics and the principles of mathematical signal processing.

The advantages of MRI over other diagnostic methods are numerous and significant. The operator easily visualises the section of the patient's body required for analysis; it can also register signals from several cross-sections simultaneously. In particular, with the necessary adjustment of the magnetic field gradients, the image can be obtained at the desired angle, which is difficult for fluoroscopy. In addition, the researcher has the ability to limit the field of observation, thereby visualising a specific organ (or part of it) with high resolution.

An additional advantage of MRI is the ability to measure the viscosity of a liquid directly at the research site. For this, a spin echo is used — a signal that is influenced by the speed with which the nuclei move in the field gradient. As a result, it becomes possible to measure the flow rate of blood or other body fluids.

By varying various parameters, for example, the length and frequency of pulses or the time during which the nuclear response accumulates, the operator can change the nature of the deviations of the nuclear magnetic moments and, thus, increase the image contrast in search of anomalies. By choosing appropriate RF coils, image resolution can be detailed down to dimensions as small as 2 μm in width and 200 μm in depth. With a suitable resolution, information on the concentration of various chemicals in the body can also be obtained.

See the Heartbeat... and Read Minds

To obtain a usable image, you must successfully overcome the most difficult problems associated with the sensitivity of the device, that is, the signal-to-noise ratio. To do this, a plurality of FID or spin echo signals are collected together. This requires a fairly long time: usually tens of minutes.

In 1977, the English physicist Peter Mansfield (in 2003, he shared the Nobel Prize with Paul Lauterbur) developed a special combination of field gradients. It does not provide particularly good images, its main quality rather being its extraordinary speed. Starting with one FID signal, it provides an image in about 590 ms! Today, even a heartbeat can be visualised with this technique (the so-called planar echo).

Finally, we will mention functional MRI techniques, which open the way to a deeper understanding of our cognitive processes. They can be used to detect active areas of the brain (activity associated with changes in blood flow).

Could the doctors of antiquity suppose that someday it would be possible to penetrate into the innermost depths of the human body and consciousness?

Chapter 28

Semiconductors and Nanophysics

Nowadays, touchscreen tablets, digital players, mobile phones and laptops are becoming more powerful, functional and miniaturised. The underlying technology is based on the use of semiconductor materials that make electrons obey the movements of our fingers.

Let us address the laws governing the nanoworld.

Miniaturisation of Technology

Over the past several decades, we have been dealing with devices that accumulate more and more functions in a tight space. First, let us recall what the technologies of our ancestors were...

Our journey begins in the 17th century — the great era of nascent technology when Christian Huygens improved the clockwork mechanism and Blaise Pascal invented the calculating machine. This mechanical device laid the foundation for the creation of more and more advanced computers. The desktop apparatus that the authors of this book encountered around 1960 was already electric; it was no less bulky than Pascal's machine (Fig. 1). More sophisticated instruments at that time already knew how to extract the square root.

The first step towards the electronic era was taken with the development of vacuum tubes. The first giant computers such as the Colossus in England and ENIAC in the USA were equipped with these devices, which strongly heated and had a short life. Built in the 1940s, these computers had thousands of vacuum tubes that needed to be replaced regularly! The most typical example is a triode, an electronic tube, developed in 1906 by the American inventor Lee de Forest (1873–1961).

Figure 1. Electromechanical calculator Olivetti “Divisumma 24”, made in Italy in the 1960s (case removed to show the mechanism and the electric motor at the back of the unit). The machine performed addition, subtraction, division and multiplication, and printed the result. It weighed almost 15 kg.

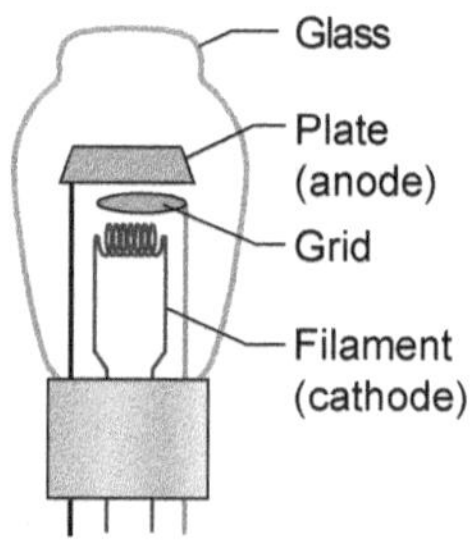

Figure 2. Until 1947, the triode was several centimetres high and had a short service life. The filament cathode emits electrons that are attracted by the anode.

The glass vacuum cylinder contains the cathode, which emits electrons when heated, and the anode, which traps these electrons insofar as an interposed grid allows them to pass (Fig. 2). The grid facilitates or prevents the passage of electrons, depending on the electric potential applied to it: in this way, it is possible to regulate, including to amplify, the potential difference between the cathode and the anode.

The triode takes advantage of the ability of electrons to travel in vacuum (and thus create an electric current in it), provided that they are emitted from a very hot cathode. It is impossible to establish a unidirectional electric current in a metal: if it can flow in one direction, then, when the sign of the potential difference is

Transistors and Moore's law

In 1965, American engineer Gordon Moore, one of the founders of Intel, formulated a law that later became famous. According to him, the number of transistors placed on a microprocessor (components that perform arithmetic and logical operations) doubles every two years.

Moore's law was later confirmed with surprising accuracy (see figure), largely because manufacturers took it as a guide in the development of their products. Obviously, this exponential growth cannot last indefinitely: the minimum size of a transistor is limited, at least by the distance between atoms, which is a fraction of a nanometre. In 2004, 100 million transistors fit into 1 cm^2, that is, one transistor occupied an area of 10^6 nm^2. According to Moore's law, atomic dimensions will be reached in the year 2004 + 2x, where x is the root of the equation $2^x = 10^6$. From this, we find $x = 6/\log 2 \approx$ 20 years. Thus, Moore's law could formally operate until 2044. In fact, it is likely that it will only be implemented until 2030.

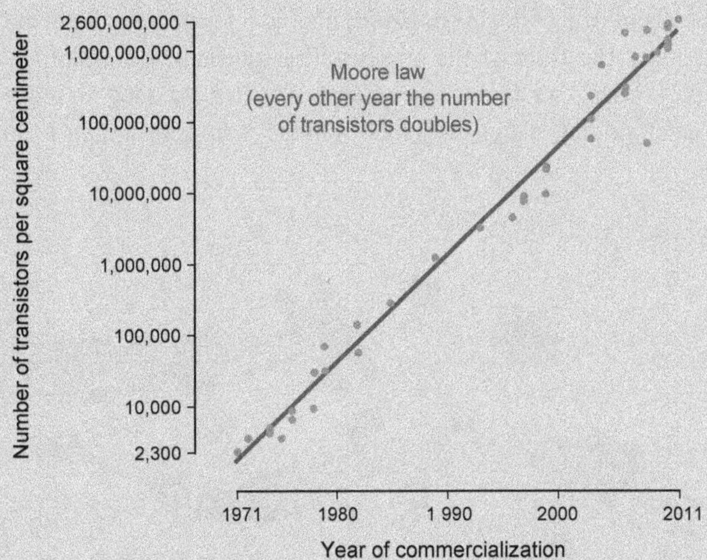

The number of transistors per 1 cm^2 of the microprocessor, depending on the year of its release. The dots correspond to the processors actually marketed, the line — Moore's law.

reversed, it can always flow in the other (see Panel on page 399). In a triode, on the contrary, the movement of electrons can be controlled. To prohibit them from coming back and to impose one-way traffic, it is enough to remove the grid: electrons will rush from the cathode to the anode, but not in the opposite direction.

From Triode to Transistor

The development of miniaturisation began in 1947 with the invention of the transistor by American physicists Walter Brattain (1902–1987), William Shockley (1910–1989), and John Bardeen (1908–1991) (whom we already met in Chapter 24). For this discovery, in 1956 they were awarded the Nobel Prize in Physics. The term "transistor" is an abbreviation of the expression "transfer resistor".

Similar to the triode, a transistor is a three-terminal device (Fig. 3). Here, the electrons, instead of flowing in the vacuum are moving in a semiconductor, namely an insulator where impurities have been added so that it can transport electric current. A terminal (called an emitter or *source*) emits electrons, another one accepts them (the *collector* or *drain*) and an intermediate element (the *base* or *grid*) modulates the flow of the current. The transistor, in a similar way to the triode of past times, can be used as an amplifier (for instance, in the radio receivers), or it can modulate a signal, as it does in the radio transmitter. It can also serve

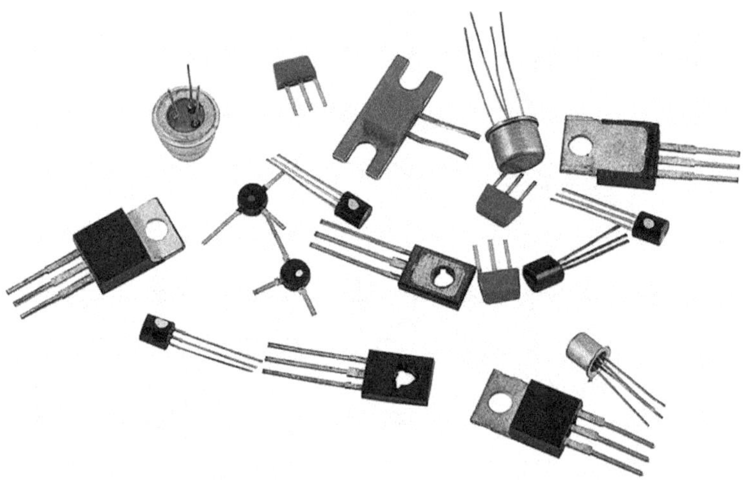

Figure 3. Transistor with individual inputs (most often, they are combined into integrated circuits). Unlike the triode, their progenitor, heating is not required for a transistor. It is also much cheaper and smaller in size.

Figure 4. Integrated circuits on a silicon plate. Hundreds of thousands of transistors form complex circuits on a crystalline substrate. The intricate design of transistors and their connections is obtained through a technological process called lithography.

as a switch in a logic circuit (in fact, computers manipulate binary digits, the bits, that are encoded by the states *up* (1) or *down* (0) of the electric voltage).

During the middle of the 20th century, the most used semiconductor was germanium, while nowadays for the electronic applications it is silicon. The impurities added to the semiconductors are of various types, and they can be classified in different ways. For this reason, different types of transistors are known. The original transistor, the one of 1947, had a size of some millimetres. The size of transistors has constantly decreased (see Panel on page 395), and nowadays transistors are collected in a huge number in so-called integrated circuits (Fig. 4). The familiar 5 cm long USB drive, with which we work quite often, can contain 4 billion transistors and stores up to 1 GB of data ($8 \cdot 10^9$ bits).[1]

[1] The data given refers to 2014 the year an earlier version of this book, *Le Kaleidoscope de la Physique*, was first published; today, the volume of USB drives can be terabits.

Controlled Electrons in Semiconductors

Let us briefly explain how semiconductors make electrons docile. In a solid consisting of a large number of atoms, the energy levels allowed for electrons are broad bands (zones) (see Panel on page 399). Electrons occupy these bands, starting with the lowest energies. The last fully filled band (called the *valence band*) and the next, at least partially, empty band (called the *conduction band*) are separated by a more or less wide band gap, which is called the *gap*.

For industrial use, silicon is doped with impurities, that is, atoms of other elements are introduced into its crystal lattice. These impurities are of two types. Impurities of the first type (for example, phosphorus atoms) willingly get rid of their valence electrons. These electrons fill the conduction band, and the semiconductor becomes a conductor — much like a metal if doped enough (see Panel on page 399). A semiconductor doped in this way is called *n*-type (from the word "negative", since the charge of electrons is negative). Impurities of the second type (for example, boron atoms), on the contrary, willingly accept free electrons in the crystal to their outer energy shells. These electrons are taken from the valence band, leaving unfilled states in it, the so-called *holes*. Under the influence of an electric field, an electron in the valence band can "jump" from its state to the vacant unfilled one, the next electron will jump in its place, and this leapfrog will continue further. What is happening can be imagined as if the hole itself carried a positive charge and moved in the direction of the electric field. As a result, in this case, the semiconductor becomes a conductor! A semiconductor doped in this way is called a *p*-type semiconductor (from the word "positive"). In both cases, the semiconductor at any point remains neutral: mobile charges are compensated by the charges of the ions.

It is convenient to represent the state of a semiconductor on a two-dimensional diagram (Fig. 5), where the *y*-axis is energy, and the *x*-axis defines the direction inside the crystal. This simplified interpretation allows to give an idea of the main processes taking place in it. The impurity levels correspond to the states inside the forbidden band, otherwise the electrons or holes introduced by them could move. The Fermi energy ε_F (also called, the Fermi level) is the border between occupied states and states at higher energy, which are empty at absolute zero. At finite, not too high temperature, some of the electrons move into the conduction band, and some of the holes move into the valence band. Their amounts are usually small without doping. With the introduction of *n*-type impurities (*n*-doping), additional electrons appear in the conduction band, so the Fermi energy, which increases with their concentration, approaches the conduction band (Fig. 5(b)). Conversely, *p*-doping shifts the Fermi energy down, closer to the valence band (Fig. 5(a)).

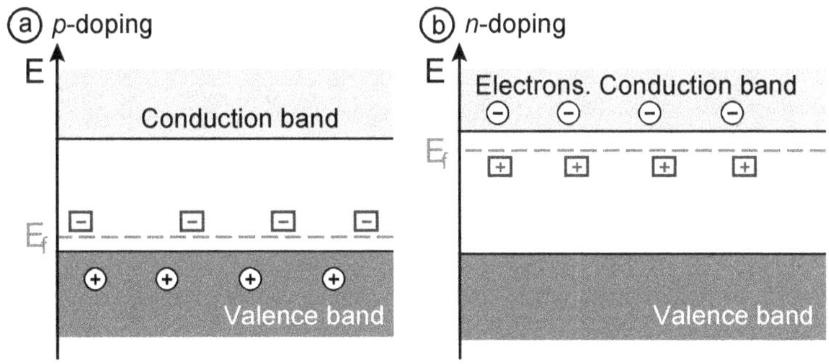

Figure 5. The bands of a p-type (a) and n-type (b) doped semiconductor. The introduced impurities (represented by squares) capture electrons that were previously in the valence band in the case of p-type doping (a). In the case of n-doping the impurity atoms donate electrons to the conduction band (b), creating additional charge carriers and thereby considerably increasing the semiconductor conductivity.

Conductor, insulator and semiconductor

Some materials, such as metals, are conducting, while others are insulators (dielectrics), do not conduct electric current (or do so very poorly). Let us consider how they differ in terms of their band structure.

In a metal, the conduction band is partially filled. If we apply a potential difference ΔU *to the ends of the metal wire, the situation becomes non-equilibrium: the energy levels shift by the value* $e\Delta U$, *and the electrons rush to where the energy is lower... just like children slide down an ice slide! Thus, a current arises that flows against the direction of the field (the electron charge* $-e$ *is negative).*

The fact that only electrons in the conduction band contribute to the electric current seems somewhat unexpected. Let us try to understand why this is so. For the current to flow, there must be more electrons moving in one direction than those moving in the opposite direction. However, symmetry requires that, in the absence of an applied field, the number of states with a positive velocity be the same as those with a negative one. Therefore, for the emergence of a current, it is necessary that the imposition of an electric field breaks this symmetry and the states occupied by electrons corresponding to a positive velocity are greater than the occupied states with a negative velocity. In the valence band, all the states are already occupied, with one electron in each (in accordance with the Pauli exclusion principle (see Chapter 22)). Therefore, here, even by applying an electric field, the symmetry cannot be broken, and the average velocity of the electrons is necessarily zero. Thus,

(Continued)

(*Continued*)

the electrons belonging to the valence band do not participate in charge transfer and do not contribute to the current.

In dielectrics, the gap between the valence band and the conduction band is large. As a result, the latter remains practically empty, and the sample does not conduct an electric current, at least at low temperatures.

There is also an intermediate category of substances located between the dielectric and the conductor. These are semiconductors, the ones that radically changed our daily life in the middle of the last century. A semiconductor is a dielectric in which the valence and conduction bands are separated by a gap narrow enough to allow the electrons to transit under the influence of temperature. At a normal temperature of about 300 K, electrons pass from the valence band to the conduction band, which, therefore, is no longer empty. Thus, the conductivity, which is absent in such substances at absolute absolute zero, becomes noticeable when the temperature rises to room temperature.

p–n Junction

The simplest semiconductor electronic device is a *p–n* junction, which consists of two connected semiconductors with different types of conductivity: electron and hole. In the contact area, such a connection becomes a place of charge accumulation (Fig. 6). Indeed, the concentration of electrons and holes in space cannot change in a discontinuous manner. Even in the case of such discontinuity, the diffusion of electrons and holes will restore continuity (similar to how thermal conductivity between two different temperature zones leads to a continuous temperature distribution, see Chapter 18). Thus, due to the separation of charges in the contact area, a strong electric field arises.

The *p–n* junction has a remarkable property: it only passes current in one direction. Suppose we need current to flow from an *n*-type semiconductor to the *p*-region (Fig. 6). In the *p*-type region, the current is carried by holes, which must move away from the contact. In the *n*-type region, the current is carried by electrons, which must also move away from the contact. If a stationary current was established in such a circuit, then there would soon be no free charges near the contact, and the current in the circuit would soon disappear. Thus electric current cannot flow from the *n*-type region to the *p*-type region[2]. On the other hand, under

[2] Generally speaking, a very weak current can still flow from the *n*-region to the *p*-region because in the former, there is a small concentration of holes, and in the latter, electrons.

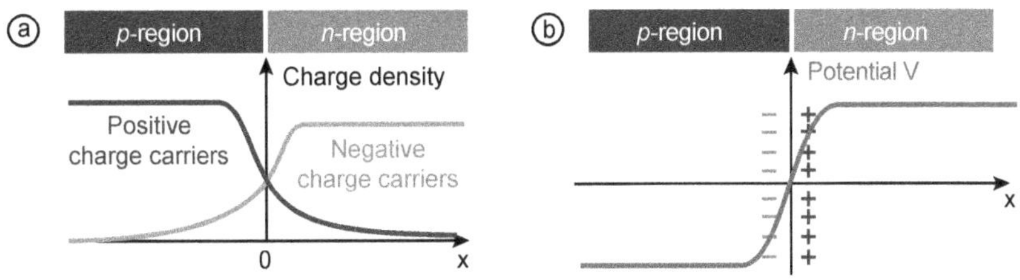

Figure 6. (a) Variation of the positive and negative mobile charge density as a function of the distance from the *p–n* junction. The requirement of a continuous variation of the density implies the formation around the junction (b) of an electrically charged region where the mobile charges do not compensate the fixed ones. As a consequence, a potential difference between the two sides of the junction arises.

the influence of the potential difference applied by the battery, the current can flow from the *p*-type region to the *n*-type region: holes in this case move to the *n*-type region, while electrons move to the *p*-type region, at the boundary of which they annihilate (*recombine*). They will be replaced by other electrons and other holes that appear in the circuit under the influence of the battery separating the charges.

Thus a *p–n* junction only passes current in one direction, such as a tube diode (a triode without a grid, which we already talked about). In order not to invent a new name, such a semiconductor device was simply called *diode*!

By choosing suitable semiconductors (e.g., gallium arsenide, (GaAs)) and doping elements, it is possible to make the recombination between holes and electrons be accompanied by strong light emission. Such *light emitting diodes* (LEDs) were recently used as indicators of device operation (Fig. 7), and today you see them everywhere in garlands, car headlights and other low-power lighting fixtures (Fig. 8). For this breakthrough in artificial light technology, the 2014 Nobel Prize in Physics was awarded to Japanese scientists Isamu Akasaki, Hiroshi Amano and Shuji Nakamura.

Photovoltaic Effect and Solar Panels

Other types of *p–n* junctions, on the contrary, instead of emitting light, are able to convert the light falling on them into an electric current — this phenomenon is called the photovoltaic effect.

This property is applied in photovoltaic cells involved, for example, in solar panels. Suppose that one photon of sufficient energy emitted by the Sun hits an *n*-type semiconductor. Its absorption leads to the formation of an "electron–hole" pair. There is the possibility that the hole, before recombining with the electron, is

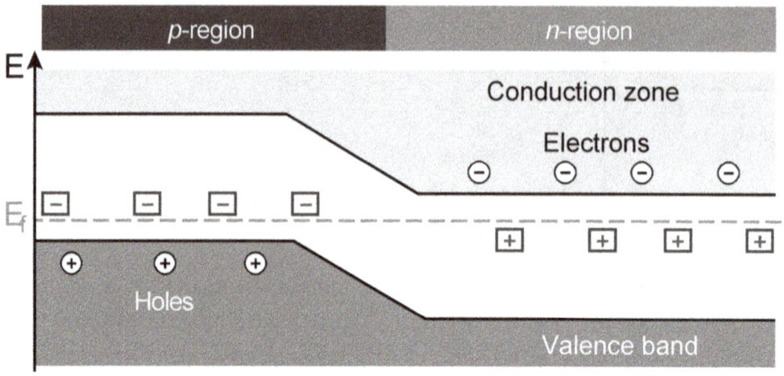

Figure 7. *p–n* junction in a state of equilibrium. Stationary charges (impurity ions that donated or received an electron) are represented by squares. The conduction band is empty, except for a few moving electrons, indicated by circles with a (–) sign. The valence band is full, except for a few moving holes, indicated by circles with a (+) sign. In the area of contact between *p*- and *n*-type semiconductors, electric charges accumulate (see Fig. 6). Thanks to these charges, the Fermi energies on both sides of the transition are equalised.

Figure 8. LEDs operating on a *p–n* junction. Diodes used for lighting, unlike incandescent lamps, emit only visible light.

carried away by the electric field ($E = -dV/dx$, in the immediate vicinity of the transition in Fig. 6) to the *p*-region, while the electron remains in the *n*-region. Similarly, if a photon creates an electron–hole pair in the *p*-region, then the electron has a good chance of going to the *n*-region, while the hole remains in the *p*-region. Thus the absorption of photons leads to charge separation: the accumulation of holes in the *p*-region, and electrons in the *n*-region. These charge carriers are just waiting for the opportunity to escape from the *p–n* junction: electrons will go in one direction, holes in the opposite direction.

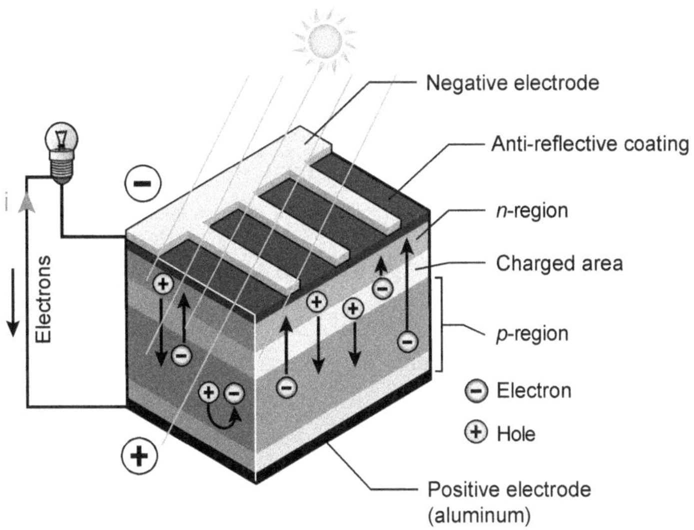

Figure 9. Principle of photocell operation. The absorbed photons lead to the formation of "electron–hole" pairs in the semiconductor. When an external electrical circuit is connected (on the left), electrons are set in motion: an electric current arises, which, for example, powers a light bulb.

The electromotive force generated by the solar cell (Fig. 9) is approximately 1V, and the current is approximately 1mA cm^{-2} of contact. Therefore, it is necessary to connect many of these elements in series in order to obtain an acceptable electromotive force, and also to connect many of these contacts in parallel in order to obtain sufficient amperage. Thus, the production of energy using solar panels uses a large surface area, and the output energy is relatively small — about 15% of the energy of the incident light. Despite these disadvantages, solar energy is an excellent alternative to fossil resources and is an inexhaustible source (see Chapter 13). Researchers estimate that 5,000 km^2 of solar panels will be sufficient to provide the current electricity needs of a large country, e.g., France (once the problem of storing the produced energy is solved, of course). That value corresponds to the area of a disc 80 km in diameter or to the roof area of 200,000 houses, 25 m^2 each.

Electrons for All Occasions

The examples above show how semiconductors are wonderful materials. They emit light, convert light into electricity, amplify signals, and respect one-way traffic. The use of semiconductors is not limited to electronics alone. They are also used in optoelectronics, which is at the interface of optics and electronics, an example of which is LEDs, and which is becoming increasingly important as fibre-optic communication develops (see Chapter 2). Semiconductors are also used in combinations of mechanics and electronics, such as Micro and Nano Electro-Mechanical systems (MEMs and NEMs), for example, accelerometers less than 1 mm in size, which are equipped with modern smartphones.

From the Computer to the Quantum Computer

The modern computer is a descendant of Pascal's computer. Its two most important, complementary properties are gigantic memory and the ability to execute programs, namely to implement complex tasks defined by a sequence of instructions. Various methods are used to store data, using semiconductors (for USB sticks), magnetism (for hard drives), or mechanical modelling (for compact discs), or a combination of all of these technologies.

Programming is a science that basically was born in the year 1936 when the English scientist Alan Turing (1912–1954) published a 16 page paper in the journal *Proceedings of the London Mathematical Society.* The article was strictly theoretical and set the main lines for the structure of a computer. Later on the theory was strongly improved by the great Hungarian-American mathematician John von Neumann (1903–1957). The main architecture of the computer involves four major elements (Fig. 10).

First, the arithmetic logic unit, or data processing unit, which performs basic operations; further, a control device responsible for the sequence of operations; then a memory containing both data and programs dictating to the control unit calculations to be performed based on these data. Finally, input and output devices enable the computer to communicate with the outside world. The memory is divided into operative (programs and data necessary during operation) and permanent (programs and data that form the basis of the device).

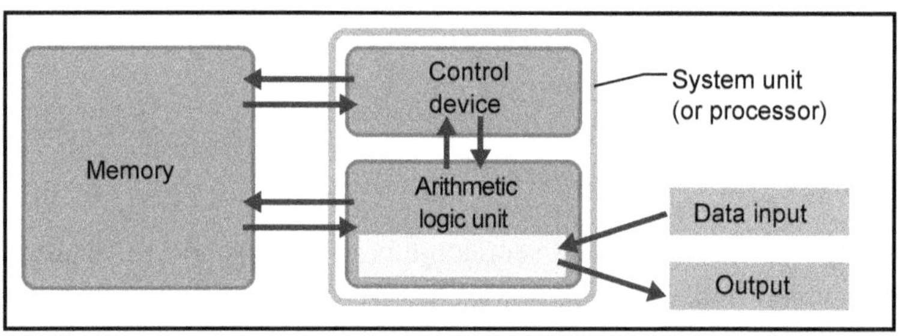

Figure 10. The von Neumann architecture.

Quantum Computer

The disadvantage of the von Neumann computer is its "sequential" nature: various stages of calculations follow one after another, and the next step is started only after the completion of the previous one. One way to save time is by introducing "parallelism". Parallel computing is already common in modern processors, but special hopes for their widespread use are associated with the development of "quantum" computers.

A quantum computer uses (or "will use", or "could use" — we are not yet sure which of these formulations to choose!) the phenomenon of *mixing* quantum states. Schrödinger's cat in his cell provides an example of such a mixture of states: the cat is both alive and dead at the same time (see Chapter 22). By the end of the 20th century, physicists realised that this phenomenon could become a valuable resource for calculations in computers of fundamentally new type. Instead of processing well-defined bits of a state, a quantum computer processes "quantum bits" (or *qubits*), the two states of which are somehow mixed up. Quantum computing performed by such a computer is a sequence of operations with quantum bits, the state of which is recorded (measured) only when required by the algorithm. Thus a quantum computer uses parallelism not by dividing computations into pieces and carrying them out by different processor cores but uses the true parallelism inherent in quantum mechanics. In a way, quantum computing allows you to consider both a dead and a living cat at the same time! By working well with quantum algorithms, a quantum computer can solve problems that are too complex for conventional sequential computers.

A typical illustrative problem is searching in the phone book for the name of a subscriber whose number is known. Since the list of subscribers is given in alphabetical order, searching without a computer takes, on average, time proportional to their number N. A sequential computer we are familiar with would also need time proportional to N, albeit with a much lower proportionality coefficient. A quantum computer, thanks to the way of operating (scientists say "algorithm") of the Indo-American computer scientist Grover, would spend time proportional to the square root of N. For large N, the time savings are significant! Another well-known quantum algorithm (Shor's algorithm) allows you to decompose numbers into prime factors. This problem is very difficult for an ordinary computer in the case of large numbers, so it underlies the RSA encryption system, which is commonly used for communication security.

Will a quantum computer work miracles anytime soon? Unfortunately, today there are only the simplest versions of quantum processors, which so far can only demonstrate the possibility of implementing the above operations, for example,

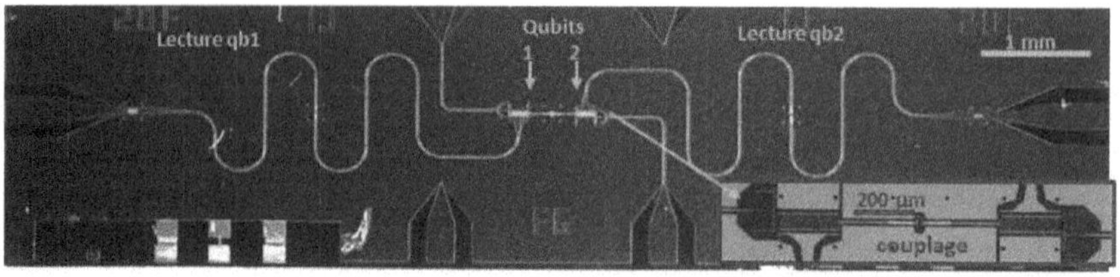

Figure 11. An example of a quantum processor with two qubits. This superconducting circuit enables Grover's algorithm to identify one item in a set of four (like a telephone book reduced to four subscribers). Courtesy of Andreas Dewes.

identify one element out of four using Grover's algorithm (Fig. 11) or factorise integers but only two-digit ones. So far, this can be done much better and more cheaply without a quantum computer.

A Look into the Nanoworld

Designing a quantum computer requires the highest level of technology. Practical creation of the nanostructures required for this implies that we know not only how to make them but also how to see them! Let us address four different devices used for this:

- *Scanning electron microscope, or SEM* (Fig. 12). This provides three-dimensional images of nano-objects with a perspective effect, such as in photography. When scanning, a beam of electrons passes over the surface of the sample, which in response reflects electrons and emits other electrons, X-rays, and light. All these particles and waves, which are carriers of information about the material and properties of the sample surface, are analysed by a microscope. SEM does not achieve atomic resolution.
- *Transmission electron microscope, or TEM* (Fig. 13). It is much bulkier than the SEM, but it is capable of achieving atomic resolution. In this case, an electron beam that has passed through the sample is analysed to obtain an image. Therefore, only thin objects can be studied with a transmission microscope. If the object is not thin enough, it would have to be cut into plates! For nano-objects, this delicate operation is not required.
- *Scanning tunnelling microscope, or STM.* This invention by German physicist Gerd Binnig and Swiss physicist Heinrich Rohrer, created in the IBM laboratory in Zurich, earned them the Nobel Prize in 1986, 5 years after its discovery. This is truly an amazing invention, as the device is capable of "feeling" atoms

Figure 12. Scanning electron microscope. You can see a sample image on the screen.

with a needle (Fig. 14). In reality, the tip does not touch the atoms: it approaches them at a distance of about 1 nm; in this case, a tunnelling current begins to flow through the gap (see Chapter 25). The distance from the tip to the atom must be kept accurate to 0.1 nm, which means, among other things, that the latter must be reliably protected from the slightest vibrations. The STM device can reach atomic resolution. On the other hand, it can provide only the image of the surface of a solid sample: the second atomic layer is not detected (Fig. 14). Finally, the material to be examined must be a conductor.

- *Scanning atomic force microscope, or AFM.* Like a tunnelling microscope, it "probes" the surfaces of solids with a needle. The distance from the tip of the needle to the surface is measured not by the magnitude of the tunneling current, but by the force with which the surface acts on the needle. The latter is determined using the deflection of the cantilever, which is recorded by a laser beam. Therefore, the substance does not have to be conductive (see the example of the resulting image for DNA molecules in Chapter 19).

Fantasy of Electrons in the Nanoworld

The nanoworld is the kingdom of strange physical laws. These laws are strictly related to quantum mechanics, particularly at low temperatures. We know that the atomic energies are quantized, namely they can take only specific values belonging to a discrete ensemble. The electric resistance R of a small circuit as well is quantized at low temperatures! It is reminded that in our world, being macroscopic and relatively hot, electrical resistance is to a great extent due to the interactions

Figure 13. Transmission electron microscope.

of the electrons with the thermal vibrations of the crystal lattice and with the impurities (see Chapter 24).

The quantisation rule is particularly simple when one refers to the conductance $1/R$: this is increasing by steps, its value being always an entire number of the quantity $2e^2/h$, where h is the Planck's constant. Let us clarify how the conductance can be varied: one uses one electrode called *grid* (by analogy with the one of the transistor) placed near an object. Which object? We will just mention two types: the quantum wires and the point contacts in semiconductors. A quantum wire is a very narrow channel of very pure conductor, with no impurities, and having a section of diameter comparable to the de Broglie wavelength of the electrons (see Chapter 22).

The tightness of the channel reveals the wave character of the electron, which implies the quantization of the motion of the electron along the transverse

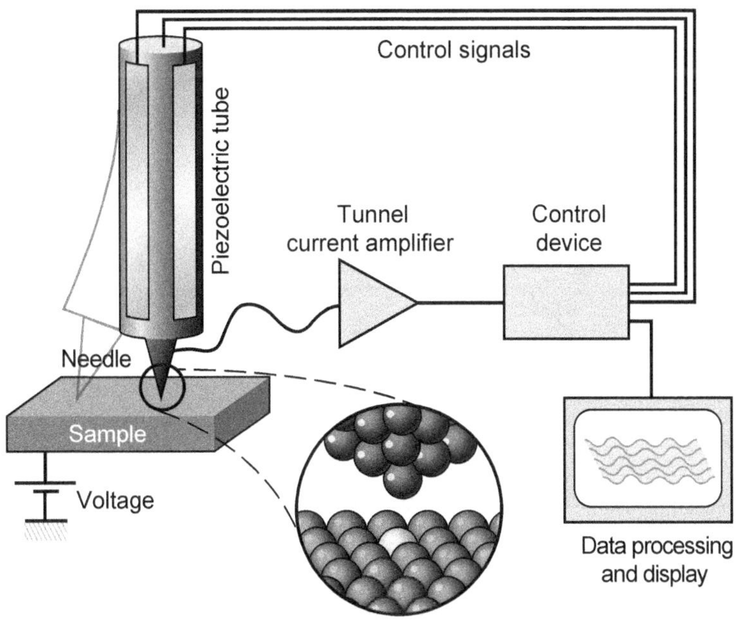

Figure 14. Scheme of operation of a tunnelling microscope. The needle is held by a piezoelectric tube at a distance of approximately 1 nm from the test sample. Tunneling current is amplified and then analysed.

directions of the wire. In ordinary conductors, due to the effect of the potential difference, the electron moves by a sequence of steps from one impurity to the other (see Fig.15(a)). The motion is said to be "diffusive". In a quantum wire, the electron propagation is said to be "ballistic" (Fig. 15(b)): it is somewhat similar to the propagation of an electromagnetic wave in a waveguide (see Chapter 2). It displays the occurrence of transverse motion schematised by a trajectory with successive reflections. In reality, the motion is quantized. Since the system is out of equilibrium, one does not speak of quantum states but rather of *modes*, as an analogy with the modes of the waveguides (see Chapter 2). By varying the electric potential of the gate (not reported in Fig. 15), it is possible to let one or two or three or more of the modes go across. Each mode contributes to the total conduction, given by the contributions of the different modes.

The point contacts between semiconductors are characterised by properties similar to quantum wires. The electrical resistance of the system is controlled by means of a "gate", and, at low temperature, one detects steps in conductance in correspondence to integer multiples of $2e^2/h$.

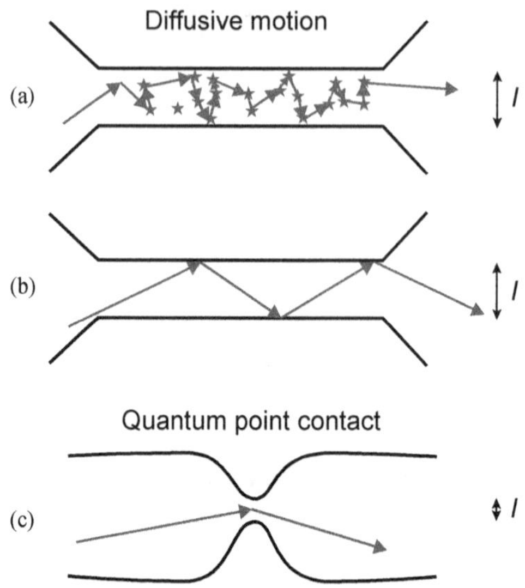

Figure 15. Motion of the electron due to the electric potential along the wire, as dependent on its width l. (a) Ordinary conductor. The trajectory of the electron can be schematised as a succession of steps from an impurity to another nearby. The motion of the electron is said to be diffusive. (b) Quantum wire. The trajectory of the electron is represented as a series of reflections against the walls. (c) Quantum point contact.

Quantum Hall Effect

The main phenomenon displaying quantized electrical resistance is the quantum Hall effect. In 1879, the young American scientist Edwin Hall (1855–1938) discovered a novel phenomenon. When an electric current is flowing in a conductor in the presence of a magnetic field $\boldsymbol{B}$ perpendicular to the electric field, the magnetic field deflects the electrons. As a consequence, besides the ordinary current along the x direction of the electric field, a further current along the y direction arises perpendicular to both the electric and the magnetic field (Fig. 16(a)). Thus along the y direction the potential difference V_y arises, the *Hall voltage*, which is related to the current I_x along the x direction by the relation $V_y = R_{xy}I_x$, where R_{xy} is the Hall resistance. This latter is proportional to the magnetic field $\boldsymbol{B}$ and to the inverse of the number of carriers per unit volume. Thus the Hall effect measures the concentration of charge carriers, which is a relevant characteristic of semiconductors.

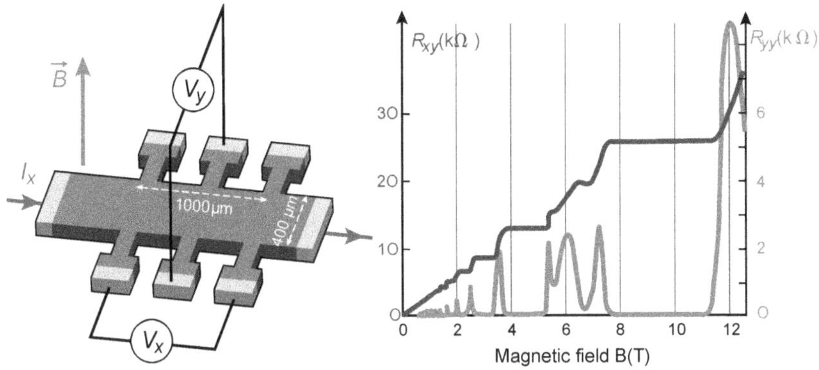

(a) (b)

Figure 16. (a) Block scheme of the experiment for the quantum Hall effect. (b) Results of the experiment. The black curve showing the steps is the Hall resistance R_{xy}. The other curve shows the ordinary resistance, which is almost null.

The Hall effect does not display any quantum property unless three conditions are verified: low temperature (a few kelvin), strong magnetic field (about 20 T); finally the electrons must form a two-dimensional gas. On increasing the magnetic field acting on the device, then one observes that the Hall resistance, instead of varying linearly, increases by steps (Fig. 16(b)). The inverse of the data, namely the value of the conductance, is integer[3] multiples of the quantity e^2/h.

How does one create such an unusual object as a two-dimensional electron gas? The method used in the early experiments for the quantum Hall effect was to apply a strong positive electric potential (by means of grid, an object always present in nanophysics) onto the surface of silicon.

The quantum Hall effect was discovered in the year 1980 by German physicist Klaus von Klitzing, thanks to the strong magnetic field provided by a magnet from Grenoble, represented a significant success of European scientific research. In 1985, von Klitzing was awarded the Nobel Prize in Physics.

An astonishing property of the quantum Hall effect is the precision in yielding the value of the ratio $e^2/h = 25{,}812.807\ \Omega$. The separation in between the steps, for instance, does not depend on the purity of the sample. After 1990, the quantum Hall effect has been used as a way to calibrate the measurements of the electrical resistance.

---

[3] In very strong fields, one can also detect steps of the conductance corresponding to fractional multiples of e^2/h.

Conclusion

In this chapter, we have recalled some technological and scientific applications of nanostructures. We have also mentioned very remarkable fundamental properties that probably will drive relevant applications in the future, but are also often studied in laboratories for their intrinsic interest. For example, flux quantization in superconductors allows one to measure extremely weak magnetic fields; quantization of the resistance in the quantum Hall effect leads to an extraordinarily precise measurement of the quantity (e^2/h).

It should be emphasized that, while making these discoveries, researchers did not even think about their possible future applications in practice. Scientists, when talking about the importance of funding fundamental science, often remind politicians of these truths. And politicians, in turn, make claims to scientists that they satisfy their curiosity without thinking in advance how future results will affect scientific and technological progress.

Further Reading

R. G. Arkhipov, Charge transport mechanism in liquid helium, *Soviet Physics Uspekhi* **9**, p. 174 (1966).

L. G. Aslamazov and A. A. Varlamov, *The Wonders of Physics*, 4th edition, World Scientific Publishing Company (2018).

K. R. Atkins, Ions in liquid helium, *Physical Review* **116**, p. 1339 (1959).

B. Audoly and S. Neukirch, Fragmentation of rods by cascading cracks: Why spaghetti does not break in half, *Physical Review Letters* **95**, p. 095505 (2005).

C. Bizon *et al.*, Patterns in 3D vertically oscillated granular layers: Simulation and experiment, *Physical Review Letters* **80**, p. 57 (1998).

A. F. Borghesani, *Ions and Electrons in Liquid Helium*, Oxford University Press (2007).

D. C. Cassidy, *Uncertainty, the Life and Science of Werner Heisenberg*, Freeman (1992).

H. Castro Neto, F. Guinea, N. M. R. Peres, K. S. Novoselov and A. K. Geim, The electronic properties of graphene, *Reviews of Modern Physics* **81**, p. 109 (2009).

A. Dewes *et al.*, Quantum speeding-up of computation demonstrated in a superconducting two-qubit processor, *Physical Review* **B85**, p. 140503 (2012).

A. Einstein, The cause of the formation of meanders in the courses of rivers and of the so-called Baer's Law (1926). Read before the Prussian Academy, January 7, 1926. Published in *Die Naturwissenschaften* **14** (English translation in *Ideas and Opinions*, by Albert Einstein, Modern Library, 1994).

J. B. Fournier and A. M. Cazabat, Tears of wine, *Europhysics Letters* **20**, p. 517 (1992).

C. Gianino, Experimental analysis of the Italian coffee pot moka, *American Journal of Physics* **75**, p. 43 (2007).

I. Grillo, Colloids and surfaces, *Physicochemical Engineering Aspects* **225**, p. 153 (2003).

R. H. Henrywood and A. Agarwal, The aeroacoustics of a steam kettle, *Physics of Fluids* **25**, p. 107101 (2013).

T. Janssen, G. Chapuis and M. de Boissieu, *Aperiodic Crystals*, Oxford University Press (2007).

D. JC MacKay, *Sustainable Energy — Without the Hot Air*, UIT Cambridge Ltd (2009).

J. Laskar, F. Joutel and P. Robutel, Stabilization of the Earth's obliquity by the Moon, *Nature* **361**, p. 615 (1993).

M. Lesieur, *Turbulence in Fluids*, Springer (2008).

T. G. Matuda, P. A. Pessoa Filho and C. C. Tadini, Experimental data and modeling of the thermodynamic properties of bread dough at refrigeration and freezing temperatures, *Journal of Cereal Science* **53**, p. 126 (2011).

M. Minnaert, On musical air-bubbles and the sound of running water, *Philosophical Magazine* **16**, p. 235 (1933).

E. Orlandini and S. G. Whittington, Statistical topology of closed curves: Some applications in polymer physics, *Reviews of Modern Physics* **79**, p. 611 (2007).

S. Parnovsky, *About the Biggest, the Smallest, and Everything Else*, World Scientific Publishing Company (2022).

S. Parnovsky and A. Parnowski, *How the Universe Works*, World Scientific Publishing Company (2018).

S. Renaud and M. de Lorgeril, Wine, alcohol, platelets, and the French paradox for coronary heart disease, *Lancet* **339**, p. 1523 (1992).

É. Reyssat, F. Chevy, A.-L. Biance, L. Petitjean and D. Quéré, Shape and instability of free-falling liquid globules, *Europhysics Letters* **80**, p. 34005 (2007).

A. Rigamonti and P. Carretta, *Structure of Matter*, Springer (2015).

A. Rigamonti and A. A. Varlamov, Superconductivity and applications, *Scientifica Acta* **5** (special issue, 2011). http://riviste.paviauniversitypress.it/index.php/sa.

T. M. Sanders, Heisenberg and the German bomb, *Contemporary Physics* **43**, p. 401 (2002).

G. D. Scott, Packing of spheres, *Nature* **188**, p. 908 (1960); *British Journal of Applied Physics* **2**, p. 863 (1969).

A. H. Shapiro, Bath-tub vortex, *Nature* **196**, p. 1080 (1962).

D. Shechtman, I. Blech, D. Gratias and J. W. Cahn, Metallic phase with long-range orientational order and no translational symmetry, *Physical Review Letters* **53**, p. 1951 (1984).

J. H. Smith and J. Woodhouse, The tribology of rosin, *Journal of the Mechanics and Physics of Solids* **48**, p. 1633 (2000).

Hervé This, *Molecular Gastronomy: Exploring the Science of Flavor* (*Arts and Traditions of the Table: Perspectives on Culinary History*), Columbia University Press (2008).

N. Vandewalle, J. F. Lentz, S. Dorbolo and F. Brisbois, Avalanches of popping bubbles in collapsing foam, *Physical Review Letters* **86**, p. 179 (2001).

M. Vollmer, K.-P. Möllmann and D. Karstädt, Microwave oven experiments with metals and light sources, *Physics Education* **39**, p. 500 (2004).

F. Wadsworth, The force required to operate the plunger on a French press, *American Journal of Physics* **75**, pp. 43–47 (2021).

Index

www.ingramcontent.com/pod-product-compliance
Lightning Source LLC
LaVergne TN
LVHW060019210926
848481LV00013B/3744
9789819833320